FOOD GELS

ELSEVIER APPLIED FOOD SCIENCE SERIES

Microstructural Principles of Food Processing and Engineering
J. M. AGUILERA and D. W. STANLEY

Biotechnology Applications in Beverage Production
C. CANTARELLI and G. LANZARINI (Editors)

Food Refrigeration Processes
ANDREW C. CLELAND

Progress in Sweeteners
T. H. GRENBY (Editor)

Food Antioxidants
B. J. F. HUDSON (Editor)

Development and Application of Immunoassay for Food Analysis
J. H. R. RITTENBURG (Editor)

Forthcoming titles in this series:

Food Irradiation
S. THORNE (Editor)

FOOD GELS

Edited by

PETER HARRIS

*Unilever Research Laboratory, Colworth House,
Sharnbrook, Bedford MK44 1LQ, UK*

ELSEVIER APPLIED SCIENCE
LONDON and NEW YORK

ELSEVIER SCIENCE PUBLISHERS LTD
Crown House, Linton Road, Barking, Essex IG11 8JU, England

Sole Distributor in the USA and Canada
ELSEVIER SCIENCE PUBLISHING CO., INC.
655 Avenue of the Americas, New York, NY 10010, USA

WITH 33 TABLES AND 164 ILLUSTRATIONS

© 1990 ELSEVIER SCIENCE PUBLISHERS LTD

British Library Cataloguing in Publication Data

Food gels.
 1. Food. Constituents. Biopolymers
 I. Harris, Peter
 641.1

ISBN 1-85166-441-6

Library of Congress Cataloging-in-Publication Data

Food gels / edited by Peter Harris.
   p. cm. — (Elsevier applied food science series)
 Includes bibliographical references.
 ISBN 1-85166-441-6
 1. Colloids. I. Harris, Peter. II. Series.
 TP453.C65F664 1990

ANDERSONIAN LIBRARY

1 0. APR 91

UNIVERSITY OF STRATHCLYDE

No responsibility is assumed by the publisher for any injury and/or damage to persons or property as a matter of products liability, negligence or otherwise, or from any use or operation of any methods, products, instructions or ideas contained in the material herein.

Special regulations for readers in the USA

This publication has been registered with the Copyright Clearance Center Inc. (CCC), Salem, Massachusetts. Information can be obtained from the CCC about conditions under which photocopies of parts of this publication may be made in the USA. All other copyright questions, including photocopying outside the USA, should be referred to the publisher.

All rights reserved. No part of this publication may be reproduced, stored in a retrieval system, or transmitted in any form or by any means, electronic, mechanical, photocopying, recording, or otherwise, without prior written permission of the publisher.

Printed in Great Britain by Galliard (Printers) Ltd, Great Yarmouth

PREFACE

The food technologist who wishes to produce a gelled product is faced with two basic options for achieving the desired effect; whether to use a protein or a polysaccharide. Although a gel can be formed by either a protein or a polysaccharide, the resultant gels have different characteristics:

- Polysaccharide gels are characterised by their fine texture and transparency which is achieved at a low polymer concentration. They can be formed by heating and cooling, pH adjustment or specific ion addition.
- Protein gels are characterised by a higher polymer concentration (5–10%) and are formed almost exclusively by heat denaturation.

Before reaching a final decision, the technologist must take a number of factors into consideration.

The purpose of this book is to help the technologist in his choice by providing fundamental practical information, in one book, on the properties of gels (and factors which influence them) for both types of biopolymer. To help the reader, each chapter is (wherever possible) organised in the same way so that, for example, information on structure will always be available in section 2. The examples in the Applications section of each chapter are not meant to be exhaustive, but to illustrate the various ways in which the particular polymer can be used to form a gelled product.

I have tried to ensure that this book forms a bridge between the academic treatise on molecular conformation and interactions of the individual biopolymers, and the technical data sheets which are readily available from the relevant ingredient suppliers. By doing this, I hope that this book will prove invaluable not only to the product developer, but also to the academic who will gain insight as to how the fundamental understanding of biopolymers is (and can be) put to practical use.

Throughout this book, the reader is urged to ensure that the chosen gelling agent is fully in solution; this advice cannot be overstressed. It is impossible to obtain 100% functionality without first having 100% of the biopolymer dissolved. To achieve this, it is essential to first disperse the gelling agent before it is allowed to dissolve.

Finally, I would like to thank my many friends and colleagues for helping me make this book possible.

PETER HARRIS

CONTENTS

Preface v

List of Contributors ix

1. Agar 1
   TETSUJIRO MATSUHASHI

2. Alginates 53
   WILMA J. SIME

3. Carrageenans 79
   NORMAN F. STANLEY

4. Casein 121
   P. F. FOX and D. M. MULVIHILL

5. Egg Protein Gels 175
   SCOTT A. WOODWARD

6. Gellan Gum 201
   G. R. SANDERSON

7. Gelatine 233
   F. A. JOHNSTON-BANKS

8. Mixed Polymer Gels 291
   EDWIN R. MORRIS

9. Muscle Proteins 361
   GRAHAM W. RODGER and PETER WILDING

10. Pectin 401
    CLAUS ROLIN and JOOP DE VRIES

11. Whey Proteins 435
    R. C. BOTTOMLEY, M. T. A. EVANS and C. J. PARKINSON

Index 467

LIST OF CONTRIBUTORS

R. C. BOTTOMLEY
> Express Foods Group Ltd, R & D Department, 430 Victoria Road, South Ruislip, Middlesex HA4 0HF, UK

JOOP DE VRIES
> Copenhagen Pectin, DK-4623 Lille Skensved, Denmark

M. T. A. EVANS
> Express Foods Group Ltd, R & D Department, 430 Victoria Road, South Ruislip, Middlesex HA4 0HF, UK

P. F. FOX
> Department of Dairy and Food Chemistry, University College, Cork, Republic of Ireland

F. A. JOHNSTON-BANKS
> Gelatine Products Limited, Sutton Weaver, Runcorn, Cheshire WA7 3EH, UK

D. W. MANNING
> Marine Colloids Division, FMC Corporation, BioProducts Department, Rockland, Maine 04841, USA

TETSUJIRO MATSUHASHI
> Nagano State Laboratory of Food Technology, Kurita, Nagano, Japan 380. Present address: Tomitake 374–15, Nagano, Japan 381

E. R. MORRIS

Department of Food Research and Technology, Cranfield Institute of Technology, Silsoe College, Silsoe, Bedford MK45 4DT, UK

D. M. MULVIHILL

Department of Dairy and Food Chemistry, University College, Cork, Republic of Ireland

C. J. PARKINSON

Express Foods Group Ltd, R & D Department, 430 Victoria Road, South Ruislip, Middlesex HA4 0HF, UK

GRAHAM R. RODGER

Imperial Chemical Industries plc, Biological Products Business, PO Box 1, Billingham, Cleveland TS23 1LB, UK

CLAUS ROLIN

Copenhagen Pectin, DK-4623 Lille Skensved, Denmark

G. R. SANDERSON

Kelco Division of Merck & Co, Inc, 8225 Aero Drive, San Diego, California 92123, USA

WILMA J. SIME

Limewood, Raunds, Northamptonshire NN9 6NG, UK

NORMAN F. STANLEY

Marine Colloids Division, FMC Corporation, BioProducts Department, Rockland, Maine 04841, USA

PETER WILDING

Unilever Research Laboratory, Colworth House, Sharnbrook, Bedford MK44 1LQ, UK

SCOTT A. WOODWARD

Poultry Science Department, Institute of Food and Agricultural Sciences, University of Florida, Gainesville, Florida 32611, USA. Present address: Sunny Fresh Foods, Egg Products, 206 W Fourth St. Box 428, Monticello, Minnesota 55362, USA.

Chapter 1

AGAR

TETSUJIRO MATSUHASHI

*Nagano State Laboratory of Food Technology,
Kurita, Nagano, Japan 380*

1 INTRODUCTION

Agar is a strongly gelling seaweed hydrocolloid composed of polysaccharides. The main structure of agar is chemically characterized by the repeating units of D-galactose and 3,6-anhydro-L-galactose,[1-5] with a few variations, as well as by a low ester sulphate content.[6-9]

Agar was accidentally discovered by an inn-keeper, Tarozaemon Minoya, in what is now called Kyoto, Japan, in the middle of the seventeenth century.[7,10,11] The documented story tells that a kind of fresh dish that was prepared from the seaweed (*Gelidium* sp.) was served to a noble feudal lord at his inn in a winter season; it happened that the residues of the dish turned to a white spongy substance, which we now know as agar. His discovery of 'agar' was subsequently developed into the frozen-and-dried agar process. Thus, this discovery could be considered as the world's first so-called 'instant food'.[12-14]

The Japanese term 'kanten' for agar was named by a Chinese Buddhist priest, Ingen (1592–1673), who was naturalized in Japan after 1654.[7,10,11] Agar gel was a favourite dish of the Zen sect Buddhists because of its plain texture and flavour.

The habitual eating of various kinds of agar gel-like seaweed extracts probably dates back to prehistoric times in many of the coastal areas,[11] and is still practised today.

As its physicochemical properties had been enigmatic, so the traditional process of agar manufacture had been mysterious and complicated until recent decades.[11]

In the past three centuries up to the 1940–1960s, Japan monopolized world agar production.[12,13] Her domestic use of agar, for instance, has been associated with a type of condensed sweet jelly containing mushed red beans. China, in the prewar period, had been a good customer of Japanese agars as well as a good international trader with Europe.

It is worth noting that a German housewife, Frau Fanny Eilshemius Hesse, suggested the idea of using agar as a solid bacterial culture medium to her physician husband in 1881.[9,11,15] Because of the work by Robert Koch (1843–1910) at the end of the nineteenth century, agar has become indispensable in any microbiological and biotechnological laboratory. Compared with food uses, laboratory uses of agar are far fewer. Most countries in the world have imported agar from Japan,[12,13] although a number of countries other than Japan are now producing agar.

The remarkable achievements on the chemical structure studies of agar by Araki[2,3] revealed a neutral polysaccharide, agarose, to be the active gelling component of agar. Thereafter, the use of agarose as well as more refined agar has been increasing in the scientific field.[16] Therefore, any scientific information on agarose will also apply to agar gels.

2 STRUCTURE

2.a Sugar Skeleton of Agar

The basic repeating unit of agar consists of alternating 1,3-linked β-D-galactopyranose and 1,4-linked 3,6-anhydro-α-L-galactopyranose units.[2–5,9,17] The disaccharide, agarobiose (Fig. 1), is the common structural unit in all of the agar polysaccharides.[5,17–20]

Araki showed that agar consists of two groups of polysaccharides:[21] agarose, a neutral polysaccharide, and agaropectin, an oversimplified term for the charged polysaccharide.[2,3] The agaropectin contains sulphuric acid,

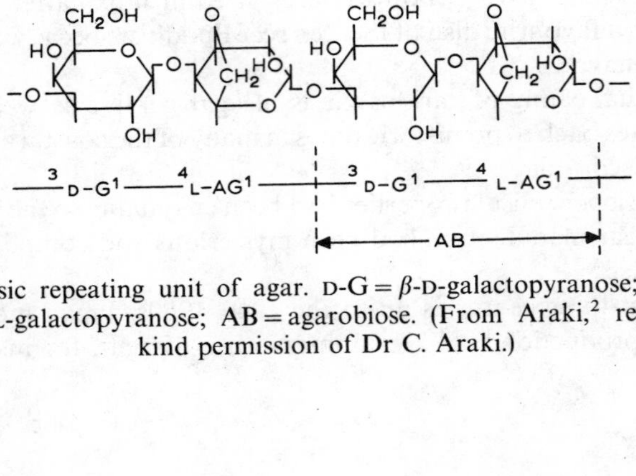

FIG. 1. Basic repeating unit of agar. D-G = β-D-galactopyranose; L-AG = 3,6-anhydro-α-L-galactopyranose; AB = agarobiose. (From Araki,[2] reproduced by kind permission of Dr C. Araki.)

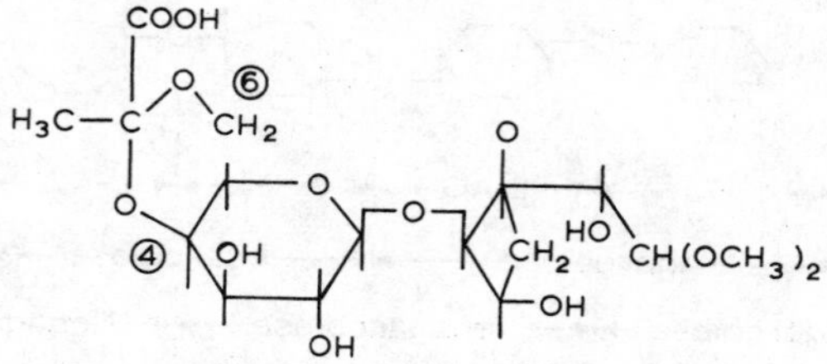

FIG. 2. One of the compositional units of agaropectin (from Araki[4]), O-4,6-O(1-carboxyethylidene)-(1 → 4)-3,6-anhydro-L-galactose dimethylacetal. A derivative of agarobiose with the bonded pyruvic acid residue from the partial methanolysis products of commercial agar made chiefly from *Gelidium amansii*. (From Hirase,[22] reproduced by kind permission of the Japanese Society of Chemistry.)

D-glucuronic and pyruvic acids (Fig. 2).[4,22] Later, Araki[17] isolated small amounts of D-xylose, L-galactose, 4-O-methyl-L-galactose and 6-O-methyl-D-galactose from agaropectin. Compared with agarose, the structure of agaropectin is more complicated and not well understood; however, the sugar skeleton is also composed of agarobiose.[5,17,21]

Yaphe and Duckworth[16] proposed that the word agar should be used to characterize the family of polysaccharides with the same backbone structure consisting of alternate (1 → 3)-linked and (1 → 4)-linked D- and L-galactose residues, respectively. Thus, the three extremes of structure in agar are neutral agarose, pyruvated agarose with little sulphation and a sulphated galactan containing little or no 3,6-anhydrogalactose or 4,6-O-(1-carboxyethylidene)-D-galactose residues.[5]

2.b Content of Agarose

The content of agarose and agaropectin will vary in commercial agar depending on the weed source of the agar[23,24] and can be further changed by the conditions used for extraction.[25]

2.c Structure of Agarose

The proposed structure of agarose[2,3] is illustrated in Fig. 3 and consists of a linear structure with no branching.

In the solid state, agarose exists as a threefold, left-handed double helix with a pitch of 1·90 nm and a central cavity along the helix axis.[26] Although this cavity is considerably smaller than in V-amylose, computer model building indicates that it can accommodate water molecules without unfavourable steric clashes. The interior of the cavity is lined with hydroxyl groups that can participate in hydrogen bonding with water.[27] Direct

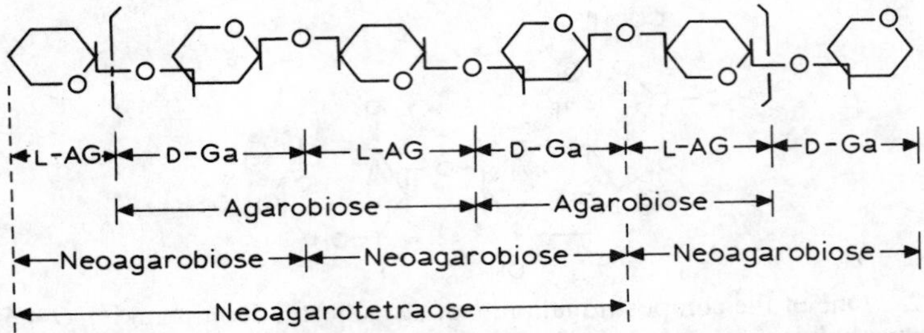

FIG. 3. Possible structure of agarose. L-AG = 3,6-anhydro-L-galactose; D-Ga = D-galactose. (From Araki,[2] reproduced by kind permission of Dr C. Araki.)

evidence of bound water within the helix has come from ^{1}H-NMR relaxation studies,[28] which showed, in addition to the expected contribution from bulk water, non-exchangeable polysaccharide protons and 'normal' bound water, the presence of a very highly bound proton species.

It has also been shown[5] that some D-galactose residues in agarose are 6-O-methylated, and the degree of methylation varies from species to species of agarophytes.

2.d Sulphate and Inorganic Constituents in Agar

Sulphate is an integral part of agar.[6,7,29,30] A study of 37 commercial agars showed the content of inorganic substances averaged 2·41% and the moisture content averaged 24·3%. Further accumulated studies show that the ash content of the refined mucilaginous substances directly processed from *Gelidium amansii* and four other typical agarophytes is around 5%; that of *Gracilaria verrucosa* is nearly 15%; and the ash content of the first-grade bar-style agar is 3·3%. For these seven samples, the SO_3 contents in ash were 19·8–39·5%, averaging 30%. The ratios of 'total SO_3 in mucilaginous substance (or agar)' to 'SO_3-in-ash' contents are 1·27 for *Gelidium amansii*, 3·14 for *Ceramium hypnaeoides* and 1·52 for the agar.[6,7]

Tanii extended the analyses of sulphates in agar.[29] For 15 commercial Japanese agars, the total SO_3 contents were 0·69–3·05%, while SO_3-in-ash contents were 0·43–1·88%. He suggested a reversed index to Yanagawa's, and the ratios of 'SO_3-in-ash' to 'total SO_3' were in a range of 0·32–1·00. However, for experimentally processed agars from eight typical species of agarophytes, Tanii analysed sulphates to produce similar results (1·10–3·60% as total SO_3 contents) to those of Yanagawa.[29]

Typical sulphate (SO_3-in-ash) contents in agar are shown in Table 1.[6,7] The significance of these sulphate levels[29,31,32] in relation to gelling ability of agar will be described in a later section.

TABLE 1
INORGANIC SUBSTANCES IN THE MUCELAGINOUS SUBSTANCES EXTRACTED FROM AGAR-BEARING RED ALGAE

Agar/Origin of substance	Ash (%)	In ash (%)					
		CaO	MgO	Na_2O	$Fe_2O_3 + Al_2O_3$	SO_3	SiO_2
Gelidium amansii	5·17	20·05	30·85	7·63	8·61	29·53	0·91
Ge. subcostatum	5·31	15·49	29·78	4·61	20·86	25·03	5·18
Ge. japonicum	5·24	32·75	8·75	1·76	8·29	36·15	8·37
Acanthopeltis japonica	5·23	17·59	9·88	2·88	11·80	19·82	32·35
Campylaephora hypnaeoides	4·51	18·48	23·13	8·31	6·99	38·83	0·99
Gracilaria verrucosa	14·66	4·60	5·42	29·91	9·60	34·73	8·62
Agar	3·30	23·65	21·24	2·74	4·42	39·50	3·55

From Yanagawa.[6,7]

The assumption by Hass[33] of the existence of ethereal sulphates bound to a carbohydrate complex in marine algae is strongly supported by the accumulated works of Yanagawa and others, who proposed that calcium is bound to agar in the following way:

$$R\begin{matrix}O \cdot SO_2 \cdot O \\ O \cdot SO_2 \cdot O\end{matrix}Ca + 2H_2O = R\begin{matrix}OH \\ OH\end{matrix} + CaSO_4 + H_2SO_4$$

where R is agar polysaccharide. This proposal has been extended for the combination of sulphate and calcium between agar and cellulose moieties in agar-bearing red algae,[34] but further work is needed for confirmation.

Agarose is defined as a neutral polysaccharide in agar, as mentioned earlier; however, agarose in practical use generally contains sulphate to some degree.[9,24,25] As illustrated in Table 2, the minimal content still shows such values as 0·3% (SO_4^{2-}).[35,36] It is noteworthy, therefore, that Duckworth and Yaphe[37] recommend as a practical definition of agarose '...that mixture of agar molecules with the lowest charge content and therefore, the greatest gelling ability, fractionated from a whole complex of molecules, called agar, all differing in the extent of masking with charged groups'.[36]

The other inorganic substances that are relatively abundant in the ash from agar are shown in Table 1.[6,7]

Boron of natural origin is one minor but noteworthy constituent of agar. It rarely exceeded a level of 600 mg (as boric acid) per 1 kg of agar in the

TABLE 2
SeaKem agarose specifications

Type[a]	Gel temperature 1·5% (°C)	EEO[b] (M)	Gel strength 1% (g/cm^2)	Sulphate (%)	Moisture (%)
LE	36 ± 1·5	0·10–0·15	≥ 1 000	≤ 0·35	≤ 10
ME	36 ± 1·5	0·16–0·19	≥ 1 000	≤ 0·35	≤ 10
HE	36 ± 1·5	0·23–0·26	≥ 650	≤ 0·35	≤ 10
HEEO	36 ± 1·5	≥ 0·30	≥ 650	≤ 0·35	≤ 10
HGT	42 ± 1·5	≤ 0·10	≥ 800	≤ 0·30	≤ 10

[a] Note: SeaKem HGT(P) agarose has been discontinued.
[b] EEO: electro endosmosis.
From FMC Bioproducts Source Book, 1985.[36] Reproduced by kind permission of Dr Guiseley of FMC.

prewar period,[7] but recent analyses have revealed larger boron contents in some commercial agars that are processed from a variety of species of raw material seaweeds throughout the world. Consequently, the present Japanese Food Sanitation Law has established a legal tolerance limit of 1000 mg of boric acid per 1 kg of dry agar.

3 SOURCE AND MANUFACTURE

3.a. Source in General

'Tengusa' in Japanese, which means *Gelidium amansii* and/or most of the genus *Gelidium* together with *Pterocladia tenuis*, etc., have been the principal source of agar traditionally.[7,10,12] The genus *Gracilaria*, especially *Gr. verrucosa*, had been a supplementary raw material[12,13] but, following the invention of Funaki and Kojima of 'alkali pretreatment of seaweeds',[31,38–40] this genus has become another important source of agar.

Gelidium species grow on rocky sea-bottoms between low-tide level and 5–20 m depth and are hand-picked by special fishing divers in Japan, whereas *Gracilaria* species grow in rather shallow sandy sea areas throughout the world. Once, Tokyo Bay was the most abundant *Gracilaria*-harvesting place in Japan and the world. The industrially useful agar-bearing red algae are listed below, on the basis of previously unpublished technological information accumulated by the author.

3.b Specific Agarophytes

The genus *Gelidium* ('Tengusa' in Japanese): *Ge. amansii* is the most abundant and useful species in Japan. *Ge. japonicum* ('Onikusa' in Japanese) bears agar of highest gelling ability among all of the known agarophytes. *Ge. pacificum* ('Obusa' in Japanese) and *Ge. linoides* ('Kinukusa' in Japanese) also yield agar with good properties, although the former plant is rigid while the latter plant is soft, which gives rise to some difference in extractability.

Ge. sesquipedale harvested along the European sea coast produces a reasonable gelling agar. *Ge. subcostatum* ('Hirakusa' in Japanese) is a popular algae among the traditional Japanese agar processors because it has a filtration-accelerating role that speeds up the filtration step, even though it yields an agar that has a much inferior gelling property to that of *Ge. amansii*.

Twenty-four species of *Gelidium* are found around Japan,[41] and 55–69 species or types of European *Gelidium* are listed by Segi,[42] all of which bear agar, but only about 10 species are processed industrially.[43]

Genus *Pterocladia*. *Pt. tenuis* ('Obakusa' in Japanese) and *Pt. densa* are often confused with *Ge. amansii*, and contain agar of similar gel properties.

Acanthopeltis japonica (common Japanese name 'Toriashi'). In spite of its curious appearance,[7,10,41] this contains agar of fairly good quality. It is usually harvested together with Ge. species.

Gelidiella acerosa ('Shima-tengusa' in Japanese). This has been imported into Japan from the south-eastern Asian countries, but generally it is a hazardous material to process because it can produce agar of variable quality.[44]

Digenea simplex. Agar having similar gelling ability to agar of *Ge. subcostatum* is obtainable from this red alga.[12,45] In Japan there is a manufacturer who specializes in using this alga for agar extraction, after chemical extraction of 'kainic acid'.

Genus *Gracilaria* Gr. verrucosa ('Ogonori' in Japanese) is probably the most abundant agarophyte in the world. It is used for both conventional agar processing and for the extraction of agarose by means of alkali treatment of the weed. Which process is used depends upon the geographical variations of the weed source as well as the particular manufacturers' decisions.[31,38–40]

In general, Japanese species,[31,39] African species,[32] Chilean species,[46]

Argentinian species[47,48] and the south-eastern Asian species produce agars that differ in both properties and extractabilities.

Gr. compressa, Gr. chorda and *Gr. gigas* are also worth mentioning, since they are similar to *Gr. verrucosa*. The mucilaginous polysaccharide, once called agaroid, of *Gr. foliifera* can be converted to agar with good properties.[49]

Ahnfeltia plicata. This red alga, which was harvested in Sakhalin and Hokkaido from early this century,[7,50] was used to develop the new industrial method that uses sodium hydroxide and mechanical processing.[50-53]

Genus *Ceramium*. *Ce. kondoi* (Yigisu' in Japanese), *Ce. boydenii* and *Campylaephora hypnaeoides* ('Egonori' in Japanese, formerly classified as *Ce. hypnaeoides*) are all characterized by their soft plant tissues and self-entangling branches. Extraction of agar from these types of algae is generally easy. Gels of these agars are viscoelastic with noticeable syneresis, and show relatively low melting temperature.[54,55] These algae usually grow attached to brown algae such as *Sarugassum fulvellum*. The practical harvesting for food and industrial uses can be unreliable owing to unexpected high tides or strong waves caused by storms.

It should be noted that *Ca. hypnaeoides* has become the most valuable red alga in recent decades and the price has become much higher than that of the finest *Ge. amansii* in Japan. The reason is that this species has long been used for ceremonial domestic cooking of gel dishes in north-eastern Japan, which faces the Japan Sea, and nowadays its cost is prohibitive for the agar industry.[56]

It is also noteworthy that *Gr. verrucosa* (once classified as *Gr. confervoides*), which has fine vegetable qualities, is used throughout Japan for a kind of salad together with raw fish ('Sashimi') in its original style with greenish colour.

Agaroid-bearing red algae. Some species of *Eucheuma* have been used for agar manufacture in limited amounts in some factories. Whether or not the mucilaginous constituents are really agar is questionable, since most of the *Eucheuma* species are recognized as raw materials for carrageenan extraction.

Genera such as *Hypnea, Gigartina* and *Grateloupia* are not agar-bearing algae and are wrongly listed in some textbooks on agar.

3.c Traditional and Conventional Extraction Methods

Cooking of refined seaweeds in an excess of water, at boiling point, is a fundamental necessity for agar extraction.[7,10,57,58] In this cooking process, a careful addition of acid, for instance 0·01–0·02% sulphuric acid or 0·05% acetic acid, is generally required to promote a good extraction.

In the traditional Japanese process, a variety of agarophytes are introduced into one cooking vessel that can hold 3–8 m^3 of aqueous contents, the formulation of which is dependent upon each processor's know-how as well as their product design.

Table 3 shows the seaweed formulae (previously unpublished, even in Japanese) of two bar-style factories. It is worth noting that the products had similar physicochemical properties!

In order to get the finest quality products for food use, reliable raw-material seaweeds from known harvesting sea coasts are used selectively.

TABLE 3
SEAWEEDS FORMULAE IN BAR-STYLE AGAR FACTORIES IN THE 1960s

K Factory		I Factory	
Species	Weight (kg)	Species	Weight (kg)
Japanese *Gelidium amansii*, A	18·75	*Gracilaria verrucosa* of Tokyo Bay, alkali-treated	187·50
Japanese *Gelidium amansii*, B	13·13	Argentinian *Gracilaria verrucosa*, untreated	33·75
Japanese *Gelidium amansii*, C	13·13	*Campylaephora hypnaeoides*	18·75
Japanese *Gelidium amansii*, D	9·37	*Ceramium kondoi* of Daikonjima	112·50
Japanese *Gelidium amansii*, E	9·37	*Ceramium* sp.	11·25
Jap. *Gelidium* spp.	13·13	*Eucheuma muricatum*	3·75
Jap. *Gelidium japonicum*	18·75	*Dignea simplex*	
Jap. *Acanthopeltis japonica*	9·37	Residual seaweeds of kainic acid extraction	22·50
Jap. *Ceramium kondoi/boydenii*, I	37·50	**Total**	**390·00**
Jap. *Ceramium kondoi/boydenii*, J	7·50		
Jap. *Campylaephora hypnaeoides*	22·50		
Jap. Red *Gracilaria verrucosa* of Daikonjima	18·75		
Argentinian *Gracilaria verrucosa*	37·50		
Indonesian *Gracilaria* sp.	18·75		
Eucheuma muricatum	3·75		
Gracilaria eucheumoides	3·75		
Total	**255·00**		

The so-called 'seaweeds formulae for agar' had been (and have been) secret *know-hows* in each of the agar factories. The data in this table were revealed to the author by the courtesy of these agar manufacturers, respectively, in the 1960s. These data have never been published. The time lag and the present situation of the Japanese agar industry will justify this first publication as some understandable examples.

Gelidium, or rigid-type seaweeds such as *Ge. amansii*, should be dominant, preferably with a minor amount of *Ceramium*, or soft-type seaweeds.

3.d Extraction under Pressure
Extraction under pressure reduces processing time and increases yield of agar, and is especially effective for rigid-type seaweeds. Generally a gauge pressure of 1–2 kg/cm^2 will be applied for 2–4 h.[32,50,57]

A comparative extraction experiment for a rigid African *Gelidiaceae* (*Ge. cartilagineum*) showed that pressurized cooking extraction gave a larger yield of agar than conventional acid-cooking extraction under atmospheric conditions.[59]

It is important to remember that both the pressure method and acid cooking are effective for agar extraction, but both of these conditions are potentially destructive for the extracted agar. (This point will be discussed further in later sections.) Therefore, the optimum extracting conditions have to be established for each kind of individual seaweed and for each mechanical operation in the factory.

3.e Alkali Treatment of Seaweeds
The so-called 'alkali treatment' method of the agar industry was first established by Kojima and Funaki,[31,38,39] at Tokyo Institute of Technology, independently of the same idea applied to carrageenan extraction.[60,61] The use of *Gracilaria verrucosa*, which was abundantly harvested in Tokyo Bay in the postwar period, was the socioeconomic background for this revolutionary development in the agar industry. The application of pressing dehydration of the agar hydrogel[50] was another promoter of the development of modern methods of agar production worldwide.

The conversion of agaroid-like mucilaginous substance in *Gracilaria* to agar substance by alkali treatment at 85–90°C is based on the following substitution reaction together with the partial elimination of sulphates:[31,39]

$$R\begin{matrix}O\cdot SO_2\cdot O\\ \\O\cdot SO_2\cdot O\end{matrix}R' + CaCl_2 + 2NaOH \rightarrow R\begin{matrix}O\cdot SO_2\cdot O\\ \\O\cdot SO_2\cdot O\end{matrix}Ca + 2NaCl + H_2O$$

$$\text{(II)} \qquad\qquad\qquad\qquad\qquad\qquad \text{(I)}$$

where R and R' are polysaccharide radicals in *Gracilaria* (or agar-bearing Rhodophiceae). It is considered that *Gelidium* species are rich in (I)-form polysaccharides that have strong gelling ability, while *Gracilaria* species (at least Japanese *Gr. verrucosa*) are rich in (**II**)-form polysaccharides that have no gelling ability. The substitution product of (**I**) in the above formula can be formed within the plants of *Gracilaria* by alkali treatment. Therefore, an aqueous solution of sodium hydroxide that contains a minor amount of ionized calcium (CaO or $CaCl_2$) is used.[31,38,39]

An interesting practical point for successful alkali treatment of *Gracilaria* (one that has never been mentioned in any agar textbook) is the use of unrefined seaweeds coated with earth and sand, which results in agars with higher gel strengths and an increased yield. The outside cell walls of the unrefined seaweed plants could be acting as protective cell membranes, thus minimizing the hydrolysis of the extracted agar.

Tagawa[35] investigated the mechanisms of alkali treatment of *Gracilaria* by which the content could be converted to agar of much higher gel strength than that of the original substance. He concluded that (a) there was an increase in agarose content and a reduction of agaropectin content in the seaweeds, and (b) the independent increase in gel strength of both agarose and agaropectin fractions was the responsible mechanism. The sulphate contents in agarose fractions were also reduced. These results suggest the conversion of L-galactose-6-sulphate moieties to 3,6-anhydro-L-galactose moieties within the alkali-treated seaweeds as the most likely reactions,[36] as in the case of carrageenan.[60,61]

3.f Extraction by Use of Polyphosphates and Others

The use of polyphosphates, or condensed phosphates, is effective for the extraction of agar from rigid-type agarophytes.[62,63] The extractability of agar is better with phosphates of higher degree of condensation. Sequestration of ionized iron in the extracts is another effect of polyphosphates in producing agar of fine whitish colour.[64,65]

Uses of cellulase as a pretreatment of seaweeds,[66,67] gamma-irradiation of agar seaweeds,[68] and an extraction method in aqueous ammonia media[10] showed some effects on extractability of agar, but are not industrially applied.

3.g A New Simplified Agar Extraction Method by 'Acid Pretreatment' of Agarophytes

This patented extraction method[34,69] is based on two experimental observations. One is that agar in an air-dried state is stable even in a strongly acidic solution of pH 1·0 provided that the temperature is kept below 20°C, preferably below 15°C.[70] The other is that agar is contained in

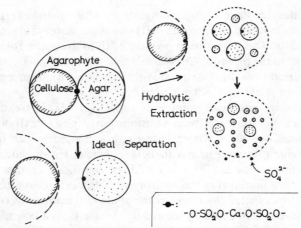

FIG. 4. A concept for extraction of agar by Matsuhashi.[34,71]

seaweed tissue, not in a free form but in a combined form with non-agar substance (see Fig. 4).

Therefore, the acid pretreatment of seaweeds could proceed as follows:[34,70,71]

$$Ra-O-SO_2-O-Ca-O-SO_2-O-Rn + 2HCl \rightarrow$$
$$Ra-O-SO_2-OH + Rn-O-SO_2-OH + CaCl_2$$

where Ra is agar and Rn is non-agar substance such as cellulose. After cleavage of bonds between agar (Ra) and non-agar (Rn) substances in seaweed tissue, only boiling in neutral aqueous solution is safe for extracting the agar. In view of this, some of the reported data on agar, and on its chemical structure and physicochemical properties, are doubtful because it is possible that the experimental materials used were degraded.

3.h Filtration of Hot Sol and Dehydration of Gel

Filtration of the hot sol extract is as important a process as agar extraction and dehydration of gel in agar manufacture. A process of filtration through roughly meshed cloth bags is commonly adopted by most of the traditional Japanese factories. It is a primitive filtering method, yet effective.[10,14] The most advanced agar factory is equipped with an automatically controlled filter press to produce physically purified agar products efficiently; filter-aid silicate is generally used.

Freezing of agar to dehydrate the hydrogel is important, historically and industrially.[72,73] Natural freezing combined with weather drying has been

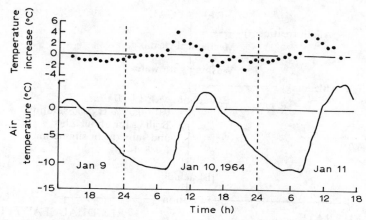

FIG. 5. The pattern of the changes of the air temperature in the process of natural slow freezing of agar gel. (From Matsuhashi.[73,74])

applied in processing 'bar-style agar' ('Kaku-kanten' in Japanese) and 'stringy agar' ('Hoso-kanten' in Japanese). Figure 5 shows a typical temperature pattern of winter weather in Chino city (Nagano Prefecture), which is located at an altitude of 800 m in the centre of the Main Island of Japan. A field experiment in Nagano[74] revealed that the proportion of water eliminated from frozen agar gel depends on sublimation for 48%, on common vaporization for 12% and on defrosted drips for 40%. The locational advantage in respect of natural heat energy for sublimation alone was roughly estimated to be equivalent to 200 kg of fuel oil per day, for one factory.[73,74] In the warmer areas, defrosted drips are quantitatively more responsible for water elimination. Mechanical freezing systems are applied to some of the more industrialized factories, where 'flakes of dry agar' are generally produced.

Freezing dehydration, whether it is mechanical or natural, is applicable to extracts of *Gelidium*-type seaweeds for which pressing dehydration is not effective. A theoretical energy of 150 kcal is required for freezing 1 kg of agar gel provided that the starting and end temperatures are 30°C and −10°C, respectively.[72]

The general flowchart for the conventional agar processing is illustrated in Fig. 6.[14,73]

Spray drying of hot agar sol has been shown to be a practical processing method,[75] but has not yet been adopted by the agar industry.

The methods for preparing agarose are well described by Guiseley.[9,36]

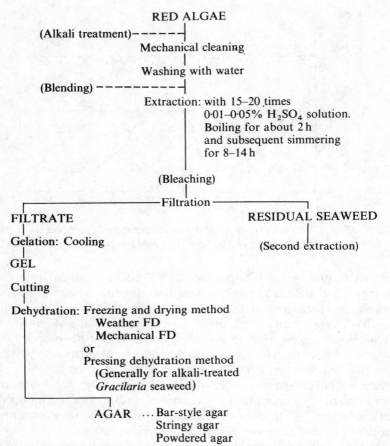

FIG. 6. Flow chart for agar manufacture by old-type processing method. (From Matsuhashi.[14,73])

4 COMMERCIAL AVAILABILITY

Japan continues to be far ahead of any other country in the production of agar. However, agar production in foreign countries is showing a gradual increase.

4.a Types of Product

Bar-style agar ('Kaku-kanten'). The weight of one piece is 7·5 g on average. Owing to its honeycomb-like structure, the bulk density is about

FIG. 7. (Left) stringy agar ('Hoso-kanten') and (right) bar-style agar ('Kaku-kanten').[13]

0.030–0.036 g/cm^3.[130] For over half a century production has been limited to the Nagano Prefecture, where peak production of 1245 t was recorded in 1939–40 agar-year.[76]

At present yearly production is maintained at the level of around 600–700 t. It is considered to be the oldest type of agar and was first produced in the Kyoto–Osaka area. Almost all of this product is for domestic use: for confectionery and for retail (home cooking), etc. Bar-style agar is traded in cuboidal packages of about 1 m^3 that contain 600 pieces of bar-style agar and weigh on average 4·5 kg. A small amount of coloured products (red or green) is also sold. For retail, one piece or a bag containing two pieces is the marketing unit in Japan.

A factory in the Philippines recently produced bar-style agar for their domestic use. In this case, mechanical freezing and sun-drying are used. It could be easily distinguished from the Japanese product, which is required to have a fine artistic shape! (Fig. 7).

Stringy agar ('Hoso-kanten'). The normal length of the agar string is 28–36 cm, although there are no definite required dimensions. The wholesale trading unit is a bundle of 15 or 30 kg net weight. Retailers sell in small quantities, as in the case of bar-style agar. For exporting overseas, more densely packed bundles are shipped.[14]

Gifu Prefecture is the biggest production district for stringy agar in Japan, although the other agar production areas, where the weather freezing-and-drying method is applied, also produce small amounts of this type of agar.

A few of these factories are trying to produce stringy agar in the summer season by using mechanical freezing equipment. The yearly stringy agar production in Japan is between 500 and 600 t.

Korea is the world's second largest producer of stringy agar using a similar process to that used in Japan, mostly from *Gelidium* species.

Agar flakes and powdered agar. 'Flakes' (or coarse powder) are normally produced by the freezing process, while most of the 'powdered agar' (or fine powder) is processed by the pressing dehydration method (or non-freezing process). In Japan and the rest of the world, the latter type of product has become dominant among the 'industrially (or chemically or mechanically) processed agars', which is the synonym of 'powder agar'.

From the viewpoint of users, it should be noted that flake agar (or agar by the freezing method) is generally processed from *Gelidium* species, while powder agar (or agar by the pressing method) is most usually processed from alkali-treated *Gracilaria* species, with very few exceptions.[12,13,57]

Since more and more countries have become agar producers, it is difficult to establish the exact total of agar production in the world. However, the following are the main countries other than Japan that produce powdered agar and/or agar flakes:

(a) Morocco, Portugal, Spain and France, all facing the Atlantic;
(b) South Africa;
(c) Chile and Argentina in the South Americas;
(d) Korea, Taiwan and India in south-east Asia;
(e) England, Australia, New Zealand and the USA are among the earliest agar-producing countries, but produce rather small quantities for their own domestic demands.

In Japan, the biggest powder-agar factory has the ability to produce 200–300 t of agar per year, but the total industrial agar producers, which once exceeded 30 in number, have been reduced to around 10.

4.b Grade and Price

In Japan, each of three types of agars are graded by the Japan Agricultural Standard (JAS), for both domestic and foreign trade. The present standards represent the historical requirements of an age of governmental economic

TABLE 4
THE JAPANESE AGRICULTURAL STANDARD FOR POWDER AGAR[a]

	Grade			
	Superior	1st	2nd	3rd
Gel strength (g/cm^3)	600 or over	350 or over	250 or over	150 or over
Insolubles in hot water (%)	Less than 0·5	Less than 2	Less than 3	Less than 4
Crude protein (%)	Less than 0·5	Less than 1·5	Less than 2	Less than 3
Crude ash (%)	Less than 4	Less than 4	Less than 4	Less than 4
Moisture (%)	Less than 22	Less than 22	Less than 22	Less than 22

Organoleptical qualities are omitted in this table.
[a] The official term is 'special type agar' rather than powder agar or agar flakes.

control (when the powdered agars were called 'special type of agar'!), and are not necessarily fitted to our current scientific understanding. Therefore, possible gaps between the required physicochemical properties in practical uses and conventional quality evaluations based on official standards should be borne in mind by potential agar users.

The Japanese Agricultural Standard for powdered agar is shown in Table 4.

Prices of agars vary considerably with time of year and with quality. The drastic changes in the international exchange rates in 1986–87 make it difficult to give realistic current agar prices. However, as an indication, the price (excluding customs duty) of agar imported into Japan from several countries, such as Chile, Morocco, Taiwan and Korea, in 1985 was around 13 $US/kg for quantities of around 380 t. This compares with 67 $US/kg (excluding customs duty) for 3-t quantities imported from the USA for bacteriological agars and agaroses for chemical use.[77]

Japan is primarily an agar-exporting country to the rest of the world; she also imports some agar, but far less than the amount she imports as agar seaweeds.

4.c Possible Contamination

It is unlikely that commercial agar will be contaminated with foreign materials; however, contamination with filter aid (silicate) may possibly occur in powdered agar. The residual levels of chemicals such as bleaching agents may be indirectly checked by measuring pH, tolerance to heating, etc.

In the old days, adulteration by mixing of other water-soluble gums was rumoured for powdered agar but never for Japanese bar-style agar and stringy agar, owing to the unique shapes of these products.

5 SOLUTION PROPERTIES

5.a Solubility

Agar is not soluble in cold water but is soluble in boiling water. In the strict sense of the word, merely heating the water, if it is kept at a temperature below boiling point, does not bring perfect dissolution. When using bar-style agar, stringy agar and agar flakes, soaking in cold water beforehand, preferably overnight, greatly assists in achieving full dissolution. Even when using so-called 'quickly dissolvable agar', at least 5–10 min soaking is recommended.

A recent report of warm-water-soluble agars (which had been modified by either freeze-drying or drum-drying)[131] needs further work in order to confirm it.

Control of pH in the course of soaking and boiling is vital to ensure that the solution always remains neutral. Failure to observe this simple precaution has long been the cause of problems for many scientists, manufacturers and users.

Agar contains small amounts of cold-water-soluble substances that are

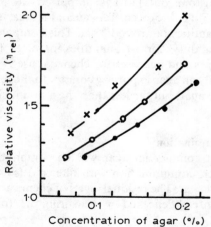

FIG. 8. The dependence of the relative viscosity of agar sol on its concentration. ●, Ogonori agar; ○, Itani agar; ×, Tengusa agar. (From Kojima et al.,[53] reproduced by kind permission of Dr Y. Kojima.)

It is interesting in the report by Forsdike at the Pharmaceutical Society of Great Britain in 1950[78] to see the general misconceptions held at that time about the identity and purity of agar.

not responsible for its gelling ability, the content varying with type and grade of agar.

It should be noted that the maximum concentration obtainable by the normal dissolution procedure will be 3–4% in water.

5.b Viscosity Profiles of Sol

Agar has the ability to form gels at very low concentrations. Therefore, viscosity measurements can really only be made on agar solutions that are above their gelling temperature, i.e. >40°C. Since agar is normally used as a gelling agent, not primarily to increase solution viscosity, its viscosity behaviour is mainly of interest in establishing processability.

The relationship between concentration and relative viscosity (η_{rel}) for three types of agar is shown in Fig. 8.[53,79] The variation between the relative viscosity of seven kinds of typical Japanese agars was shown to be 1·2–3·5 (η_{rel}) at a concentration of around 0·2%.[79] Arrhenius plots (Fig. 9) show a linear relationship between $\log \eta_{rel}$ and concentrations in the range of 0·06–0·2% sol. Therefore, for $\log \eta_{rel} = KC$, the values of K were 0·90 for Ogonori-agar, 1·12 for Itani-agar and 1·22 for Tengusa-agar, respectively.[53]

The viscosity of high concentrations of agar, e.g. 0·8–4·1%, can be measured using a rotational viscometer such as a Brookfield.

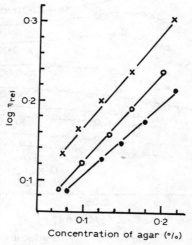

FIG. 9. The relation between the logarithm of relative viscosity and the concentration of agar. ●, Ogonori agar; ○, Itani agar; ×, Tengusa agar. (From Kojima et al.,[53] reproduced by kind permission of Dr Y. Kojima.)

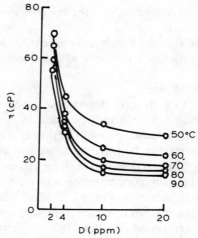

FIG. 10. Relative viscosity of agar sol (1·2% or 1·6%) versus rotation velocity (D) at various temperatures. (From Fuse and Goto,[80] reproduced by kind permission of the Japanese Society of Agricultural Chemistry.)

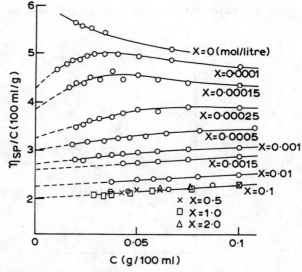

FIG. 11. Reduced viscosity curves of agar in aqueous potassium chloride at different ionic strengths (X = concentration of added KCl in moles (litres)). (From Sakamoto and Kishimoto,[81] reproduced by kind permission of the Japanese Society of Scientific Fisheries.)

The effect of shear rate and temperature on the viscosity of a typical agar is shown in Fig. 10.[80]

For the same agar there is a correlation between the melting point of 1·5% gel and the relative viscosity of a 0·2% sol at 50°C.[55,79] Also, the relative viscosity correlated with Tanii's C_c/C_r value (which expresses a gel property of agar) for a number of different agars (provided that each of the indexes were measured at the same concentration levels). This means that agar solutions of higher viscosity (at the same concentration) generally make firmer gels.

The presence of ions tends to reduce the viscosity of an agar solution, the magnitude of the effect depending on the level and type of ion used, see, e.g., Fig. 11.[81]

5.c Setting Point of Sol

The setting point of sol (or temperature of gelation) increases with increasing agar concentration.[53,79] At a constant concentration, the higher

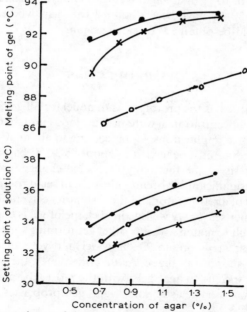

FIG. 12. The dependence of the melting point of gel and the setting point of sol on the concentration for three different agars. ●, Ogonori agar; ○, Itani agar; ×, Tengusa agar. (From Kojima *et al.*,[53] reproduced by kind permission of Dr Y. Kojima.)

the setting point of the sol, the higher the rigidity coefficient of the gel will be.[79] For 15 kinds of agar, the setting point of a 1·5% sol was between 28·3 and 37·3°C (Fig. 12).

The gelling temperature of agarose becomes higher with increase in methoxyl content.[82] Gelation also occurs at higher temperatures when slower cooling rates are used, for example an agarose with a dynamic (e.g. cooling at a rate of 0·5°C/min) gelling temperature of 36°C exhibits a static gelling temperature of 40–42°C.[36]

5.d Colloid Titration of Agar Sol

Indirect colloid titration of agar developed by Senju and Tanii[83–85] is a useful analytical method for establishing quickly the presence of agar. Five millilitres of about 0·05% agar sol is enough for the titration. The reagents used are the potassium salt of polyvinyl sulphate (PVSK) and N-methylated glycol chitosan iodide (MGC), and the indicator is a 0·1% solution of toluidine blue (TB).

A PVSK value of 10^5/g of dry agar is equivalent to a so-called 'E' value of 1.[83] E values of 30–90 have been obtained for agar, whereas a value of around 300 would be obtained for carrageenan.[49,83,86]

6 GEL PROPERTIES

Agar can be regarded as the prototype and model for all gelling systems.[60] Typically, 1/100 concentration in water will be good enough to make a rigid gel suitable for most applications. Upon cooling to about 30–40°C, the sol sets to a firm gel that must be heated to about 85–95°C before it will melt. Thus, the hysteresis of the thermally reversible changes from gel to sol phases of agar is significant,[136] because the sol can set at room temperature without the need of a refrigerator and the resultant gel is stable even during the hottest summer season. A well-defined shape of gel is characteristic of agar that is difficult to match with any other gel-forming substance. A gel of good quality agar may contain 99·9% water and yet still be capable of retaining its self-supporting shape. As described later, the term 'brittle' is not an adjective specific to agar gels; with the most advanced technology an agar that has very elastic and rigid gel properties can also be obtained.[11,29,54,55]

It is worth reminding the reader what an agar gel is. Confectioners usually have a good understanding of how to prepare an agar gel, but novices in a scientific laboratory may have quite wrong ideas. The preparation of a 'gel' for 'gel chromatography' by merely mixing with water will only make a 'false gel' of the polymer. It is essential to boil the agar at neutral pH in order to make a 'true gel'. Rapid cooling of the hot sol and moving of hot sol vessels should be avoided in order to produce firm gels.

The dynamic properties of agar gels can be expressed popularly in three different ways: 'gel strength', 'apparent gel strength' and 'strength of gel'.

6.a Gel Strength

The meaning of 'gel strength' varies between countries, between research workers, and with the type of gel-forming substances used. Fortunately, the gel strength of Marine Colloids[87] and the gel strength of Meer Corp.[88] are similar to one another and to the Japanese gel strength defined by JAS, although there are minor differences among them that should be borne in mind.

6.a.1 Gel Strength of Marine Colloids[87]

Gel strength in the case of agarose is defined as that force, expressed in grams per square centimetre, that must be applied to a 1% agarose gel to cause it to fracture. Controlled conditions are used to determine this value. After preparation of the gel, formed in a standard 50×70 mm crystallizing dish, it is held for 2 h at 10°C; then the break force is measured using a constant-speed drive gel tester. The Marine Colloids' Gel Tester (Springfield, New Jersey) consists of a spring balance at the bottom and a motor driving a downward plunger fixed overhead. Gel strengths of 200–400 g/cm^2 (1% gel) are quite common for agar, whereas values for good-quality agarose are generally 600–1000 g/cm^2.[87]

6.a.2 Gel Strength of Meer Corp.[88]

The tester is a modification of a two-pan balance. Gel (about 200 g) in a weighed 400-ml beaker is placed on one of the pans. A 1-litre beaker is placed on the other pan, and weights are added to counterbalance the weight of the gel sample. A controlled rate of water running into the 1-litre beaker will force the gel to be broken by a statically held plunger of 1 cm^2 area, which is set against the top surface of the gel. In this measuring system, 3·00 g of agar is first dispersed in 230 ml of water and boiled to dissolve it. The solution-containing beaker is kept in a 25°C water bath for 22 h, to make a gel of around 200 g net weight.

6.a.3 Japanese Agar Gel Strength[10,62]

Gel strength is the maximum cracking load (g/cm^2), after 20 s duration, on the top surface of the tested gel. The gel is in a 200-ml mould of about

38 mm depth and is measured using the Japanese Agar Gel Tester (Kiyaseisakusho, Tokyo), which is equipped with a cylindrical plunger having a flat bottom area of 1 cm². In the commercial evaluation of agar quality, gel strength of a 1·5% agar gel as defined by the Japan Agricultural Standard is widely used.

Preparation of gel based on JAS. Weighed agar (3·00 or 9·00 g) is put into a 1-litre flask, moistened well with a given amount of water (197 or 591 ml) under a reflux condenser and boiled for 30 min for dissolution. The hot solution is poured into a stainless-steel mould (200 or 600 ml), which is kept in a 20°C thermostatic cabinet for 15 h. According to the JAS specification, the concentration of gel should be 1·5% on a dry-matter basis.[8,9] Weighing a definite amount of absolutely dried agar is not very practical. If necessary, the moisture in the agar sample can be measured, or use can be made of the relationship between 'apparent gel strength' and concentration (see later).[90] For most testing it is normal to disregard the moisture content of agar, which is normally about 20%, or at least 10%,

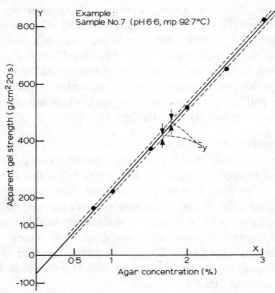

FIG. 13. The relation between 'apparent gel strength' (Y) and agar concentration (X): regression line of Y on X. $S_y^2 = s^2(1 - r^2)$ (S_y = error by estimation, based on the regression line; s = standard deviation of y; r = correlation coefficient). $a = 291$; $b = -67$; $-b/a = 0.23$; $-100(a/b) = 430$; $S_y = 8.1$; $r = 0.99_9$. (From Matsuhashi.[90])

and also to round the water volume used to 200 or 600 ml. Correction tables can be used to allow for the effects of water evaporation and temperature of gel between 6 and 32°C.[91]

Significance of gel strength. The Japanese Agar Gel Tester is inexpensive and the test is simple and reliable, with an accuracy of 10 g unit. The tester is based on the traditional finger test on the agar gel surface developed in the old prosperous days of the Japanese agar industry. The time duration of 20 s is unique for agar gel and consequently the gel strength should strictly be expressed as (g/cm^2 . 20 s). However, in normal commerce, gel strength is often given simply in g/cm^2 without causing confusion.

Instruments such as the Curd Meter (Iio-denki, Tokyo[24,92]), GF Texturometer (Zenken, Tokyo[49,93,94]) and Rheometer (Fudo-kogyo, Tokyo[86]) can be used in place of the conventional agar gel tester.[95]

6.b Apparent Gel Strength

'Apparent gel strength'[90] means a value at a non-defined concentration. It is differentiated from 'gel strength of agar', which is defined only for 1·5% gel. Linear relationships between apparent gel strength (Y) and concentration of gel (X) were confirmed in a wide range of concentrations for a number of agars.[53]

Figure 13 shows a typical linear regression line, which is expressed by the following equation:

$$Y = aX - b$$

where a and b are the constants for individual agars. The value of $-b/a$ should theoretically correspond to the minimum concentration (in per cent) of agar that forms a gel.[7,90]

6.c Strength of Gel

'Gel strength' is a very popular indicator and defines the quality of agar gel in one respect. But it does not cover all of the physical properties and gelling properties of agar. The concept of 'strength of gel', established by Tanii[29] in 1957, consists of two parameters for the physical properties of agar gel.

In the first step, Tanii measured the rigidity coefficient (T/θ) and the breaking (or cracking) load (T_c) of agar gels by use of an improved torsion dynamometer developed by Sheppard *et al.* around 1920 for gelatine gel. The notation used is T = twisting moment in grams, θ = angle of twist in degrees and T_c = breaking load in grams that caused a crack of about 2 mm in the moulded gel. The gel mould used by Tanii was a cylinder having an inner diameter of 20 mm, a total height of 70 mm and a movable part of 20 mm height.

In the second step of his study, by using a numerical relationship between each of the physical properties and the concentration of gel, the properties of any sampled gel were expressed relative to those of the standard agar as follows: C_r related to rigidity coefficient and C_c related to cracking load. $C_{r.s}$ related to the rigidity coefficient of a 1% gel. From his data Tanii proposed that:

(i) C_c/C_r was independent of concentration; when the gel is viscoelastic, the value of C_c/C_r should be high, whereas a low value means a more brittle gel.

(ii) When the purity of the agar gel is high, the value of $C_{r.s}$ will also be high.

For 16 commercial agars, two of the parameters showed values of 1·00 to 1·50 for C_c/C_r and 0·64 to 1·32 for $C_{r.s}$. For experimentally processed agars from 37 kinds of seaweed, C_c/C_r and $C_{r.s}$ values were 1·63 ± 0·12 and 0·92 ± 0·04 for *Gelidium* agars (average of 22 species), 2·28 and 0·84 for a *Ceramium* agar, 1·72 and 0·57 for a *Gracilaria* agar, and 2·70 and 0·11 for a *Eucheuma* extract, respectively.

Tanii also claimed that the C_c/C_r and $C_{r.s}$ of agar gel could be deduced by estimation of the melting points at several concentration levels.

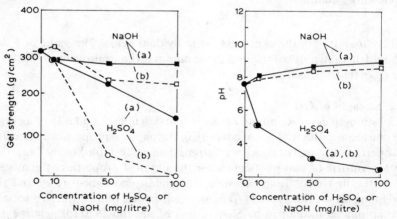

FIG. 14. Effects of sulphuric acid and sodium hydroxide on gel strength and pH: (a) acid or alkali was added into hot agar solution immediately after cessation of boiling; (b) agars were dissolved in boiling water containing acid or alkali. (From Matsuhashi.[11])

6.d Effects of Acid and Alkali on Gel Strength

Figure 14 shows the effects of sulphuric acid and sodium hydroxide on gel strength of agar.[11] Sulphuric acid acts on agar more vigorously than does sodium hydroxide. Furthermore, dissolution of agar in boiling aqueous acid solution will cause significant degradation. Figure 15 shows the effects of sodium carbonate on gel strength of agar.[11] The addition of 10 mg/litre Na_2CO_3 to the slightly acidic agar improved gel strength.

The gel strength (or gel property) of agar is most stable at about pH 7·0 (or at a pH a little over 7·0). Therefore, if any kind of acidic substance is required to be added to the gel preparation, it should be added after the agar is dissolved and at a temperature lower than boiling point.

The relationship between amount of digested agar (D, per cent) and temperature of solution (T, Kelvin) for a 15-min soaking in 1N HCl is approximately as follows:[70]

$$\log_{10} D = (1/25)T - 11 \cdot 52$$

For example, $D = 100$ at 338 K (65°C) and $D = 0.01$ at 288 K (15°C).

Both reducing and oxidizing bleaching chemicals such as sodium dithionite and sodium hypochlorite, which may be used in the processes of agar manufacture, affect gel strength in a similar way to acid.

6.e Melting Point of Gel

To measure the melting point, 5 ml of agar solution is gelled in a glass test tube of 10 mm inside diameter and 100 mm length. After holding at 20°C overnight, the glass tubes are turned over and placed in a water bath, which

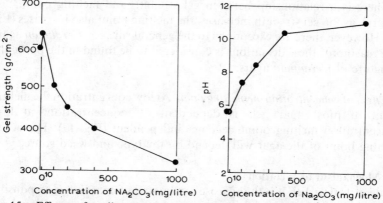

FIG. 15. Effects of sodium carbonate on gel strength and pH. (Agars were dissolved in boiling water containing sodium carbonate.) (From Matsuhashi.[11])

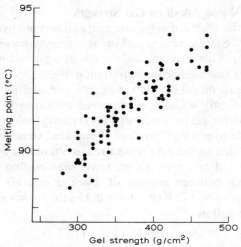

FIG. 16. Correlation between gel strength and melting point of gel for agars from one factory ($n = 78$, $r = 0.78$). (From Matsuhashi.[96])

is heated to boiling point at a rate of 0·5–1·0°C rise per minute. The rate of heating should be reduced to about 0·2°C/min when the temperature of the bath reaches about 5°C below the expected melting point, in order to obtain thermal equilibrium. The temperature at which the gel in the glass tube suddenly drops is recorded as the melting point of the gel at that concentration.[62] This measurement is time consuming; however, it is accurate (generally $\pm 0.2°C$) and reproducible.[29]

There is a relationship between gel strength and melting point.[96] In general, as the gel strength increases, the melting point also increases (Fig. 16). However, there are exceptions to the general rule, e.g. *Ceramium* agars. The reason for these deviations is considered to be found in the particular structure of *Ceramium* agars.[14,17,97]

Effect of concentration on melting point. At low concentration the melting point of most agar gels is dependent on concentration; at high concentration melting point becomes independent (Fig. 17). The actual melting point of the agar will depend on the type and weed source.[54]

6.f Mechanism of Gelation

As described earlier, it is the agarose component of agar that is responsible for gelation. At elevated temperatures agarose exists as a disordered 'random coil', but on cooling it forms strong gels at low polymer

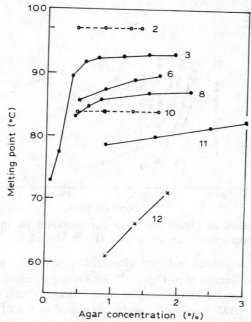

Fig. 17. Dependence of melting point of gel on agar concentration for various kinds of agars. (From Matsuhashi.[54])

concentration. The gelation process involves the adoption of the ordered double-helix state.[27,28,98,99]

Formation of a gel network rather than a solid precipitate can be traced[100] to the presence in native agarose of occasional 1,4-linked residues in which the anhydride bridge is absent. This allows the sugar ring to adopt the normal 'chair' shape rather than the alternative higher-energy chair form necessary for the closure of the anhydride ring. This consequently changes the overall residue geometry to a shape that cannot be incorporated within the double-helix structure. The occurrence of such 'kinking' residues thus terminates helix formation, and allows each chain to participate in further interchain functions with other agarose molecules and thus build up a three-dimensional gel network. Each ordered junction zone involves between 7 and 11 double helices.[101,102]

6.g Heat Energy Required to Melt the Junction Zones

This method was first developed by Eldridge and Ferry[103] for gelatine gels and applied to agar gels by Tanii.[29] The curves of dependence of melting

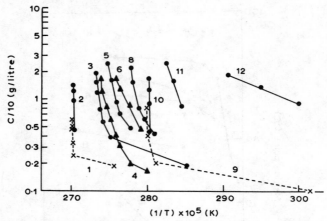

Fig. 18. Dependence of logarithmic agar concentration on inverse value of absolute temperature of melting. (From Matsuhashi.[54])

point on concentration in Fig. 17 are converted to curves of $\log_{10} C$ (g/litre) vs $1000/T$ (K^{-1}), as illustrated in Fig. 18.[54] According to Eldridge and Tanii, the 'heat energy required to dissociate cross-linkage of gel', or to melt the junction zones ($-\Delta H°$ kcal/mol), can be calculated by use of the following equation, where R is the gas constant:

$$\log_{10} C = \frac{\Delta H°}{2 \cdot 303 RT} + \text{const}$$

which is further converted as follows:

$$\Delta H° = \frac{k \log_{10}(C_1/C_2)}{(1/T_1) - (1/T_2)} \qquad k = 2 \cdot 303 R$$

The $-\Delta H°$ values are obtained from the linear parts of the individual slopes, which straighten up at the relatively higher concentration region in Fig. 18. The values of $-\Delta H°$ for six kinds of common commercial agars were 79–351 kcal/mol of cross-links, which agree with Tanii's data of 70–260 kcal/mol for seven kinds of typical agars,[29] and Tagawa's data of 214 kcal/mol for his experimentally processed agarose.[35] The corresponding value for a kappa-carrageenan was smaller than 1/30 of that for an agarose.[132]

It is worth remembering that the organoleptic perception of firmness of gel does not correlate with gel strength alone; it does correlate with the $-\Delta H°$, such that the larger the value, the firmer the sensory perception. The

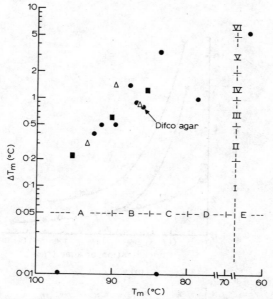

FIG. 19. T_m–ΔT_m diagram: a new proposed classification of gel property of agar. (From Matsuhashi.[11])

$-\Delta H°$ value is dependent on the processing or extraction conditions as well as the seaweed source.[54] The highest $-\Delta H°$ value for agar ever obtained by the author himself exceeds 2000 kcal/mol.

6.h T_m–ΔT_m Diagram: A New Practical Evaluation of Gelling Property

It has been found that the melting point difference between 1·0% and 1·5% concentrations (ΔT_m) is inversely correlated with the calculated heat energy to dissociate cross-linkage of gel ($-\Delta H°$). Also, there is a correlation between gel strength and melting point (T_m) of gel for agars of the same origin.[96]

On the basis of these experimental facts, a new simplified evaluation for agar gel has been developed,[11,104,105] the '$T_m - \Delta T_m$ diagram', which is illustrated in Fig. 19. It assumes that the value of T_m (abscissa) depends on species and types of agarophytes, and the value of ΔT_m (ordinate) depends on processing conditions of agar in which the extraction process is the most decisive factor.

Each of the axes can be subdivided to make a practical classification of the gelling property of agar. For example, the Difco agar was ranked as

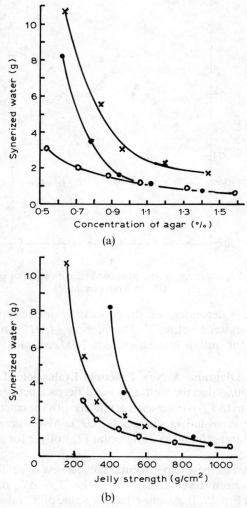

FIG. 20. Syneresis of agar gel: (a) amount of synerized water versus concentration of gel; (b) amount of synerized water versus apparent gel strength. ●, Ogonori agar; ○, Itani agar; ×, Tengusa agar. (From Kojima et al.,[53] reproduced by kind permission of Dr Y. Kojima.)

B-III class, that of Japanese stringy agar as A-II, that of *Gelidium* agar as A-I, that of *Ceramium* agar as C-I, and that of bar-style agar of Argentinian *Gracilaria* as D-IV.

6.i Syneresis of Gel

An agar gel in a closed vessel will eliminate a certain amount of water at its surface with time. This phenomenon is called syneresis and results in volumetric shrinkage of the gel. The phenomenon will be encountered with food gels, bacterial media and intermediate products of agar during processing. Where food gels are concerned, this phenomenon is nicknamed 'Naki' in Japanese ('tear' in English) or 'Ase' (corresponding to English 'sweat') among confectioners and other users.

The basic factors that influence syneresis of agar gels are listed below.[53,79]

(i) *Concentration of gel.* Low concentration gels eliminate the larger amount of synerized water. Very roughly, the syneresis of agar gels becomes significant at around 1% or lower concentrations (Fig. 20(a)). Approximately, the amount of synerized water is inversely proportional to the square of the concentration for most practical concentrations.[79]

(ii) *Holding time of gel.* In an experiment using Ogonori (*Gracilaria*) agar gel, syneresis almost reached an equilibrium state in 72 h at 30°C; 144 h is recognized as the time to reach practical equilibrium[53] (Fig. 21).

(iii) *Apparent gel strength.* The higher the apparent gel strength, the lower the syneresis (Fig. 20(b)).

(iv) *Rigidity coefficient* ($C_{r.s}$). For a variety of agars the amount of synerized water was inversely proportional to the rigidity coefficient ($C_{r.s}$).[79]

(v) *Pressurization.* For most agar gels the application of pressure increases syneresis; however, exceptions have been reported, even though the gels showed the same rigidity coefficient.[79] The conditions used for the extraction of agar are important in influencing whether pressure affects syneresis.

(vi) *Total sulphate content.* The amount of synerized water is generally inversely proportional to the total sulphate content. This is one reason why the pressing dehydration step is applicable to the alkali-treated seaweeds in agar manufacture; gels from agarophyte not processed in alkaline medium previous to extraction hardly eliminate water even under pressure.[12,13,57,79]

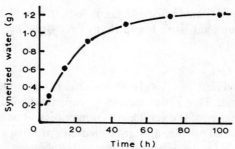

FIG. 21. Syneresis of agar gel: effect of time. (From Kojima et al.,[53] reproduced by kind permission of Dr Y. Kojima.)

6.j Turbidity (or Clarity) of Gel

Agarose sols may be absolutely colourless and perfectly clear. In contrast, agar sols are usually somewhat hazy and may be yellowish, and become cloudier upon setting.[106] Agar gels are usually opaque and hazy, while agarose gels are usually much clearer. As pointed out by Guiseley, it is generally possible to read six-point type (six-point type) through 5 cm of a 1% agarose gel, a fact seldom possible with a 1% agar gel.[9]

6.k Irreversibility of Agar Gel by Freezing

Gel and sol phases of agar are thermally reversible at temperatures above 0°C, or freezing point. Once frozen, however, the agar gel collapses, and does not recover its gel phase when thawed. This is one of the fundamental characteristics of agar, which contrasts with carrageenan. However, if boiled with water to dissolve and cooled once again, a gel can be obtained with little or no changes in gel properties.[12,72]

Owing to its higher water content, 98–99% or more, the freezing point of agar gels in general use is very close to 0°C.

When food-seasoning substances such as sugar and salt are added, or when other mucilaginous substances such as carrageenan and other agaroids are mixed with the agars, a gel phase may be obtained after thawing. In such a case the recovered gel phase can differ from the original gel owing to syneresis and a volumetric shrinkage of gel.

With a larger sulphate content, and a larger $-\Delta H°$ value, agar will show more carrageenan-like behaviour in the freezing and thawing processes.

6.l Effects of Addition of Other Materials on Gel Properties

6.l.1 Effects of Inorganic Salts

Iwase[107] investigated the effects of 1N inorganic salts on the time required to set 0·3–0·5% agar sols. Among 11 kinds of salts, potassium sulphate accelerated gelation most remarkably. The time for gelation of a 0·5% sol was 5·0 min without any additives, whereas it was shortened to 1·25 min with K_2SO_4 and to 3·25 min with NaCl. In summary, the accelerating effects of the individual ions on gelation of agar were in the following orders:

$K^+ > Na^+$ and $SO_4^{2-} > CH_3COO^- > C_4H_4O_6^{2-} > Cl^- > Br^- > NO_3^- > I^-$

where Br^- and the other two anions written to the right rather retarded gelation.

In another report, Fe^{2+} and Fe^{3+} at 1 ppm concentration level in water were found to reduce transparency of 1% agar gel.[108]

For practical purposes, 0·5–5% sodium chloride (or table salt) when added to 1% agar sol gives only a slight increase in 'gel strength' over that of the control. Even at the most effective peak point of 2% salt concentration, gel strength remained less than 110% relative to that of the control.[109]

6.1.2 Agar–Sugar Gel

The effect of added sucrose on the rheological properties of 1% agar gel was investigated,[110] using sugar concentration in the range 0–75%. Commercial powder agar was used, and all of the gels were prepared to ensure exactly 1% concentration of agar as well as to maintain the selected sugar concentrations. Each of the rheological parameters increased with increasing concentration of sugar to reach maxima. Hardness (N/m^2) by plastometer and gel strength (N/m^2) by curdmeter showed a maximum at a sugar concentration of 60%, while elasticity (per cent) by plastometer showed a maximum at a sugar concentration of 65%. The value of gel strength at the maximum was over twice that of the control.

The effects of glycerin and sorbitol are similar to that of sucrose.[110,111]

The addition of sucrose noticeably increases the transparency of agar gels.[106]

6.1.3 Effects of Milk, Fat and Protein

Milk, soy milk, butter, soy oil and casein can be mixed at reasonable concentrations with agar sols to form gels. Gels containing high levels of any of these materials will be weakened in gel strength, but, in a typical formulation, the effects should be negligible.[109]

6.1.4 Compatibility with Other Macromolecular Gums

Locust-bean gum (LBG) is one of very few exceptional substances that are capable of strengthening an agar gel.[88,109,112] For example, a formulation of 10% LBG and 90% agar had an increased gel strength of about 10%.[109]

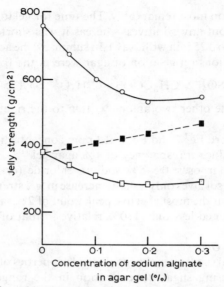

FIG. 22. Effect of combined use of sodium alginate and calcium chloride on gel strength of cubic agar gel (15 mm^3). ○, Agar gels of standard size; ■, after soaking in 0·1M calcium chloride solution; □, no treatment. (From Kojima et al.,[113] reproduced by kind permission of Dr Y. Kojima.)

However, LBG does not always increase firmness of agar gel according to the author's observations.

Kojima et al.[113] developed a technique to reinforce agar gel in canned 'mitsumame', in which 0·1–0·2% sodium alginate was added to 1·5% agar gel and soaked in 0·1M calcium chloride solution at 25°C for 20 min (Fig. 22). The use of carboxymethylcellulose (CMC) was also investigated by these authors.

The appearance of agar gel mixed with other gums does not tend to change much; however, the physical strength of the agar gel will generally be reduced, a fact that could be exploited in new food applications of agar gels.

In agarose–gelatine gels, the two hydrocolloids do not form a combined network structure, and at relatively higher concentrations they interfere with each other.[88,109,112,114]

The scanning electron microscopic observations by Watase and Nishinari[115] for mixed gels composed of 2% agarose and 5% gelatine support the mutual interference of phases in such network structures of gel. The rheological data from measurements of stress relaxation and dynamic viscoelasticity by the same workers[115,116] should be treated with caution by food technologists because of the abnormally concentrated gels of '2% agarose and 15 (or 30)% gelatine' gels used, which would not normally be encountered within the food industry.

6.l.5 Agarose–Agaropectin Gel

In a study on artificial agar composed of experimentally processed agarose (AG) and agaropectin (AP), Tagawa reported two facts.[35] In the compositional ranges of AG 100%–AP 0% to AG 20%–AP 80%, gel strength decreased proportionally to AG content, and when the gels contained less than 10% AG they showed no recognizable gel strength. Melting points of the gels remained almost constant over the range from AG 100%–AP 0% to AG 30%–AP 70%, but decreased by 6–10°C at the composition level AG 10%–AP 90%. The $-\Delta H°$ value of that AG was 214 kcal/mol.

6.m Other Physicochemical Properties of Gels

6.m.1 Rheological Properties of Agarose Gel with Respect to Molecular Weight

The relationship between rheological properties of gel and molecular weight of agarose has been studied.[133] In the experiment, molecular weights of agarose from the same source (*Gelidium* sp.) were indirectly differentiated into four fractions by means of intrinsic viscosity (100 ml/g) $[\eta]$, which ranged from 2·0 to 5·4. It was shown that

(1) When Alfrey's approximation was applied to the relaxation spectra, the 'wedge-type' spectra at a shorter relaxation time (i.e. for 10^{-1}–10 h) are all alike for the four varieties of agarose, but the 'box-type' spectra at the longer time (i.e. for 10^2–10^6 h) are more prominent for the higher-molecular-weight samples.

(2) The maximum strain at breaking point for 2·4% (w/w) gel at 25°C was as large as 20% (Fig. 23), when measured by use of a chainomatic balance relaxometer.[134]

(3) The dynamic Young's modulus, E', was measured by use of Nishinari's dynamic viscoelasticity measuring apparatus,[135] and the relation between the ($\sqrt{E'/c}$) value and the value of $[\eta]$ was found to be linear, as in the case with gelatine,[103] for gels at concentration levels between 1·5% and 2·5% (w/w), and for the temperature range between 15°C and 65°C. The E' value of the agarose ($[\eta] = 5·4$) showed maxima at 30–35°C for 1·5–4% (w/w) concentration gels in a temperature range from 5°C to 65°C.[132]

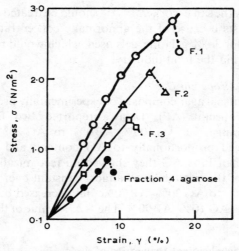

FIG. 23. Stress–strain relation of 2·4% (w/w) agarose gels at 25°C. (From Watase and Nishinari,[133] reproduced by kind permission of Dr M. Watase and Dr K. Nishinari.)

6.m.2 Thermal Expansion of Agarose Gel

The apparent specific volume change in the transition at 0°C for water in agarose gel is less than 10% of that for ordinary water.[117] Figure 24 shows the hysteresis in thermal expansion of agarose gel, the AG concentration of which was equivalent to 2·6%; it is noteworthy that the specific volume changed sharply at about −10°C with decrease in temperature. In the dilatometric studies of the properties of water in macromolecular gels, it was reported that 'the agarose gel with a lower water content ($w = 0.29$) showed neither hysteresis in thermal expansion nor any anomalous change in the specific volume around 0°C'; the agarose concentration of that gel was claimed to be equivalent to 71%.[118]

6.m.3 PMR and NMR Properties of Gels

A high-resolution and broad-line PMR (proton magnetic resonance) analysis was applied to agarose gels over the temperature range from +50°C to −60°C with respect to three states of water in the gel. It was found that the transition temperature of a quantity of water was distributed in the range from 0°C to −20°C.[119]

The anomalous temperature dependence of the NMR (nuclear magnetic resonance) line width was more marked in agarose gels than in other

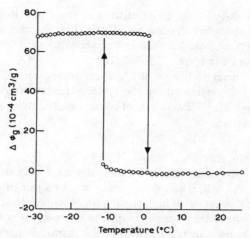

FIG. 24. Hysteresis in thermal expansion of agarose gel ($w = 0.974$). $\Delta \Phi_g$ is the specific volume change in the gel. (From Aizawa and Suzuki,[117] reproduced by kind permission of the Japanese Society of Chemistry.)

polysaccharide gels.[120] The close relationship between the motional state of water and the rheological properties of a gel was established using NMR analyses[121] (see Fig. 25(a) and (b)).

A change in the concentration dependence of the observed ^{1}H spin–spin relaxation rate, $1/T_2$ (s^{-1}), of an agarose gel at 0.5–1% concentration, in comparison with that of suspension at a similar or lower concentration, was

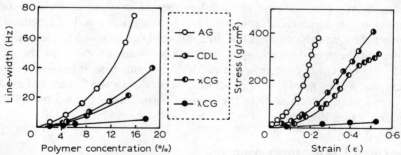

FIG. 25. Relationship between the motional state of water and the rheological properties of polysaccharide gels: (a) NMR analyses; (b) rheological analyses. AG = agarose; CDL = curdlan-type polysaccharide; CG = carrageenan. (From Aizawa et al.,[121] reproduced by kind permission of the Japanese Society of Chemistry.)

interpreted as signifying the onset of helix aggregation (Ref. 28; cf. 2.c). A graph of the temperature dependence of the ^{1}H spin–spin relaxation time, T_2 (ms), of a 0·6% agarose gel also showed the minima at 298–323 K, in the observed temperature range 273–343 K.[28]

The thermal reversibility of the gel–sol transition in agarose–water system was well investigated by the PMR experiments for a 2% (w/w) agarose–H_2O system.[136] Two kinds of commercial agarose were used as the experimental materials.

6.m.4 Film-forming Ability
Agar sol or gel can be turned into a film by drying. For example, a 0·5% gel of 1 mm thickness will, when dried, make a thin sheet of clear agar film about 5 μm thick. Agar gel of high rigidity (in other words of low ΔT_m value) will more easily produce stronger film from more dilute solutions of agar. This property is used in the laboratory for infrared spectrophotometric analysis.[35,49,55] Agar film is not readily soluble in water.

Agarose molecules are believed to have the same conformation both in gel and film on the basis of NMR analysis.[28]

Non-spinability of sol is also another rheological characteristic of agar.[122]

7 APPLICATIONS

In any application of agar more and better uses can be expected if the characteristic physical and chemical properties of agar as well as the variations in commercial products are better understood. Remember, neither the agar in Koch's age nor the agarose of the present day necessarily represent the most desirable agar for food use. 'Many people think there is only one texture of agar, a brittle gel'.[16] This is not true. It is possible by changing weed source and/or processing to produce a whole range of agar textures from the classical brittle to very elastic, as described before in Section 6 (see Fig. 26).

7.a Direct Use of Stringy Agar
Well-swollen stringy agar in cold water and cut short (i.e. 3 cm or up to 1·5 in) is used in a summer-style menu mixed with other appropriate materials. In this kind of use a stringy agar of medium or lower gel strength will be suitable. Variations may be seen in Chinese dishes.

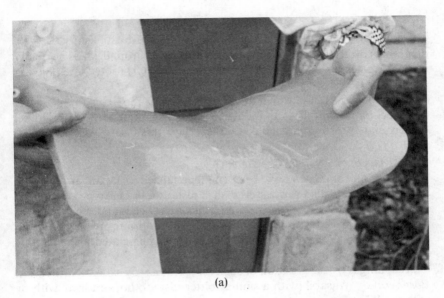

(a)

(b)

FIG. 26. (a) A rubber-like high-elasticity agar gel; (b) bending of a gel of *Gelidium japonicum* extract. (Absolute agar concentration, 0.77%. See Sample No. 2 in Figs 17 and 18.)[11]

7.b Plain Gel

Gel classified as A-I or C-I (ΔT_m less than 0·2°C) in Matsuhashi's T_m–ΔT_m diagram is preferred,[104,105] and a concentration of around 1% is normally used. Plain gel is one of the traditional Japanese-style foods, often having unique shapes of the gel.

The important thing, which most people do not realize, is that agar gel must be bland in taste and have a soft texture that is rubber-like. Unfortunately, few commercial products have this desired property.

7.b.1 'Tokoroten' (Noodle-like Agar Gel)

Two types of Tokoroten have been commercialized in Japan, as a packed family dessert and also as a dessert dish in restaurants. One type is prepared directly from the fresh extracts from *Gelidium* seaweeds and the other type is prepared from commercial agar having the appropriate properties. In the case of the former, the natural seaweed flavour is appreciated, although this is an acquired taste![53,123]

Seasoning is by the addition of such materials as soy sauce, 'Nori' (dried *Porphyra*) or 'Wasabi' (with a similar bitter taste to horseradish), with or without Japanese vinegar. Instead of soy sauce seasoning, liqueur is very good with plain agar gel.

7.b.2 'Mitsumame'

This is a sweet dessert popular among women and children in Japan. Cubic-style agar gels (ca. 1 × 1 × 1 cm), which have enough rigidity to keep straight edges, are soaked in syrup, together with sweetened peas ('mame') and pieces of sweetened fruits. For canned 'mitsumame', agar having a sufficiently high gel melting point, above 85°C, is required.[113]

7.c Fabricated Gel (Jelly)

7.c.1 Home-made-type Jelly

Agar jellies mixed with fruits, fruit extract–agar jelly, sweet jelly with mushed beans, wine–agar jelly, etc., are popular home-made-type agar jellies, and some of them are served in restaurants or sold in supermarkets. The following is an example of a formulation for 'custard-type agar jelly'.[124]

 Agar: 1 block of bar-style agar (ca. 7·5 g), or 4–5 g of powdered agar
 Water: 1–1·5 cups
 Milk: 2 bottles (ca. 360 ml)
 Sugar: $\frac{3}{4}$ cup

Egg yolk: 2 eggs
Vanilla essence: to taste
Sweetened fruits: 1 small can of mixed fruits

7.c.2 Confectionery-type Condensed Jelly

'Yokan' is the most traditional Japanese agar jelly, which is composed of sugar, mushed azuki beans, agar and water.[125] It is a commercial product, and is consumed in the tea ceremony, at home as a dessert for visitors and for hiking as a preserved food. Compared with the other types of agar jelly, its consumption has been declining. Chestnut in place of azuki beans is used for the most expensive product.

There are several kinds of condensed sweet agar jellies other than 'yokan'. An example of a formulation is as follows.

Agar	50 g
Water	2000–3000 g
Sugar	1000–2000 g
(invert sugar)	(1500–3000) g
(citric acid)	(2) g
Flavour and colour substances, if desired	

The agar is dissolved in boiling water, to which sugar and probably invert sugar is then added. The solution is cooked to about 106°C for a number of hours to reduce the volume. After a brief cooling, fruit juice and/or synthetic organic acid, flavouring agents and colour additives are added, and the mixture is allowed to gel at room temperature overnight. At this stage the solids content of the gel will be 70–75%. Further drying is carried out by placing the gel in a 55°C oven for 15–30 h to reduce the moisture content to 16–18%. The gel is cut into small pieces and then manually wrapped, first in 'Oblate' (see below) then cellophane.[126–129]

7.d Stickiness-preventing Agent

In the manufacture of processed cheese, a trace addition of agar is effective in machine process control. In confectionery, a similar use of agar added to sweetened mushed bean paste makes the paste smoother and easier to handle.

7.e Edible Paper

'Oblate' is a registered trade name for an edible paper made from starch and agar. Its thickness is 10–15 μm. It is used for wrapping jelly and powdered medicines.[88]

ACKNOWLEDGEMENTS

The author would like to thank Dr Peter Harris for his encouragement in the writing of this chapter. Thanks are also due to Dr Katsuyoshi Nishinari for his help in reviewing the recent literature.

REFERENCES

1. Araki, C. & Arai, K., Chemical studies on agar—8. 3,6-Anhydro-L-galactose. *J. Chem. Japan*, **61** (1940) 503.
2. Araki, C., Structure of the agarose constituent of agar–agar. *Bull. Chem. Soc. Japan*, **29** (1956) 543.
3. Araki, C., Structure of agarose, a main polysaccharide of agar–agar. *Memoirs, Faculty of Indust. Arts, Kyoto Technical Univ., Sci. Technol.*, **5** (1956) 21.
4. Araki, C., Carbohydrates of agar. In *Jikken Kagaku Koza*, Vol. 22. Chemical Society of Japan, Tokyo, Japan, 1958, pp. 468–87.
5. Araki, C., Some recent studies on the polysaccharides of agarophytes. In *Proc. Fifth Int. Seaweed Symp.*, ed. E. G. Young & J. L. McLachlan. Pergamon Press, Oxford, UK, 1966, p. 3.
6. Yanagawa, T., Studies on agar—4. Processing and chemical composition of agar substances of seaweeds. *Repts Imp. Res. Inst., Osaka, Japan*, **14**(5) (1933) 1.
7. Yanagawa, T., *Kanten*. Kogyo-tosho, Tokyo, 1942, pp. 1–352.
8. Araki, C., Chemical studies on agar—1. Agar substances from *Gelidium* seaweeds. *J. Chem. Japan*, **58** (1937) 1214.
9. Guiseley, K. B., Seaweed colloids. In *Encyclopedia of Chemical Technology*, Vol. 17, 2nd edn, ed. Kirk-Othmer. John Wiley and Sons, Inc., New York, 1968, p. 763.
10. Hayashi, K. & Okazaki, A., *'Kanten' Handbook*. Korin-shoin, Tokyo, 1970, pp. 1–534.
11. Matsuhashi, T., Fundamental studies on the manufacture of agar. Dissertation to Tokyo University of Agriculture, 1978.
12. Matsuhashi, T., Industrial chemistry of agar manufacture and physical properties of agar (1). *Chem. Times (Kantokagaku, Tokyo)*, **58** (1970) 1000.
13. Matsuhashi, T., Industrial chemistry of agar manufacture and physical properties of agar (2). *Chem. Times (Kantokagaku, Tokyo)*, **59** (1971) 1017.
14. Matsuhashi, T., Kanten. In *Food Industries*, ed. M. Fujimaki *et al.* Koseisya-koseikaku, Tokyo, 1985, p. 886.
15. Chapman, V. J., *Seaweeds and Their Uses*, 2nd edn. Methuen, London, 1970, p. 151.
16. Yaphe, W. & Duckworth, M., The relationship between structures and biological properties of agars. In *Proc. Seventh Int. Seaweed Symp.*, ed. T. Nishizawa. University of Tokyo Press, Japan, 1972, pp. 15–22.
17. Araki, C., Studies on sugar skeleton of agaropectin. *Mem. Shijonawate-gakuen Women's Coll.*, **3** (1969) 1.
18. Araki, C., Chemical studies on agar—13(1). Separation of agarobiose by partial hydrolysis of *Gelidium* agar substance. *J. Chem. Japan*, **65** (1944) 533.

19. Araki, C., Chemical studies on agar—13(2). Chemical structure of agarobiose. *J. Chem. Japan*, **65** (1944) 627.
20. Araki, C. & Hirase, S., Studies on the chemical constitution of agar–agar—17. Isolation of crystalline agarobiose dimethylacetal by partial methanolysis of agar–agar. *Bull. Chem. Soc. Japan*, **27** (1954) 109.
21. Araki, C., Chemical studies on agar—3. Acetylation of *Gelidium* agar substance. *J. Chem. Japan*, **58** (1937) 1338.
22. Hirase, S., Studies on the chemical constitution of agar–agar—19. Pyruvic acid as a constituent of agar–agar (3). Structure of the pyruvic acid-linking disaccharide derivative isolated from methanolysis products of agar. *Bull. Chem. Soc. Japan*, **30** (1957) 75.
23. Araki, C., Seaweed polysaccharide. In *Proc. Fourth Int. Congress Biochem., Vienna*. Pergamon Press, London, UK, 1958, pp. 15–30.
24. Tagawa, S., Separation of agar–agar by dimethylsulfoxide into agarose and agaropectin. *J. Shimonoseki Univ. Fish.*, **14**(3) (1966) 165.
25. Tagawa, S. & Kojima, Y., On agarose and agaropectin in agar prepared from *Gracilaria* harvested in different places. *J. Shimonoseki Univ. Fish.*, **15** (1966) 11.
26. Arnott, S., Fulmer, A., Scott, W. E., Dea, I. C. M., Moorhouse, R. & Rees, D. A., The agarose double helix and its function in agarose gel structure. *J. Mol. Biol.*., **90** (1974) 269.
27. Morris, E. R. & Norton, I. T., Polysaccharide aggregation in solutions and gels. In *Aggregation Processes in Solution*, ed. E. Wyn-Jones & J. Gormally. Elsevier, Amsterdam, 1983.
28. Ablett, S., Lillford, P. J., Baghdadi, S. M. A. & Derbyshire, W. *J. Colloid Interface Sci.*, **67** (1978) 355.
29. Tanii, K., Study on agar. *Bull. Tohoku Reg. Fish. Res. Lab.*, **9** (1957) 1.
30. Neuberg, C. & Ohle, H., Über einen Schwefelgehalt des Agars. *Biochem. Z.*, **125** (1921) 311.
31. Kojima, Y. & Funaki, K., Studies on the preparation of agar–agar from *Gracilaria confervoides*—2. *Bull. Jap. Soc. Sci. Fish.*, **16** (1951) 405.
32. Tagawa, S., Tateyama, Y. & Kojima, Y., On the agar–agar prepared from *Gracilaria verrucosa* in Africa. *J. Shimonoseki Coll. Fish.*, **11** (1961) 71.
33. Hass, P., On carrageen (*Chondrus crispus*)—2. On the occurrence of ethereal sulphates in the plant. *Biochem. J.*, **15** (1921) 469.
34. Matsuhashi, T., Acid pretreatment of agarophytes provides improvement in agar extraction. *J. Food Sci.*, **42** (1977) 1396.
35. Tagawa, S., Chemical studies on manufacture of agar–agar. *J. Shimonoseki Univ. Fish.*, **17**(2) (1968) 35.
36. *FMC Source Book*. FMC Bioproducts, Rockland, Maine, 1985, pp. 12–14.
37. Duckworth, M. & Yaphe, W., The structure of agar. Part 1. Fractionation of a complex mixture of polysaccharides. *Carbohydrate Res.*, **16** (1971) 189.
38. Funaki, K., Manufacturing method of agar from seaweeds. Japanese Patent No. 175290 (1947).
39. Funaki, K. & Kojima, Y., Studies on the preparation of agar–agar from *Gracilaria confervoides*—1. *Bull. Jap. Soc. Sci. Fish.*, **16** (1951) 401.
40. Kojima, Y., Kusakabe, S. & Funaki, K., Studies on the preparation of agar–agar from *Gracilaria confervoides*—5. Report on the industrialization

for manufacturing crude agar from *Gr. conf. Bull. Jap. Soc. Sci. Fish.*, **18** (1952) 245.
41. Segawa, S., *Genshoku Nippon Kaiso Zukan* (Natural-colored Picture Book of Marine Seaweeds). Hoikusha, Tokyo, Japan, 1965.
42. Segi, T., The type or authentic specimens of *Gelidium* in Europe. *Rep. Fac. Fish., Pref. Univ. of Mie*, **4** (1963) 509.
43. Tagawa, S., Nozawa, S. & Kojima, Y., Chemical studies on agar from imported agar seaweeds—4. On the agar prepared from *Gelidium* species from Egypt and Portugal. *J. Shimonoseki Univ. Fish.*, **14**(1) (1965) 15.
44. Tagawa, S. Goto, K. & Kojima, Y., Chemical studies on agar preparation from imported agar seaweeds—3. On the agar prepared from *Gelidiella acerosa* from Philippines. *J. Shimonoseki Univ. Fish.*, **14**(1) (1965) 9.
45. Humm, H. J., Agar from *Digenia simplex*. *Fla State Univ. Studies*, **1** (1950) 1.
46. Tagawa, S., Ogata, T. & Kojima, Y., Chemical studies on agar preparation from imported seaweeds—2. On the agar prepared from *Gracilaria verrucosa* collected from Chile. *J. Shimonoseki Univ. Fish.*, **13** (1963) 13.
47. Tagawa, S. & Kojima, Y., Chemical studies on agar preparation from imported seaweeds—5. On the agar prepared from *Gracilaria verrucosa* from Argentina. *J. Shimonoseki Univ. Fish.*, **14**(13) (1966) 157.
48. Matsuhashi, T., Kitazawa, T. & Takahashi, B., Gelling properties of agar manufactured in Nagano—3. The distributions of melting point and gelation temperature of the two groups of commercial agar processed from the raw material dominant with either Argentina *Gracilaria* or the Japanese *Gelidium* species. *J. Jap. Soc. Food Sci. Technol.*, **18** (1971) 291.
49. Matsuhashi, T. & Hayashi, K., Agar processed from *Gracilaria foliifera* of Florida. *Agric. Biol. Chem.*, **36** (1972) 1543.
50. Kojima, Y., Fukushima, Y. & Kono, M., Studies on the new method of preparation of agar–agar from *Ahnfeltia plicata* (1). *J. Shimonoseki Coll. Fish.*, **9** (1960) 43.
51. Kojima, Y., Tagawa, S. & Kono, M., Studies on the new method of preparation of agar–agar from *Ahnfeltia plicata*. (2) On the decolorization and dehydration of crude agar gel. (3) On the industrialization for manufacturing Itani agar. *J. Shimonoseki Coll. Fish.*, **9** (1960) 317, 323.
52. Tagawa, S., Kojima, Y. & Kono, M., Studies on the new method of preparation of agar–agar from *Ahnfeltia plicata* (4). On the constituents of *Ah. plicata* and agar–agar prepared from it. *J. Shimonoseki Univ. Fish.*, **10** (1960) 35.
53. Kojima, Y., Tagawa, S. & Yamada, Y., Studies on the new method of preparation of agar–agar from *Ahnfeltia plicata* (5). On the properties of Itani agar. *J. Shimonoseki Univ. Fish.*, **10** (1960) 43.
54. Matsuhashi, T., Firmness of agar gel, in respect to heat energy required to dissociate cross linkage of gel. In *Proc. Seventh Int. Seaweed Symp.*, ed. T. Nishizawa. University of Tokyo Press, Tokyo, 1972, p. 460.
55. Matsuhashi, T., Studies on freezing and drying of agar gel—5. *Refrigeration (Tokyo)*, **49** (1974) 958.
56. Itoh, T., Mita, K. & Hirota, N., Studies on cooking properties of 'egonori'. 1. *Bull. Nagano Pref. Jr. Coll.*, **30** (1975) 5.
57. Matsuhashi, T., Industrial chemistry of agar manufacture and physical

properties of agar (3). *Chem. Times (Kanto-kagaku, Tokyo)*, **60** (1971) 1043; idem. (4), *ibid.*, **61** (1971) 1059.
58. Matsuhashi, T., Extraction of agar by the traditional cooking process from seaweeds. *J. Jap. Soc. Food Sci. Technol.*, **16** (1969) 522.
59. Matsuhashi, T. & Takahashi, B., Trial extractions and discussions for a few characteristic agarophytes—5. African *gelidiaceae*. *Nagano State Res. Lab. for Agar–Agar Indust., Res. Report*, **1** (1964) 33.
60. Rees, D. A., Structure, conformation, and mechanism in the formation of polysaccharide gels and networks. In *Advances in Carbohydrate Chemistry and Biochemistry*, ed. M. L. Wolform & R. S. Tipson. Academic Press, New York, 1969, p. 267.
61. Guiseley, K. B., Stanley, N. F. & Whitehouse, P. A., Carrageenan. In *Handbook of Water-Soluble Gums and Resins*, ed. R. L. Davidson. McGraw-Hill, New York, 1980, p. 5-1.
62. Matsuhashi, T., Effects of polyphosphates on extractability of agar in cooking process of seaweeds. 1. Relationship between variety of polyphosphates and extractability of agar. *Bull. Jap. Soc. Sci. Fish.*, **37** (1971) 441.
63. Matsuhashi, T., Effect of polyphosphates on extractability of agar in cooking process of seaweeds. 2. Effects of concentration of polyphosphates and cooking period on extractability of agar and gel properties. *Bull. Jap. Soc. Sci. Fish.*, **37** (1971) 449.
64. Matsuhashi, T., Use of polyphosphates in agar processing. 1. Chelating action of polyphosphates with iron and their effects on agar decolorization. *J. Jap. Soc. Food Sci. Technol.*, **15** (1968) 557.
65. Matsuhashi, T., Use of polyphosphates in agar processing. 3. Residual phosphate in agar reproduced through freezing process. *J. Jap. Soc. Food Sci. Technol.*, **16** (1969) 1.
66. Hachiga, M. & Hayashi, K., Studies on the extraction of agar from the red seaweeds by the cellulase treatment. *J. Ferment. Technol. (Japan)*, **42** (1964) 207.
67. Tagawa, S. & Kojima, Y., Effects of the treatment of agar-seaweeds by cellulase on the agar–agar preparation. *J. Shimonoseki Univ. Fish.*, **14**(2) (1965) 9.
68. Matsuhashi, T. & Itoh, H. (a) Carrageenan extracted from the irradiated seaweeds. *Food Irradiation, Japan*, **21** (1986) 43. (b) Effects of gamma irradiation on 'melting point of gel' of agar and carrageenan. *Ibid.*, **21** (1986) 29.
69. Matsuhashi, T., Processing method of Tokoroten and agar from seaweeds. Japanese Patent No. 739750 (1974).
70. Matsuhashi, T., Effect of acid treatment on air dried agar. *Bull. Jap. Soc. Sci. Fish.*, **43** (1977) 831.
71. Matsuhashi, T., Migration of chemical components in a slowly frozen agar gel: a study in relation to agar extracting method from seaweeds. In *Proc. Seaweed Conf. Gifu*, Japanese Bureau of Fisheries, Research Division, Tokyo, 1979, pp. 29–32.
72. Matsuhashi, T., Kanten. In *New Edition Handbook of Refrigeration and Air-Conditioning*, Vol. *Applications*. Japanese Association of Refrigeration, Tokyo, 1981, p. 946.

73. Matsuhashi, T., Processings of agar and frozen-and-dried waxy rice cake. *Refrigeration (Tokyo)*, **58** (1983) 1127.
74. Matsuhashi, T., Studies on freezing and drying of agar gel. 6. The mechanism of weather freezing-and-drying of agar hydrogel. (1) The functional proportion of dehydration by sublimation. *Refrigeration (Tokyo)*, **49** (1974) 966.
75. Matsuhashi, T., Takahashi, B., Kitazawa, T., Nakajima, G. & Kuramochi, A., Spray drying of aqueous agar solution. *J. Jap. Soc. Food Sci. Technol.*, **16** (1970) 1.
76. Gomi, N. & Matsuhashi, T., Statistics of agar production. *Nagano State Res. Lab. for Agar–Agar Indust., Res. Report*, **1** (1964) 77–79.
77. Japanese Marine Products Importers' Association, *Japanese Imports of Marine Products: Statistics, Calendar Year 1985*.
78. Forsdike, J. L., A comparative study of agars from various geographical sources. *J. Pharmacy and Pharmacology*, **2** (1950) 796.
79. Tanii, K., Study on agar. 2. *Bull. Tohoku Reg. Fish. Res. Lab.*, **15** (1959) 67.
80. Fuse, T. & Goto, F., Studies on utilization of agar. 8. Viscosity of comparatively concentrated solution of agar. *Agric. Chem. (Japan)*, **43** (1969) 694.
81. Sakamoto, M. & Kishimoto, A., Viscosities of agar in aqueous potassium chloride. *Bull. Jap. Soc. Sci. Fish.*, **26** (1960) 25.
82. Guiseley, K. B., The relationship between methoxyl content and gelling temperature of agarose. *Carbohydrate Res.*, **13** (1970) 247.
83. Tanii, K., Indirect colloid titration of agar sol. *Bull. Tohoku Reg. Fish. Res. Lab.*, **17** (1960) 80.
84. Senju, R., Colloid titration. 1. *Wako-junyaku Jiho (Osaka)*, **37**(1) (1968) 4.
85. Senju, R., *Colloid Titration*. Nankodo, Tokyo, 1969, p. 104.
86. Matsuhashi, T. & Katsumata, T., Gelling properties of mucilaginous polysaccharides extracted from *Eucheuma gelatinae* of Okinawa. *Bull. Jap. Soc. Sci. Fish.*, **53**(9) (1987) 1593.
87. Guiseley, K. B. & Renn, D. W., Agarose: purification, properties, and biochemical applications. *Marine Colloids*. FMC Bioproducts, Rockland, Maine, 1977.
88. Meer, W., Agar. In *Handbook of Water-Soluble Gums and Resins*, ed. R. L. Davidson. McGraw-Hill, New York, 1980, p. 5-1.
89. Hayashi, K., Nagata, Y. & Harada, N., Chemical studies on agar–agar manufacture in Gifu Prefecture. 3. *Res. Bull. Faculty Agric., Gifu Univ.*, **6** (1956) 141.
90. Matsuhashi, T., Regression line of 'apparent jelly strength' of agar gel on agar concentration. 1; 2. *J. Jap. Soc. Food Sci. Technol.*, **17** (1970) 29, 32.
91. Hayashi, K. & Yasumoto, N., Chemical studies on agar–agar manufacture in Gifu Prefecture. 23. On the jelly strength of agar gel. *Res. Bull. Faculty Agric., Gifu Univ.*, **22** (1966) 131.
92. Hayashi, K. & Hiramitsu, T., Estimation of load deformation of agar gel by curdmeter. *J. Jap. Soc. Food Sci. Technol.*, **16** (1969) 160.
93. Gifu Pref. Agar Res. Lab., A study on processing of agar from agarophytes other than *Gelidium* species. Special Report of GPARL for 1968–70 fiscal years, 1971.

94. Gifu Pref. Agar Res. Lab., Studies on preservation of agar and agar seaweeds. Special Report of GPARL for 1977 fiscal year, 1978.
95. Matsuhashi, T., Textural measurements and the scientific quality evaluation of solid state foods. *Shokuhin-kogyo (Korin, Tokyo)*, **14**(10) (1971) 17.
96. Matsuhashi, T., Correlation between melting point and jelly strength of agar gel. *J. Jap. Soc. Food Sci. Technol.*, **16** (1969) 520.
97. Hayashi, K. & Hiramitsu, T., Physical properties of agar prepared from Amikusa (*Ceramium boydenii* Gepp.). *J. Jap. Soc. Food Sci. Technol.*, **16** (1969) 162.
98. Dea, I. C. M., McKinnon, A. A. & Rees, D. A., Tertiary and quaternary structure in aqueous polysaccharide systems which model cell wall cohesion: Reversible changes in conformation and association of agarose, carrageenan and galactomannans. *J. Mol. Biol.*, **68** (1972) 153.
99. Hayashi, A., Kinoshita, K., Kuwano, M. & Nose, A., Studies of the agarose gelling systems by the fluorescence polarization method II. *Polymer J.*, **10** (1978) 5.
100. Morris, E. R., Rees, D. A., Thom, D. & Welsh, E. J., Conformation and intermolecular interactions of carbohydrate chains. *J. Supermolec. Struct.*, **6** (1977) 259.
101. Florry, P. J. *Proc. Roy. Soc. Ser. A*, **50** (1956) 234.
102. Laurent, T. C., Determination of the structure of agarose gels by gel chromatography. *Biochem. Biophys. Acta*, **136** (1967) 199.
103. Eldridge, J. E. & Ferry, J. D., Studies of the cross-linking process in gelatin gels. 3. Dependence of melting point on concentration and molecular weight. *J. Phys. Chem.*, **58** (1954) 992.
104. Matsuhashi, T., Properties and qualities of agar, with respect to gelling property. *New Food Industry (Shokuhin-shizai Kenkyukai, Tokyo)*, **17**(2) (1975) 14.
105. Matsuhashi, T., Genuine property of 'tokoroten' (agar gel). *New Flavor (Tokyo)*, **14**(6) (1980) 6.
106. Nakahama, N., Setting point and transparence of the agar–agar gel. *J. Home Economics, Japan*, **17** (1966) 203.
107. Iwase, E., Experimental methods in colloid chemistry. In *Jikken-kagaku-koza*, Vol. 7, ed. H. Nanjo. Kyoritsusha, Tokyo, 1938, p. 90.
108. Tazaki, Y. & Nagasawa, S., The properties of agar gel prepared with inorganic salt solutions. *J. Jap. Soc. Food Sci. Technol.*, **9** (1962) 274.
109. Takegawa, O., Strength of agar gel. *Nippon Kaiso-kogyo Res. Lab. Report*, No. 2, 1963.
110. (a) Nakahama, N., Rheological studies of the agar–agar gel. *J. Home Economics, Japan*, **17** (1966) 197. (b) Isozaki, H., Akabane, H. & Nakahama, N., Viscoelasticity of hydrogel of agar–agar: analysis of creep and stress relaxation. *J. Agric. Chem., Japan*, **50** (1976) 265.
111. Nakahama, N., Maeda, F. & Kujira, S., Rheological studies on the egg-white gel. 1. The effects of sugar, sorbitol and glycerin on the gel. *J. Home Economics, Japan*, **18** (1967) 365.
112. Selby, H. H. & Selby, T. A., Agar. In *Industrial Gums*, ed. R. L. Whistler. Academic Press, New York, 1959, pp. 15–49. [Translated by T. Yabuki & S. Suzuki in *NKRL Report*, No. 4, 1966.]

113. Kojima, Y., Inamasu, Y. & Shiraishi, T., Studies on the agar gel in canned 'Mitsumame'. 1. On the gel reinforcing agents. *J. Shimonoseki Coll. Fish.*, **8** (1959) 161.
114. Moritaka, H., Nishinari, K., Horiuchi, H. & Watase, M., Rheological properties of aqueous agarose–gelatin gels. *J. Texture Studies*, **11** (1980) 257.
115. Watase, M. & Nishinari, K., Effect of coexistence of gelatin on gelation of agarose. *J. Jap. Soc. Food Sci. Technol.*, **30** (1983) 368.
116. Watase, M. & Nishinari, K., Rheological properties of agarose–gelatin gels. *Rheol. Acta*, **19** (1980) 220.
117. Aizawa, M. & Suzuki, S., Properties of water in macromolecular gels. 3. Dilatometric studies of the properties of water in macromolecular gels. *Bull. Chem. Soc. Japan*, **44** (1971) 2967.
118. Mizuguchi, J., Takahashi, M. & Aizawa, M., Electric conductivity of natural macromolecule gels. *J. Chem., Japan*, **91** (1970) 723.
119. Aizawa, M., Mizuguchi, J., Suzuki, S., Hayashi, S., Suzuki, T., Mitomo, N. & Toyama, H., Properties of water in macromolecular gels. 4. Proton magnetic resonance studies of water in macromolecular gels. *Bull. Chem. Soc. Japan*, **45** (1972) 3031.
120. Aizawa, M., Suzuki, S., Suzuki, T. & Toyama, H., Properties of water in macromolecular gels. 5. Anomalous temperature dependence of the nuclear magnetic resonance line-width of water in macromolecular gels. *Bull. Chem. Soc. Japan*, **46** (1973) 116.
121. Aizawa, M., Suzuki, S., Suzuki, T. & Toyama, H., The properties of water in macromolecular gels. 6. The relationship between the rheological properties and the state of water in macromolecular gels. *Bull. Chem. Soc. Japan*, **46** (1973) 1638.
122. Nakagawa, T., *Rheology.* Iwanami, Tokyo, Japan, 1960, p. 154, p. 245.
123. Matsuhashi, T., Consumer preference and texture of 'tokoroten'. *New Flavor*, **5**(3) (1971) 26.
124. Matsuhashi, T., Applications of agar to jelly-type desserts. *New Food Indust. (Tokyo)*, **17**(9) (1975) 15.
125. Matsuhashi, T. & Shimada, T., Texture studies on Yokan, a Japanese sugar-mushed beans-agar jelly-like confectionery. *J. Jap. Soc. Food Sci. Technol.*, **18** (1971) 370.
126. Ishida, K., Okada, Y. & Koyama, Y., Properties of beet sugar and cane sugar as the raw materials of dried (condensed) agar jelly. *J. Jap. Soc. Food Sci. Technol.*, **15**(6) (1968) 109.
127. Goto, F., The comparison of block-agar gel and powder-agar gel. 2. Study on some characteristic tastes of 'Mizu-yokan' by sensory test and objective methods. *Home Economics, Japan*, **29** (1978) 67.
128. Goto, F., Kaneko, E., Yoshimatsu, F. & Matsumoto, F., A study on cooking of agar–agar: textural properties of Kingyokuto. *Home Economics, Japan*, **29** (1978) 67.
129. Goto, F., Yoshimatsu, F. & Matsumoto, F., A study on cooking of agar–agar: decomposition of sucrose and agar–agar in Kingyokuto. *Home Economics, Japan*, **35** (1984) 459.
130. Matsuhashi, T., Three kinds of the traditional frozen-and-dried food products

in Nagano Prefecture: the industrial background and the technology transfer. *Quest of Traditional Foods (Japan)*, **5** (1987) 22.
131. Pappas, G., Rao, V. N. M. & Smit, C. J. B., Development and characteristics of modified agar gels. *J. Food Sci.*, **52** (1987) 467.
132. Watase, M. & Nishinari, K., Rheology, DSC and volume or weight change induced by immersion in solvents for agarose and kappa-carrageenan gels. *Polymer J.*, **18** (1986) 1017.
133. Watase, M. & Nishinari, K., Rheological properties of agarose gels with different molecular weights. *Rheol. Acta*, **22** (1983) 580.
134. Watase, M. & Nishinari, K., Effect of sodium hydroxide pretreatment on the relaxation spectrum of concentrated agar–agar gel. *Rheol. Acta*, **20** (1981) 155.
135. Nishinari, K., Horiuchi, H., Ishida, K., Ikeda, K., Date, M. & Fukada, E., A new apparatus for rapid and easy measurement of dynamic viscoelasticity for gel-like foods. *Nippon Shokuhin Kogyo Gakkaishi*, **27** (1980) 227.
136. Indovina, P. L., Tettamanti, E., Micciancio-Giammarinaro, M. S. & Palma, M. U., Thermal hysteresis and reversibility of gel–sol transition in agarose–water systems. *J. Chem. Phys.*, **70** (1979) 2841.

Chapter 2

ALGINATES

WILMA J. SIME
Limewood, Raunds, Northamptonshire NN9 6NG, UK

1 INTRODUCTION

The term 'alginate' (or algin) refers to a group of naturally occurring polysaccharides that are extracted from the brown seaweeds (*Phaeophycea*), a variety of seaweed that grows in the shallower waters of the world's temperate zones. Alginates should be distinguished from the other seaweed extracts agar and carrageenan, which are obtained from red seaweeds. Both the chemical composition and properties of alginates differ significantly from those of agar and carrageenan.

Although alginate was first prepared by Stanford in the 1880s[1] and isolated pure by Krefting in 1896,[2] it was not until the 1930s that alginate was first exploited commercially.

There are many different species of brown seaweed, and although all contain alginates only a few are sufficiently abundant and conveniently located for commercial production. The most widely used species are *Laminaria hyperborea*, *Macrocystis pyrifera* and *Ascophyllum nodosum*.

In its natural environment, alginate exists in the cell wall as the mixed calcium/sodium/potassium salt of alginic acid. It is available commercially principally as the sodium salt, although alginic acid, other metal salts and organic derivatives can be obtained. Sodium alginate is soluble in water, producing viscous solutions, and is used in both food and non-food systems as a thickening and stabilizing agent. The special property of alginate, however, is its ability to form gels in the presence of calcium ions, a property that has successfully been utilized in a wide variety of applications.

2 STRUCTURE

Alginic acid is a linear copolymer composed of two monomeric units, D-mannuronic acid and L-guluronic acid.[3] The monomers occur as regions made up exclusively of one unit or the other, referred to as M blocks or G blocks, or as regions in which the monomers approximate an alternating sequence—MG blocks.[4-6] D-Mannuronic acid exists in the C1 configuration (Fig. 1(a)) and is connected in the alginate polymer to its neighbouring units through the 1 and 4 positions. L-Guluronic acid has the 1C configuration (Fig. 1(b)) and is 1,4-linked in the polymer.[7]

Owing to the particular shapes of the monomers and their mode of linkage in the polymer, the geometries of the G-block regions, M-block regions and the alternating regions are substantially different (Fig. 2). Specifically, the G blocks are buckled while the M blocks have a shape

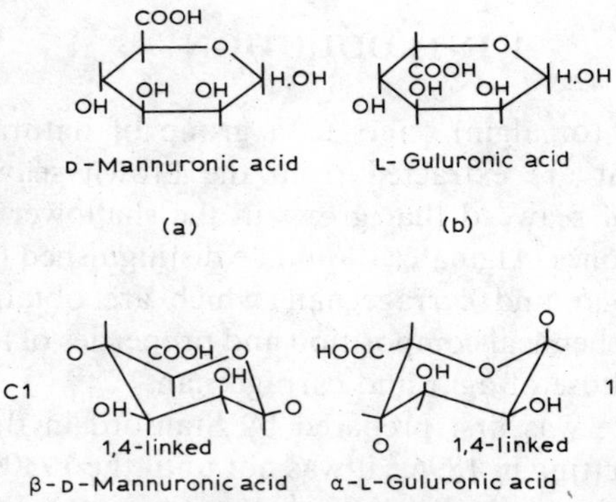

FIG. 1. Structures of the alginate monomers D-mannuronic and L-guluronic acids.

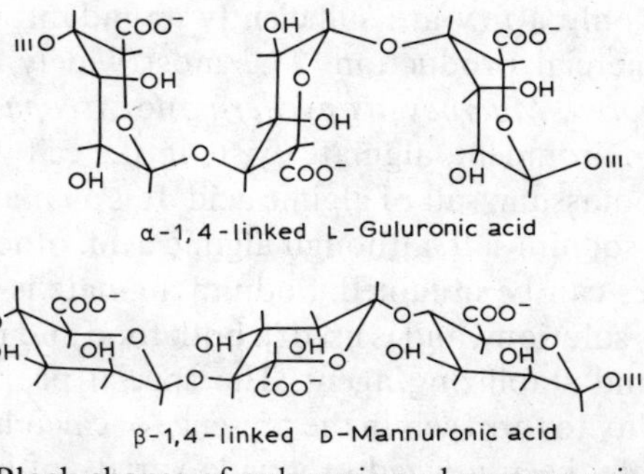

FIG. 2. Block shapes of guluronic and mannuronic acids.

TABLE 1
COMPOSITION OF COMMERCIAL BROWN SEAWEEDS[9,10]

	Macrocystis pyrifera	Ascophyllum nodosum	Laminaria hyperborea
Mannuronic acid (%)	61	65	31
Guluronic acid (%)	39	35	69
M blocks (%)	40·6	38·4	12·7
G blocks (%)	17·7	20·7	60·5
Alternating segments (%)	41·7	41·0	26·8

referred to as an extended ribbon.[8] The behaviour of a particular alginate depends both on its molecular weight (i.e. the length of the polymer chains) and the proportion and arrangement of mannuronic to guluronic acid units present. The latter is determined by the variety of seaweed from which the alginate is derived. Different weeds yield alginates that differ in monomeric composition and block structure. Table 1 shows the ratios of mannuronic to guluronic acid found in the more commonly processed weeds. Owing to the differing monomer ratios, alginates are commonly referred to as being 'high-M' or 'high-G' alginates. *Macrocystis pyrifera* and *Ascophyllum nodosum* are of the high-M type, whereas *Laminaria hyperborea* gives a high-G alginate.

3 SOURCE AND PRODUCTION

3.a Raw Materials

As mentioned above, only a few of the many species of brown seaweed are suitable, owing to abundance and location, for commercial production. The weeds most commonly processed are:

- *Laminaria hyperborea*, found along the North Atlantic coastal regions where the storm-cast weed, particularly the stipes (stems), is collected from the beaches;
- *Macrocystis pyrifera*, mechanically harvested off the west coast of America;
- *Ascophyllum nodosum*, which grows between the high- and low-tide levels in Northern Europe and Canada. The weed is cut by hand at low tide and floats at high tide, where it is collected in nets and towed to the processing or drying plant.

The seaweed, once collected, can either be dried before processing or processed wet.

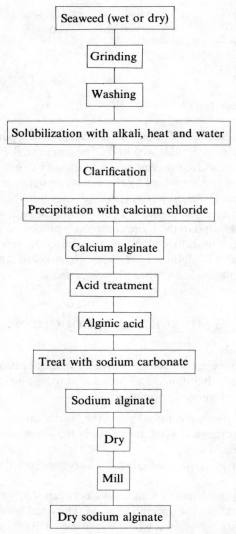

FIG. 3. Flow sheet for the manufacture of sodium alginate.

3.b Process

The various steps in the production of a typical sodium alginate are shown in Fig. 3. Alginate extraction is essentially an ion-exchange process. The seaweed is milled and washed, and alginate is brought into solution as sodium alginate by treatment with a strong alkali and heat. Calcium chloride is added to the viscous solution to precipitate the alginate as calcium alginate, which is then treated with acid to form alginic acid. The alginic acid may then be treated in a variety of ways to produce the desired commercial product, e.g.:

(1) Treated with sodium carbonate or other bases to produce alginate salts.
(2) Reacted with propylene oxide to produce propylene glycol alginate.

This sequence of events ensures production of a relatively pure, colourless, odourless product.

3.c Production of Propylene Glycol Alginate (PGA)

PGA is prepared by reacting moist alginic acid with propylene oxide—depending on reaction conditions, up to 90% of the carboxyl groups become esterified, the remaining groups being free or neutralized with sodium or calcium (Fig. 4). Different grades of propylene glycol alginate with a range of degrees of esterification are available to suit different applications.

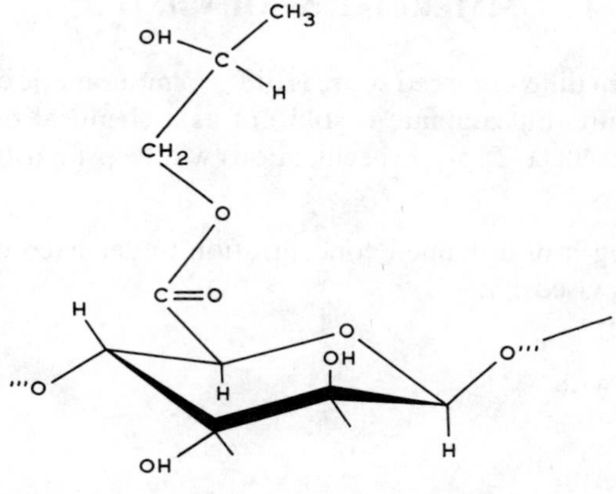

FIG. 4. Structure of propylene glycol alginate.

3.d Bacterial Alginates

The production of alginate is not unique to the brown algal seaweeds. Alginate-like polymers are produced as an exocellular material by several bacteria.[11] A partially acetylated alginate containing variable amounts of M and G has been isolated from *Pseudomonas aeruginosa*,[12] a secondary pathogen of patients with cystic fibrosis.[13] A similar polymer produced by the soil bacterium *Azotobacter vinelandii*[14] was shown to have a similar block structure to that of the seaweed alginates.[15] Recently, alginate production has been reported in three non-pathogenic species of *Pseudomonas*, including *P. mendocina*, *P. putida* and *P. fluorescens*.[16]

O-acetyl groups invariably seem to be present in bacterial alginates and are the most characteristic feature distinguishing them from the seaweed alginates.[17] To date, there have been no reports of commercial bacterial alginate production.

3.e Enzyme Modification of Alginates

It is possible to 'tailor' alginates by enzymatic modification *in vitro*. High-molecular-weight alginates having a variety of initial composition and sequential structures can be modified with a mannuronan-C-5 epimerase from *Azotobacter vinelandii* to yield polymers with a high content of guluronic acid and, hence, an enhanced ability to form gels with calcium ions.[18] At present this technique appears restricted to the laboratory and is not commercially exploited.

4 COMMERCIAL AVAILABILITY

The alginates from different weed sources differ in monomeric composition and block structure, but alginate is sold not as a chemical but rather to produce a desired effect. Typical specifications will be primarily concerned with

- the gel strength of a defined concentration under fixed conditions;
- the solution viscosity;
- the salt type;

and secondarily with

- particle size;
- particular form;
- powder (and solution) colour.

The levels of mannuronic and guluronic acid residues are not usually measured by the supplier but merely assumed from their knowledge of the weed source. Weeds from different sources will often be blended together to produce a desired effect, be it gel strength, solution viscosity, etc.

Calcium, potassium and ammonium alginates are also available commercially, as is alginic acid (all are food-approved materials), although the available range of these products is not as wide as that for sodium alginate, owing to the greater number of applications for the latter.

Alginates can be obtained as blends with, for example, sugar or other polysaccharides for specific applications.

5 SOLUTION PROPERTIES

5.a Solubility and Preparation of Solutions

Alginic acid and calcium alginate are insoluble in water and will not form smooth solutions. They do, however, absorb large amounts of water, swelling to form paste-like globules. Sodium alginate, on the other hand, will form smooth solutions in water provided that the level of calcium in the water is low.

When making a solution of alginate, it is essential that the alginate is dispersed in the water before it starts to dissolve.

To obtain a solution within a short time (5–10 min), care should be taken over the method of addition of the alginate to water. Two suitable methods for the rapid dissolution of sodium alginate are as follows.

(1) HIGH-SHEAR STIRRING

Water should be vigorously stirred to form a vortex and the alginate gradually sprinkled into this vortex. When using this method important points to note are:

(a) The alginate should be added sufficiently slowly and gradually so that the individual particles remain separate and are rapidly dispersed, but all of the alginate should be added before the solution starts to visibly thicken and the vortex disappears.
(b) The mixer blades should be sufficiently well submerged so as not to incorporate air into the solution.

(2) DRY MIXING WITH OTHER FORMULATION INGREDIENTS

The alginate can either be dry-mixed with other powders such as starch or sugar or dispersed in a non-aqueous liquid, for example vegetable oil or glycerol, before addition to water. Pre-mixing in this way ensures separation of the alginate particles, which then disperse and hydrate more easily. Some agitation will still be required but there is no longer the need for high-shear mixing.

If neither of these methods is possible or practical then a specially treated sodium alginate, designed for good dispersion, can be used; the drawbacks are that the alginate is more expensive and solution time is much longer.

If sodium alginate is not properly dispersed it will form clumps, which become swollen on the outside, preventing contact between water and the centre of the clump, a phenomenon known as 'blinding'. Should this occur, the mixture should be subjected to very high-shear mixing, e.g. by use of a Silverson mixer. Care must be taken to ensure that the mixture does not overheat; the combination of heat plus high shear will almost certainly degrade the alginate, resulting in a decrease in solution viscosity. Alternatively, the mixture can be left to stand, with agitation at regular intervals, in which case dissolution may take up to 24 h depending on the severity of the problem.

Wherever possible, the use of soft water is recommended for preparing alginate solutions so as to avoid reaction with the calcium ions present in hard water. Depending on the hardness of the water, the use of hard water will at best slow down dissolution and lead to a 'false' viscosity, due to some polymer cross-linking, and at worst prevent dissolution altogether. The problems of hard water can, however, be overcome by the use of sequestrants—see Section 5.c.5.

5.b Measurement of Viscosity

The viscosity of an alginate solution should only be measured using an instrument suitable for non-Newtonian liquids, e.g. a rotational viscometer. Two popular rotational viscometers in common use are the Brookfield Synchro-Lectric and Haake Rotovisco, both of which are available with a wide range of attachments, including facilities for temperature control.

Many alginate suppliers quote the viscosity of their various grades of alginate as $x-y$ mPa s^{-1} using a 1% solution at 25°C and a Brookfield LVF Viscometer at 60 rpm.

As a general 'rule of thumb', doubling the alginate concentration will increase the viscosity of the solution by a factor of 10.

5.c Factors Affecting Solution Properties

When dissolved in distilled water, pure alginates form smooth solutions having long flow properties. The solution properties are dependent on both physical and chemical variables. The physical variables which affect the flow characteristics of alginate solutions include polymer size (molecular weight), temperature, shear rate, concentration in solution and the presence of miscible solvents. The chemical variables affecting alginate solutions include pH, sequestrants, monovalent salts, polyvalent cations and quaternary ammonium compounds.

5.c.1 Molecular Weight

Solution viscosity ideally should be related to molecular weight,[19] but this can be influenced significantly by the levels of residual calcium from manufacture. Most commercial high-viscosity alginates have a molecular weight of over 150 000 corresponding to a degree of polymerization (DP) of about 750. Others are prepared with a range of values of DP down to about 80. It should be remembered that a particular viscosity can be achieved either by (a) all of the molecules in solution being of the same molecular weight or (b) some of the molecules being of low molecular weight and the rest of high molecular weight. Therefore, although both solutions have the same viscosity, it is conceivable that they could perform differently in a product situation.

5.c.2 Temperature

Alginate solutions, like most other polysaccharides, decrease in viscosity with an increase in temperature. Over a limited range the viscosity of an alginate solution decreases by approximately 12% for each 5·5°C increase in temperature. Heating of sodium alginate solutions results in some thermal depolymerization and the amount of depolymerization is dependent on time, temperature and pH. The lower the pH, the faster the degradation at elevated temperatures.

Temperature reduction causes a viscosity increase in an alginate solution but does not result in gel formation. A sodium alginate solution that has been frozen and then thawed will not have its appearance or viscosity changed.

5.c.3 Water-miscible Solvents

The addition of non-aqueous water-miscible solvents (alcohols, glycols, etc.) to an alginate solution results in viscosity increases and eventual precipitation, although the levels of added solvent can be quite high before precipitation occurs. A 1% sodium alginate solution can tolerate up to 20% ethanol and up to 70% glycerol. The source of the alginate, the degree of polymerization, the cation present and the concentration in solution all affect the solvent tolerance of the alginate solution.

5.c.4 Effect of pH

The lower the molecular weight of the alginate, the more stable the solution is at low pH. Although sodium alginate solutions appear to tolerate high pH conditions, long-term stability is poor above pH 10. At high pH, beta elimination[20] and hydrolysis result in depolymerization, with an accompanying viscosity loss.

Propylene glycol alginates are more stable at acid pH values than sodium alginate and therefore can be used to thicken and stabilize acidic products such as salad dressings. Sodium alginate is precipitated at a pH below 3·5, whilst propylene glycol alginate, with a degree of esterification of 60%, will remain in solution at pH values down to 2. Acid stability of PGA increases with degree of esterification, but in strong acid the alginate chains depolymerize, with a resultant loss of viscosity.

5.c.5 Effect of Sequestrants

Sequestrants are used in alginate solutions either to prevent the alginate from reacting with polyvalent ions in the solution or to sequester the calcium inherent in the alginate. The use of sequestrants as a protective device is common in most applications of alginates. Polyvalent ion contaminants can occur in water, chemicals, pigments or almost any material of natural origin.

The viscosity of an alginate solution is dependent on molecular weight and the level of residual calcium. Sequestrants can be used to establish how much of the 'viscosity' is due to the presence of this residual calcium (Table 2).

5.c.6 Effect of Monovalent Salts

The viscosity of a dilute sodium alginate solution is depressed by the addition of monovalent salts. The alginate polymer contracts as the ionic strength of the solution is increased. The maximum viscosity effect is attained at a salt level of 0·1N. As the concentration of the alginate is increased the electrolyte effect is decreased except for alginates high in calcium. As the concentration of the salt is increased, the viscosity of the solution may increase. This effect is most evident after prolonged storage. The salt effects vary with the source of alginate and the degree of polymerization, the concentration in solution and the type of monovalent salt used. Calcium present in the alginate will possibly be replaced by

TABLE 2
THE EFFECT OF SEQUESTRANT ON THE VISCOSITIES OF HIGH-M AND HIGH-G COMMERCIAL ALGINATES

Alginate	Type	Claimed viscosity $(mPa\,s^{-1})^a$	Viscosity in sequestrant $(mPa\,s^{-1})^b$
A	M	250	55·5
B	M	175	47·5
C	M	25	14·5
D	M	4	6·5
E	G	800–1 000	217·5
F	G	400–600	153
G	G	200–300	81·5
H	G	100–150	65
I	G	50–60	26·5
J	G	5–10	9·5

^{a} Manufacturer's claim of viscosity of a 1% solution using a Brookfield LV Synchro-Lectric Viscometer (60 rpm, spindle 2, 20 °C).
^{b} Viscosity measured as in a but in this instance using a solvent of 0·12% sodium phosphate and 0·03% sodium carbonate.

sodium in more concentrated sodium salt solutions and this interchange in itself will have an effect on the ultimate viscosity of the solution.

6 GEL PROPERTIES

6.a Mechanism of Gelation
6.a.1 Sodium Alginate Gels
Under controlled conditions, alginates will form gels with a number of divalent cations (and borax).[20] By far the most important gels, particularly in food applications, are the gels formed with calcium ions.

Until comparatively recently it was considered that calcium alginate gels were formed by simple ionic bridging of two carboxyl groups on adjacent polymer chains with calcium ions. Although these bonds may play a role in the gelation mechanism, they are not considered sufficiently energetically favourable to account for the gelation of alginate.[21] The poly-L-guluronate segments show an enhanced binding of calcium above a chain length of 20 residues, suggesting that a cooperative mechanism is involved in the

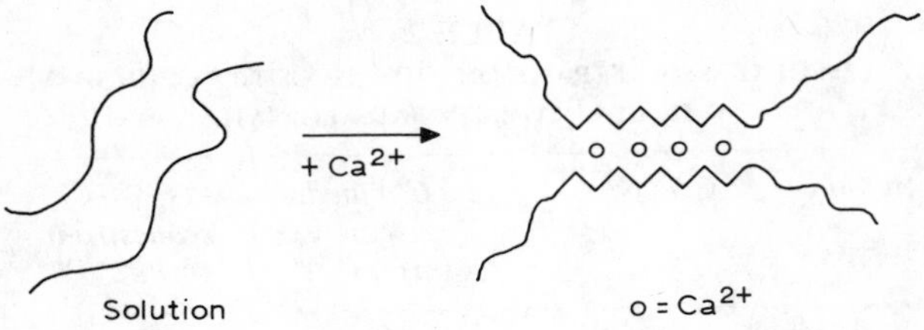

FIG. 5. 'Egg-box' model for alginate gelation.[24]

gelation process in which binding sites exist in an ordered array and binding of one ion facilitates binding of the next. These effects are not observed for poly-D-mannuronic acid or alternating sequences.[22] Therefore, the gel-forming ability is related mainly to the content of guluronic acid.[23]

Two contiguous, diaxially linked G residues form a cavity that acts as a binding site for calcium ions. Long sequences of such sites form cross-links with similar sequences in other alginate molecules, giving rise to junctions in the gel network. For obvious reasons, this is often referred to as the 'egg-box' model of alginate gelation[24] (Fig. 5). The calcium ion interacts not only with the carboxyl groups but also with the electronegative oxygen atoms of the hydroxyl groups. The primary gelation event involves chain dimerization, although secondary aggregation of the dimers may then occur. Thus it can be seen that the proportions of G blocks is the main structural feature contributing to gel formation[25] (see also Chapter 10, Section 6).

6.a.2 Propylene Alginate Gels
Propylene glycol alginate does not form a very good gel. The propylene glycol groups hinder aggregation of the alginate chains and fewer groups are available for interaction with cations. Reaction with calcium depends on the degree of esterification—at 60% or less, soft gels are formed; at 85% there is no noticeable change in viscosity in the presence of calcium.

The extent of change brought about by calcium is dependent on the pH of the solution and is much less marked at pH 3 than at pH 5. This is the consequence of ion-exchange equilibrium between hydrogen and calcium ions.

6.b Preparation of a Gel
For the practical food technologist, formation of an alginate gel is somewhat more complex than the simple addition of calcium ions to an

alginate solution. Alginates are capable of producing gels of differing strengths and textures suitable for a wide range of applications, but correct formulation requires consideration of the following areas:

(1) type of alginate used
(2) degree of conversion to calcium alginate
(3) source of calcium ions and method of preparation

6.b.1 Type of Alginate Used

High-G alginates form strong, brittle gels with a tendency to synerise, whereas high-M alginates form weaker more elastic gels that are less prone to syneresis.

6.b.2 Degree of Conversion to Calcium Alginate

The term calcium conversion refers to the ratio of calcium ions to sodium alginate. A molar ratio of 0·5 (where theoretically there is sufficient calcium to totally replace the sodium) is expressed as a calcium conversion of 100%.

The introduction of a small amount of calcium brings about an increase in viscosity. As the proportion of calcium is increased, the solution becomes more viscous and develops some gel structure on standing. Further amounts of calcium result in an increase in gel strength, but up to a certain point the gel remains thixotropic, breaking down to a fluid on shearing and resetting when left undisturbed. Addition of further calcium produces

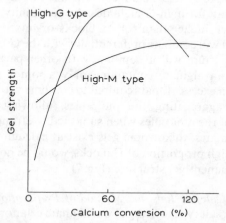

FIG. 6. Relative gel strengths of high-G and high-M alginates at different calcium conversions.

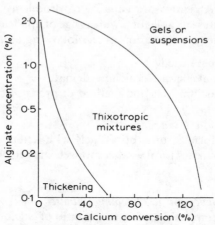

FIG. 7. Influence of calcium ion level on alginate rheology.

irreversible gels, suspensions or flocs depending on the method of addition. A continuous gel formed from these higher levels of calcium will, on shearing, break down into a suspension or paste and the gel structure does not reform on standing (Fig. 6).

The ratio of calcium to alginate over which thixotropic gels are formed depends on the alginate type, the pH and the solids content of the system. Both a decrease in pH and an increase in solids result in a shift of the thixotropic region to lower calcium conversions. For high-M alginates the range is wider than for high-G alginates. As the affinity of calcium for G blocks is so much higher than for M blocks or mixed MG blocks, the available calcium will combine preferentially with the G blocks, and only if there is excess calcium will it combine with other parts of the alginate molecule. With an alginate containing only a minor proportion of G blocks, calcium in excess of that required to combine with them will not be so effective in aggregating the molecules, allowing separation and rearrangement of the molecules when subjected to strong shearing forces. For high-M alginates, thixotropic gels exist at calcium levels that, in an alginate with a high proportion of G blocks, would be holding the alginate chains in a permanent gel structure (Fig. 7).

6.b.3 Source of Calcium Ions and Method of Preparation

The best calcium salt type to bring about alginate gelation will depend both on the method of preparation and the pH of the final product. Alginate gels can be prepared by either diffusion or internal setting.

6.b.3.1 DIFFUSION SETTING

Diffusion setting is the simplest technique and, as the name implies, the gel is set by allowing calcium ions to diffuse into an alginate solution. Since the diffusion process is slow, this process can only be utilized effectively to produce thin strips or small spheres, or to provide a thin gelled coating on the surface of a food product. The most common source of calcium ions for diffusion is calcium chloride. The diffusion rate can be increased by increasing the calcium concentration in the setting bath. This does, however, have limitations, since calcium chloride imparts an unpleasant bitter taste to foodstuffs when used at high levels. Calcium lactate and calcium acetate, although not as soluble as calcium chloride, can be used effectively as diffusion setting agents without any taste problems.

When an alginate solution is extruded into, and gelled in, a calcium-containing bath, the choice of calcium salt is unaffected by the pH of the alginate system. However, for acidic systems, diffusion gelling can also be brought about by depositing an alginate (containing a calcium salt soluble only under acidic conditions) into an acid bath. Diffusion of acid into the mixture results in solubilization of the calcium salt and subsequent gelation. A suitable calcium salt is calcium hydrogen orthophosphate.

It is important to remember that diffusion is a two-way process, i.e. calcium ions are diffusing into the alginate system at the same time as soluble material is diffusing out of the alginate into the setting bath. Therefore, if, for example, you wish to prepare an alginate gel containing a high level of sugar, then the setting bath should contain an equivalent concentration of sugar as well as the calcium.

6.b.3.2 INTERNAL OR BULK SETTING

In internal, or bulk, setting, which is normally carried out at room temperature, the calcium is released under controlled conditions from within the system. Calcium sulphate, calcium carbonate and calcium hydrogen orthophosphate are the sources of calcium most commonly used. The rate at which the calcium is made available to the alginate molecules depends primarily on pH, and the amount, particle size and solubility characteristics of the calcium salt. Small particle size and low pH favour rapid release of calcium.

In almost all situations in internal setting, calcium release during mixing of the ingredients is so rapid that a calcium sequestrant is required to control the reaction by competing with the alginate for the calcium ions. Typical food-approved sequestrants are sodium hexametaphosphate, tetrasodium pyrophosphate and sodium citrate. Disodium hydrogen orthophosphate, although it has little affinity for calcium at pH values below 5, is sometimes used to remove calcium ions (as insoluble calcium phosphate) from tap water. Removal of the calcium permits more efficient hydration and solubilization of the alginate.

For a given level of alginate and calcium salt, an increase in the level of sequestrant causes a decrease in the setting rate of the gel and results in a progressively weaker final gel because the ultimate distribution of the calcium ions between the alginate and the sequestrant increasingly favours the latter. Control of the gelling reaction with sequestrant is only necessary during mixing to prevent gelation starting during the mixing process and the subsequent irreversible breakdown of the gel structure. With highly efficient and rapid mixing equipment, only small amounts of sequestrant are required, since only a small proportion of the calcium salt has the opportunity to dissolve during the mixing process.

6.c Measurement of Gel Properties

Measurements of gel properties are widely used to establish the differences between different gels, and a common measurement is that of gel strength. This parameter is often used as an indicator of the quality of a particular alginate, and is obtained using a variety of instruments such as an Instron Universal Materials Tester or a Marine Colloids' Gel Tester. It is important to remember that although a number of instruments are claimed to measure gel strength, they are not always measuring the same parameter. Therefore, comparisons of gel strengths should only be performed using the same instrument under controlled conditions.

Gel strength can be a measure of the 'hardness' of a gel, that is the maximum force (in newtons) needed to break the gel, or the resistance to deformation (compression modulus measured in newtons per square metre). Samples of alginate for measurement are usually prepared as gelled discs or cylinders.

For compression modulus, the sample is compressed by a fixed amount at a fixed rate. The result is calculated from the initial slope of the force–deformation curve (usually calculated at 10–20% total deformation, i.e. small deformation). Hardness is calculated as the force required to break the gel.

It cannot be stressed too strongly that reproducible results will only be obtained if the instrument and plungers are maintained in first-class order and the samples are prepared and presented to the instrument in a consistent manner.

For really accurate measurements of gel properties, it is necessary to prepare gel cylinders with very close dimensional tolerances. These can be made by first placing the solution of sodium alginate in a flat-bottomed cylindrical mould. Calcium alginate gels are then formed by mixing into the sample D-glucono-δ-lactone and Ca-EDTA,[26] slow hydrolysis of which yields a mixed H^+,Ca^{2+} gel (pH 3·5). These gels are then fully converted into the calcium form by removal from the mould and dialysis against 0·34M calcium chloride solution.[18]

Calcium alginate gels prepared in this way have a perfectly symmetrical geometry, with very small variations in mass or dimensions. This is in contrast to gels obtained by dialysing sodium alginate against calcium chloride in the casting cylinders.

6.d Factors Affecting Gel Properties

6.d.1 Effect of Temperature on Gel Formation

The rate of setting of internally set alginate gels can also be controlled by dissolving in *hot* water the alginate, calcium salt and sequestrant, and allowing the solution to set by cooling. Unlike gelatine gels, however, alginate gels are not thermally reversible and once set will not melt on reheating. The calcium salts and sequestrants used in this system are the same as those already mentioned for internal setting. Although the calcium ions are already in solution with the alginate, setting does not occur at elevated temperatures because the alginate chains have too much thermal energy to permit alignment. It is only when the solution cools that calcium-induced interchain associations can occur.

An interesting feature of gels prepared in this way is their stability to syneresis, i.e. water loss from the network is minimal. This stability is due to the fact that the calcium required for gel formation is available to all of the alginate molecules at the same time, allowing formation of a thermodynamically stable network. By contrast, in diffusion setting, the alginate molecules closest to the calcium ions in the setting bath react first; also in internal setting, at room temperature the molecules closest to the macroscopic particles of dissolving calcium salt react first. In the latter two systems, not all of the alginate molecules have the opportunity to align at the same time and the resulting gel network has a certain amount of inherent instability, which gives rise to greater gel shrinkage and syneresis.

The temperature required to prevent gelation depends both on the alginate concentration and the amount of calcium present. Higher alginate concentrations and calcium levels require higher temperatures. At 0·6% sodium alginate, and sufficient calcium to give 15% conversion, the system is still fluid at room temperature but will gel on freezing. At the same alginate concentration but with sufficient calcium to give 60% conversion, a temperature of 80°C is required to prevent gelation. Hence, this method is more suitable for the formation of weaker, more elastic gels than for strong, brittle gels.

6.d.2 Effects of pH

It is possible, with care, to produce a gel of alginic acid, i.e. without the use of calcium. This is best achieved by diffusion setting using acid (below pH 3·5) in the setting bath; the main problem is the tendency for the alginate to precipitate rather than form a three-dimensional gel network. Alginic acid gels tend to be grainy with a high level of syneresis.

The pH of a sodium alginate solution can effect the amount of calcium required to form a gel. For example, an alginate solution at pH 4 requires half the level of calcium to form a gel of that of the same alginate solution at pH 6.

6.d.3 Effects of Proteins

Under acid conditions, proteins interact with sodium alginate.[27] The interaction is considered to be electrostatic in nature as non-ionic polysaccharides do not show any pH-dependent changes in behaviour with proteins.[28] The strength of the interaction between proteins and alginate increases to a maximum as the pH is reduced.[29] When heated at acid pH, alginate destabilizes proteins so that they denature at a lower temperature than does the protein alone, and stable, soluble high-molecular-weight complexes are produced.[29]

This means that caution should be exercised when heating solutions of alginate with proteins, particularly at acid pH.

6.d.4 Effects of Neutral Polymers

Neutral polymers, at reasonable concentrations, have very little effect on the formation of alginate gels. However, their presence in the formed gel may lead to alterations of the rheological characteristics as compared to those of the 'pure' alginate gel.

7 APPLICATIONS

Over the years, numerous applications of alginates in food systems have been reported (see, e.g., Ref. 30). Therefore, it is of limited value to repeat them all again in this chapter. What will be done is to give examples of the various ways that alginate can be gelled, and give indications of the more recent attempts at exploiting alginate gel technology.

7.a Internally Set Gels
7.a.1 Structured Fruit
Internal setting at room temperature is used to prepare fruit analogues (such as apple, peach, pear and apricot) that have a close to uniform texture. A two-mix process involving rapid mixing is employed. One mix contains the alginate and the calcium ion source (anhydrous dicalcium phosphate), and the other mix contains fruit purée, sequestrant and acid (Table 3).

Note in Table 3 that part of the total sugar is in the alginate mix; this has been dry-blended with the alginate to assist in its dissolution.

Although the calcium source is in the alginate solution, no reaction occurs, because anhydrous dicalcium phosphate (DCP) is essentially insoluble at neutral pH. The structured fruit is prepared by pumping the two mixes through a suitable in-line high-shear mixer (e.g. Oakes) and allowing the final mixture to set under shear-free conditions. The gelling reaction is brought about by the calcium ions released from the DCP, which dissolves as the pH is lowered on contact with the purée phase. If the reaction is too rapid and gelation occurs in the mixer, the gel structure will be broken down irreversibly. Also, since the gel reaction takes place very rapidly after the two mixes come into contact, it is important that they remain separate until they reach the mixing head.

A convenient method of preparing structured fruit on a continuous basis is to extrude the mix as a slab on to a moving conveyor (Fig. 8). Once the slab reaches the end of the belt, setting has occurred and it can be diced or cut into fruit-like shapes.

TABLE 3
FORMULATION FOR INTERNALLY SET PEACH

Alginate	(%)	Fruit mix	(%)
Sodium alginate (high-G)	0·85	Fresh peach purée	33·55
Dicalcium phosphate anhydrous	0·30	Sucrose	10·00
		Glucose (dextrose)	5·00
Disodium hydrogen ortho-phosphate dodecahydrate	0·07	Citric acid (anhydrous)	0·80
		Sodium citrate dihydrate	0·65
Glucose (dextrose)	5·00		
Sucrose	5·00		
Water	38·78		
	50·00		50·00

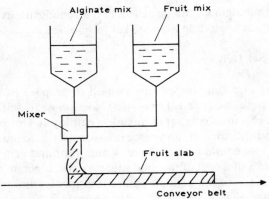

Fig. 8. Schematic layout of a process to produce internally set structured fruit.

7.a.2 Shear-reversible Dessert Gel

A shear-reversible alginate gel can be formulated by choosing alginate concentration and percentage calcium conversion such that the system lies in the thixotropic region (see Section 6.b.2), taking into account both pH and solids content.

A typical formulation for such a gel is shown in Table 4. A high-M alginate is chosen because a non-brittle, soft-textured gel with good freeze/thaw stability is normally required, and the wider range of calcium conversion over which shear-reversible gels can be formulated with a high-M alginate allows for greater recipe tolerance.

The product is prepared by first rapidly mixing the dry ingredients in

TABLE 4
FORMULATION FOR SHEAR-REVERSIBLE DESSERT GEL

	(%)
Part A	
Sodium alginate (high-M)	0·70
Calcium hydrogen orthophosphate (dihydrate)	0·07
Sodium hexametaphosphate	0·12
Sugar	37·60
Water	56·36
Part B	
Citric acid (anhydrous)	0·15
Water	5·00

Part A into the water. This mixture is then heated to 70°C and the acid (Part B) is dissolved in the small amount of water (to aid more rapid and even dispersion in the final mixture) and added with vigorous stirring to Part A. The system is then allowed to cool. Addition of the acid results in the release of calcium ions from the calcium hydrogen orthophosphate dihydrate, but the high temperature prevents setting occurring until the system has cooled (see Section 6.d.1). Sodium hexametaphosphate is present both to sequester any calcium that is in the water and to control the rate of release of calcium when the acid is added. Too rapid release of calcium can result in localized gelation even at high temperature.

Shear-reversible alginate gels are ideal for the production of layered desserts. The gel can be aseptically prepared in bulk and then pumped into individual containers. As the gel is deposited cold, and regains a large proportion of its gel strength almost immediately after the shearing action is removed, other layers can be deposited on top without delay. Heat-sensitive materials such as whipped cream or ice-cream are unaffected by the gel, and, since it does not melt, the addition of hot layers such as custard can be tolerated.

7.b Diffusion-set Gels

Diffusion setting is ideally suited to the preparation of fruits with an outer skin and a liquid centre such as blackcurrants and blueberries. An alginate solution and a fruit purée mix are used, but they are kept separate and fed through nozzles consisting of two coaxial tubes, as shown in Fig. 9. This process is sometimes referred to as co-extrusion. A key element is the maintenance of a continuous flow of alginate solution coupled with intermittent pulsing of the central fruit purée stream. As each pulse of purée breaks away from the nozzle, it is coated with a thin uniform coating of alginate solution. The extruded berry capsule then drops into the calcium setting bath, where it rapidly acquires a resilient calcium alginate skin.

The distance of the co-extrusion nozzle above the setting bath is important; if it is too small, the berries do not have a chance to separate and therefore enter the setting bath like a 'string of beads'. If the distance is too great, the berries hit the setting bath with such a force that the berries either break up and form small berries, or they set misshapen. It is also critical to ensure that the viscosities of the fruit purée and the alginate are similar. This can be achieved by adjusting the level of CMC in the example fruit purée formulation.

A formulation for restructured blackcurrants is shown in Table 5. A high-M alginate is used because a flexible gelled alginate film is required.

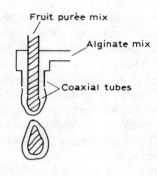
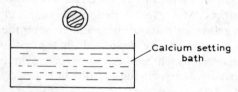

FIG. 9. Schematic diagram of a co-extrusion system for preparing reformed blackcurrants. (Unilever patent, BP 1 484 562.)

TABLE 5
FORMULATION FOR RESTRUCTURED BLACKCURRANTS

	(%)
Alginate mix	
Sodium alginate (high-M)	2·00
Water	98·00
Fruit purée mix	
Blackcurrant purée	41·0
Citric acid	0·2
Calcium lactate (pentahydrate)	1·0
Sugar	12·7
Cross-linked potato starch	1·7
Carboxymethyl cellulose	0·5
Water	42·8
Setting bath	
Calcium lactate (pentahydrate)	3·0
Sugar	12·7
Water	84·3

For taste reasons, calcium lactate is used in the setting bath. The presence of calcium lactate in the fruit purée results in gelation of the inner surface of contact between the alginate solution and the fruit purée, thus minimizing the time required in the setting bath.

7.c Recent Applications of Alginate Gel Technology

It seems incredible that, although alginates have been commercially available for more than 50 years, new and exciting uses for alginate are still being discovered. The following are just a few of the more recent applications that have been reported.

7.c.1 Use in the Extrusion of Proteins

It has been reported that alginate can improve the performance of a cooker extruding proteins, and can influence the texture of the extruded protein.[31]

7.c.2 Extending the Storage Life of Potatoes

Alginate has been shown to extend the storage time of potatoes, before budding commences, by up to 50%.[32]

7.c.3 Immobilization of Banana Enzymes

Use of alginate has helped in the elucidation of the pathways for starch metabolism in bananas. Immobilizing the whole banana pulp in a calcium alginate gel enabled the authors[33] successfully to assay the activity of the starch-degrading enzymes in the ripe banana pulp.

7.c.4 Use in Minced Fish Patties

The addition of alginate has been shown to improve the texture of patties made either from a mixture of fish, soy-flour and soy-protein concentrate,[34] or from pure minced 'sheepshead' fish.[35]

7.c.5 Use in Meat Products

Alginate gels have been used to produce structured beef steaks that bind not only in the cooked state but also in the raw, refrigerated state. It is claimed that this alginate gel technology will allow structured meat products to be marketed retail with other fresh meats. The technology will also allow structured beef to be marketed to the hotel, restaurant and institutional industries through existing fresh-meat channels. Because freezing of this product is not required, energy savings can be realized during marketing and final cooking from the raw as compared to the frozen state. This application of alginate gel technology could prove useful in the development of microwaveable structured beef steaks. The authors believe that this technology could represent a new generation of formed meat products that are void of some of the problems associated with salt and phosphate products.[36-38]

These applications are shown to illustrate the many ways in which alginate can be utilized to formulate gelled food products. It must be stressed that for any successful application care must be taken over formulation control. If the factors described in this chapter are taken into consideration, then the range of products that can be prepared from alginates is limited only by the imagination of the food technologist.

REFERENCES

1. Stanford, E. C. C. *J. Soc. Chem. Ind.*, **4** (1885) 518.
2. Krefting, A., An improved method of treating seaweed to obtain valuable products therefrom. British Patent 11 538 (1896).
3. Fischer, F. G. & Dorfel, H., Die Polyuronsauren der Braunalgen Kolenhydrate der Algen. Part 1. *Z. Physiol. Chem.*, **302** (1955) 186.
4. Haug, A., Larsen, B. & Smidsrod, O., A study of the constitution of alginic acid by partial acid hydrolysis. *Acta Chem. Scand.*, **20** (1966) 183.
5. Haug, A., Larsen, B. & Smidsrod, O., Studies on the sequence of uronic acid residues in alginic acid. *Acta Chem. Scand.*, **21** (1967) 691.
6. Haug, A., Myklestad, S., Larsen, B. & Smidsrod, O., Correlation between chemical structure and physical properties of alginates. *Acta Chem. Scand.*, **21** (1967) 768.
7. Atkins, E. D. T., Mackie, W. & Smolko, E. E., Crystalline structure of alginic acids. *Nature (London)*, **225** (1970) 626.
8. Atkins, E. D. T., Mackie, W., Parker, K. D. & Smolko, E. E., Crystalline structures of poly-D-mannuronic and poly-L-guluronic acids. *J. Polymer Sci., Part B: Polymer Lett.*, **9** (1971) 311.
9. Haug, A. & Larsen, B., Quantitative determination of the uronic acid content of alginates. *Acta Chem. Scand.*, **16** (1962) 1908.
10. Haug, A., Composition and properties of alginates. Rept No. 30, Norwegian Institute Seaweed Research, Trondheim, Norway (1964).
11. Sutherland, I. W., *Surface Carbohydrates of the Prokariotic Cell*. Academic Press, London, pp. 41–2.
12. Linker, A. & Jones, L. R., A new polysaccharide resembling alginic acid isolated from *Pseudomonas*. *J. Biol. Chem.*, **291** (1966) 3845–51.
13. Doggett, R. C. & Harrison, G. M., in *Proc. 5th Int. Cystic Fibrosis Conf.*, ed. D. Lawson. Cambridge University Press, Cambridge, 1969, pp. 175–8.
14. Gorin, P. A. J. & Spencer, J. F. T., Exocellular alginic acid from *Azotobacter vinelandii*. *Can. J. Chem.*, **44** (1966) 993–8.

15. Larsen, B. & Haug, A., Biosynthesis of alginates. *Carbohydrate Res.*, **17** (1971) 287–96.
16. Govan, J. R. W., Fyfe, J. A. M. & Jarman, T. R., Isolation of alginate-producing mutants of *Pseudomonas fluorescens*, *Pseudomonas putida* and *Pseudomonas mendocina*. *J. Gen. Microbiol.*, **125** (1981) 217–20.
17. Skjak-Braek, G., Grasdalen, H. & Larsen, B., Monomer sequence and acetylation pattern in some bacterial alginates. *Carbohydrate Res.*, **154** (1986) 239–50.
18. Skjak-Braek, G., Smidsrod, O. & Larsen, B. *Int. J. Biol. Macromol.*, **8** (1986) 330.
19. Donnan, F. G. & Rose, R. C. *Can. J. Res.*, **28** (1950) 105.
20. McDowell, R. H., *Properties of Alginates*. Kelco International.
21. Rees, D. A., Structure conformation and mechanism in the formation of polysaccharide gels and networks. In *Advances in Carbohydrate Chemistry and Biochemistry*, Vol. 24, ed. M. L. Wollfrom & R. S. Tipson. Academic Press, New York, 1969, p. 267.
22. Morris, E. R., Rees, D. A. & Thom, D., Characterization of polysaccharide structure and interactions by circular dichroism. Order–disorder transition in the calcium alginate system. *J. Chem. Soc. Chem. Commun.* (1973) 245.
23. Smidsrod, O. & Haug, A. *Acta Chem. Scand.*, **22** (1968) 1989.
24. Grant, G. T., Morris, E. R., Rees, D. A., Smith, P. J. C. & Thom, D., Biological interactions between polysaccharides and divalent cations: the egg-box model. *FEBS Lett.*, **32** (1973) 195.
25. Smidsrod, O., Haug, A. & Whittington, S. *Acta Chem. Scand.*, **26** (1972) 2563.
26. Arnott, S., Fulmer, A., Scott, W. E., Dea, I. C. M., Moorhouse, R. & Rees, D. A., The agarose double helix and its function in agarose gel structure. *J. Mol. Biol.*, **90** (1974) 269.
27. Imeson, A. P., in *Gums and Stabilisers for the Food Industry 2*, ed. G. O. Phillips, D. J. Wedlock & P. A. Williams. Pergamon Press, Oxford, 1984, p. 189.
28. Ganz, A. J., How cellulose gums react with protein. *Food Engineering*, **46** (1974) 67.
29. Imeson, A. P., Ledward, D. A. & Mitchell, J. R., On the nature of the interaction between some anionic polysaccharides and proteins. *J. Sci. Food Agric.*, **28** (1977) 661.
30. Cottrell, I. W. & Kovacs, P., in *Handbook of Water-Soluble Gums and Resins*, ed. R. L. Davidson. McGraw-Hill, New York, 1980, p. 2-1.
31. Oates, C. G., Ledward, D. A. & Mitchell, J. R., 7th World Congress on Food Science and Technology, Singapore, 1987.
32. An-qi, Xu, 7th World Congress on Food Science and Technology, Singapore, 1987.
33. Glass, R. W. & Rand, A. G. Jr, Alginate immobilisation of banana pulp enzymes for starch hydrolysis and sucrose interconversion. *J. Food Sci.*, **47** (1982) 1836.
34. Rockower, R. K., Deng, J. C., Otwell, W. S. & Cornell, J. A., Effect of soy flour, soy protein concentrate and sodium alginate on the textural attributes of minced fish patties. *J. Food Sci.*, **48** (1983) 1048.
35. Ahmed, E. M., Cornell, J. A., Tomaszewski, F. B. & Deng, J. C., Effects of tripolyphosphate and sodium alginate on the texture and flavour of fish patties prepared from minced sheepshead. *J. Food Sci.*, **48** (1983) 1078.

36. Means, W. J. & Schmidt, G. R., Algin/calcium gel as a raw and cooked binder in structured beef steaks. *J. Food Sci.*, **51** (1986) 60.
37. Means, W. J., Clarke, A. D., Sofos, J. N. & Schmidt, G. R., Binding, sensory and storage properties of algin/calcium structured beef steaks. *J. Food Sci.*, **52** (1987) 252.
38. Means, W. J. & Schmidt, G. R., Process for preparing algin/calcium gel structured meat products. US Patent 4 603 054 (1986).

Chapter 3

CARRAGEENANS

Norman F. Stanley
*Marine Colloids Division, FMC Corporation, BioProducts Department,
Rockland, Maine 04841, USA*

1 INTRODUCTION

The term 'carrageenan' is used to name a class of galactan polysaccharides that occur as intercellular matrix material in numerous species of red seaweeds (marine algae of the class Rhodophyta).

Carrageenans serve a function in the structure of the plant analogous to, but differing from, that of cellulose in land plants. Whereas land plants require a rigid structure capable of withstanding the constant pull of gravity, marine plants need a more flexible structure to accommodate the varying stresses of currents and wave motion. They have adapted accordingly by developing hydrophilic, gelatinous structural materials having the necessary compliancy—the carrageenans are important classes of this type of material.

Humanity, with its aptitude for turning nature's inventions to its own use, has put carrageenans to use as thickening and gelling agents in a host of food and industrial applications. Carrageenans have been used to thicken foods for centuries, the earliest recorded use being in Ireland,[1] where the red alga *Chondrus crispus* was boiled with milk to give a thickened product; hence the name Irish Moss for *Chondrus crispus*. Since then carrageenans have been used commercially in a wide range of thickening, suspending and gelling applications.

2 STRUCTURE

Carrageenans are linear polysaccharides made up of alternating β-1,3- and α-1,4-linked galactose residues. Thus, the repeating units are disaccharides.

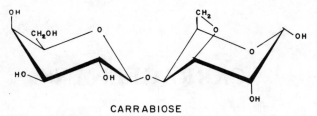

CARRABIOSE

FIG. 1. Basic repeating unit of carrageenans.

The 1,4-linked residues are commonly, but not invariably, present as the 3,6-anhydride. Carrageenans differ from agars in that the 1,4-linked residue in agars is the L-enantiomer, whereas in carrageenans it is the D-enantiomer; the 1,3-linked residues are D-galactose in both agars and carrageenans. Figure 1 shows the general structure of the repeating unit, termed carrabiose—4-β-D-pyranosyl-3,6-anhydro-α-D-galactopyranose. Variants on this basic structure result from substitutions on the hydroxyl groups of the sugar residues and from the absence of the 3,6-ether linkage. Substituents may be either anionic (sulphate, pyruvate) or non-ionic (methoxyl).

In contrast to agars (see Chapter 1 for more details), carrageenans characteristically are highly sulphated. The 1,3-linked D-galactose residues occur as the 2- and 4-sulphate, or are occasionally unsulphated, while the 1,4-linked residues occur as the 2-sulphate, the 6-sulphate, the 2,6-disulphate, the 3,6-anhydride and the 3,6-anhydride 2-sulphate. Sulphate at C-3 apparently never occurs. Pyruvate has been reported to be present in the carrageenans from some *Gigartina* species; these carrageenans have been termed 'π-carrageenan.[2-4] Methoxyl groups occur in sulphated galactans from the *Grateloupiaceae* family.[5-7] There is some question, though, as to whether these have the alternating structure characteristic of carrageenans.[8]

In their original work on fractionation of the carrageenan from *Chondrus crispus* with potassium chloride, Smith and Cook[9] isolated two fractions, which they named κ- and λ-carrageenan; κ was defined as the fraction that was precipitated by potassium chloride, while λ was the fraction that remained in solution. Chemical studies of these fractions revealed that nearly half of the sugar units in κ were 3,6-anhydro-D-galactose, while λ contained little or none of this sugar.[10]

Owing largely to the investigations of Rees and his co-workers,[11-22] carrageenans are now defined in terms of chemical structure. While it is true that more or less of a continuous spectrum of carrageenans exists,[23] it is

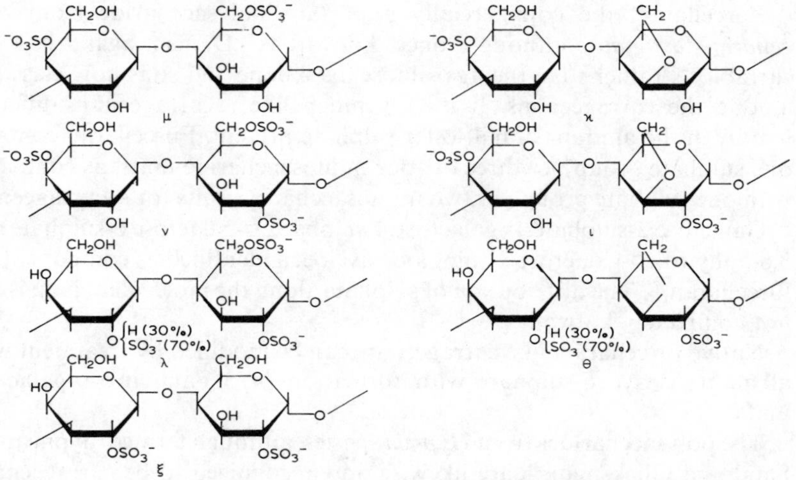

FIG. 2. Repeating units of limit carrageenans. (Reproduced with permission from R. L. Davidson (ed.), *Handbook of Water-Soluble Gums*, McGraw-Hill, New York, 1980.)

nevertheless possible to distinguish a small number of ideal or limit polysaccharides. The names μ, κ, ν, ι, λ, θ and ξ are presently applied to these limit carrageenans. Figure 2 shows the ideal structures of these galactans. μ and ν are believed to be precursors in the biosynthesis of κ and ι respectively,[5,24] the transformations being accomplished in the alga by a desulphatase, 'dekinkase',[25] or, in industrial processing, by the base-catalysed $S_N 2$ elimination of 6-sulphate.[26] λ likewise can be at least partially converted to θ-carrageenan by this reaction, but θ has yet to be identified as occurring naturally.

ξ-Carrageenan, which constitutes the KCl-soluble fraction of some *Gigartina* species (*G. chamissoi* and *G. canaliculata*), has not been completely characterised but seems to differ from λ in that the 1,3-linked residues are completely sulphated at C-2, while at least some of the 1,4-linked residues are unsubstituted at C-6.[19]

A new family of carrageenans for which the 1,3-linked residues are unsulphated and the 1,4-linked units lack sulphate at C-2 has recently been reported as the polysaccharide of *Eucheuma gelatinae*. This family comprises β-carrageenan, analogous to κ but lacking sulphate on the 1,3-linked residues, and its precursor, γ-carrageenan, analogous to μ.[27] β-Carrageenan occupies a position in the carrageenan system of polysaccharides analogous to that of agarose in the agar system.

Furcellaran, the commercially important polysaccharide from *Furcellaria fastigiata*, although once known as 'Danish agar', has the carrabiose, rather than the agarobiose, backbone and so is now classified among the carrageenans. It is very much like κ-carrageenan, differing mainly in the amount of half-ester sulphate present. Furcellaran contains one sulphate group per three or four monosaccharide units, as compared with one sulphate group per two monosaccharide units for κ-carrageenan. D-Galactose 2-sulphate, D-galactose 4-sulphate, D-galactose 6-sulphate and 3,6-anhydro-D-galactose 2-sulphate have been identified as components of furcellaran.[28] The distribution of sulphate along the molecular chain is still not completely known.

Native furcellaran, like carrageenan, can be modified by treatment with alkali to cleave 6-sulphate with formation of 3,6-anhydro-D-galactose units.

The polysaccharides from *Hypnea* species, although termed 'hypnean' by Sand and Glicksman,[29] are likewise now recognized to be carrageenans. The earlier distinction from carrageenan was based on the observation that 'hypnean' gels can be dewatered by a freeze–thaw process, whereas

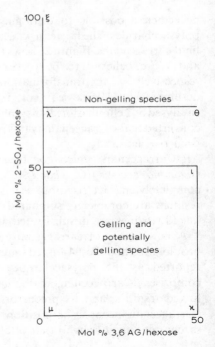

FIG. 3. Composition of carrageenans. (Reproduced with permission from R. L. Davidson (ed.), *Handbook of Water-Soluble Gums*, McGraw-Hill, New York, 1980.)

carrageenan gels cannot. This also led early on to the erroneous characterization of *Hypnea* extractives as agars. Furneaux and Miller[30] characterized the polysaccharide from *H. spicifera* as κ-carrageenan. Laserna *et al.*[31] reported a low half-ester sulphate content (17·72%) for the polysaccharide from Philippine *H. cervicornis*, which would make it more like furcellaran. It is likely that the carrageenans from different *Hypnea* species differ in composition. Greer *et al.*,[32] using enzymic degradation and ^{13}C NMR spectroscopy, determined that the carrageenan from *H. musciformis* consisted mainly of κ-type repeating units with minor proportions of ι-type units and other components.

Native carrageenans from different algae may therefore be regarded as varying mixtures of the limit polysaccharides and intermediate hybrids ranging in degree of anhydridation and 2-sulphation of the 1,4-linked residues. This is shown graphically in Fig. 3. This diagram divides the carrageenans into two general groups. One consists of μ, ν, κ, ι and their hybrids. Furcellaran and the *Hypnea* galactans also fit into this scheme. Carrageenans in this group form solutions that are gelled by potassium ions, or can be made to do so by treatment with alkali (see Section 6.c for more details); they are characterized by having their 1,3-linked residues either unsulphated or sulphated only at C-4. The other group comprises λ, ξ, θ and their hybrids; these do not gel either before or after alkaline treatment and characteristically have both their 1,4- and 1,3-linked units sulphated at C-2, though occasionally the latter are unsulphated.[33]

The complex fine structure of the carrageenans is still an active field of research. Enzymic, immunological and ^{13}C NMR techniques have proved to be powerful tools for these investigations.

3 SOURCE AND PRODUCTION

3.a Source
Carrageenan-yielding species (carrageenophytes) have been reported as occurring in seven different families: Solieriaceae, Gigartinaceae, Furcellariaceae, Phyllophoraceae, Hypneaceae, Rhabdoniaceae and, most recently, Rhodophyllidaceae.[34] Table 1 lists carrageenophytes that have been exploited commercially.

3.b Production
Carrageenans, which exist in the algae as gels at the temperature of the natural environment, are recovered from the algae by extraction with water

TABLE 1
ALGAE YIELDING CARRAGEENANS

Name	Location	Type
Eucheuma cottonii	Philippines, Indonesia	κ
Eucheuma spinosum	Philippines, Indonesia	ι
Chondrus crispus	Canada (Maritimes), USA (New England), France	$\lambda,^a \kappa^b$
Chondrus ocellatus	Korea	$\lambda,^a \kappa^b$
Gigartina stellata	France, Spain	$\lambda,^a \kappa^b$
Gigartina acicularis	France, Spain, Portugal, Morocco	λ
Gigartina pistillata	France, Spain, Portugal, Morocco	λ
Gigartina canaliculata	Mexico	$\lambda,^a \kappa^b$
Gigartina chamissoi	Mexico	$\lambda,^a \kappa^b$
Gigartina radula	Chile	$\lambda,^a \kappa^a$
Gigartina skottsbergii	Chile	$\lambda,^a \kappa^b$
Furcellaria fastigiata	Denmark, Canada (Maritimes)	Furcellaran
Hypnea musciformis	North Carolina, Brazil	κ

[a] From diploid plants (tetrasporophytes).
[b] From haploid plants (gametophytes).

at temperatures above the melting points of their gels. Extraction is usually conducted under alkaline conditions. The alkali functions to assist extraction by macerating the algae, to retard acid-catalysed depolymerization of the galactan, and to modify the structure of the galactan chain to enhance the gelling properties of the extract. Although extraction is simple in principle, sophisticated extraction technologies have been developed by the producers of carrageenans, with details generally guarded as trade secrets.

Three processes are in common use. These are:

- alcohol precipitation
- freeze–thaw
- gel press

Typical steps in the manufacture of carrageenan by the alcohol process are as follows.

(1) *Extraction.* The selected seaweed is first washed to remove sand, stones, salt and other impurities. The carrageenan is then extracted from the weed with dilute alkali. Extraction takes from 1–24 h to complete.

(2) *Purification.* The seaweed residues are separated from the extract by centrifugation or filtration and the extract is purified by filtration

through porous silica or activated charcoal to yield a carrageenan solution of 1–2%.
(3) *Concentration.* Multistage vacuum evaporators are used to concentrate the carrageenan extract to 2–3%.
(4) *Precipitation.* The carrageenan is precipitated from solution by the addition of isopropyl alcohol to yield a fibrous coagulum, which is separated and pressed to remove excess liquid.
(5) *Drying.* The recovered carrageenan is dried using either steam-heated drum driers or vacuum or inert gas-filled drying cabinets. The dried product is then ground to the required particle size.

The freeze–thaw and gel press processes differ from the alcohol process principally in the manner of recovering the carrageenan from solution (step 4).

The freeze–thaw process employs the same principle of dewatering a frozen gel that is used for agar recovery (see Chapter 1 for more details). Originally developed for furcellaran production, it is also employed for κ-carrageenan. Following concentration by evaporation, the extract is extruded through spinnerets into a cold solution of potassium chloride. The resulting gelled threads are further dewatered by subsequent potassium chloride washes followed by pressing. The gel is then frozen, thawed, chopped, again washed with fresh KCl solution and air-dried.

A limitation of the freeze–thaw process as applied to carrageenans is that it is applicable only to furcellaran and κ-carrageenan, which are the only types whose gels made with potassium ion exhibit marked syneresis. Moreover, the requirement that potassium be present precludes making products wherein sodium is the only counter-ion.

The gel press process, used by some minor producers of carrageenans, likewise relies on pressure to dewater the gel, but omits the freeze–thaw cycle.

4 COMMERCIAL AVAILABILITY

The functionality of carrageenans in various applications depends largely on their rheological properties. Unlike agars, which for most functional purposes can be considered neutral polymers, carrageenans typically carry a high density of half-ester sulphate groups, which are strongly anionic, being comparable to inorganic sulphate in this respect. The free acid is unstable, and commercial carrageenans are available as stable sodium, potassium and calcium salts or, most commonly, as a mixture of these. The associated cations, together with the conformation of the sugar units in the polymer chain, determine the solution and gelling properties of the carrageenans.

4.a Types Available

Commercial carrageenans are usually sold as κ, ι or λ forms. These commercial carrageenans are not pure κ, ι or λ, but will contain a varying amount of the other types, the exact amount depending on the weed source and extraction procedure.

Carrageenans specifically tailored for water-thickening applications are usually λ types or the sodium salts of mixed λ and κ. They dissolve in either cold or hot water to form viscous solutions. For this purpose, high water viscosities are desirable, and the high molecular weight and hydrophilicity of λ contribute to this. For gelling applications, a low viscosity in hot solution is usually desirable for ease in handling, and, fortunately, high-gel-strength carrageenans (mixed calcium and potassium salts of κ or ι) fulfil this requirement because of their lower hydrophilicity and the effect of the calcium ions.

Seaweed flour, in which the κ-carrageenan has been alkali-modified without extraction from the weeds, is available for petfood applications (see Section 7.b for more details).

4.b Molecular Weight

Both synthetic polymers and naturally occurring polysaccharides are polydisperse, meaning that in general they do not have sharply defined molecular weights, but rather have average molecular weights representing a distribution of molecular species identical in structure but of varying chain length. Various types of averages, such as number-average ($\bar{M}_n$) and weight-average ($\bar{M}_w$), are defined.

Carrageenans, being polysaccharides, are polydisperse. Commercial food-grade carrageenans typically have number-average molecular weights in the region of 200 000–400 000 daltons.

The functionality of carrageenans in most applications depends on molecular weight, and is largely lost at molecular weights under 100 000 daltons. Figure 4 shows cumulative molecular weight distributions for carrageenans of various number-average molecular weights. As this shows, food-grade carrageenans have little material in this region.

4.c Standardization

Since the seaweed stock that is used for the raw material undergoes natural variations, the carrageenan that results from the extraction process will

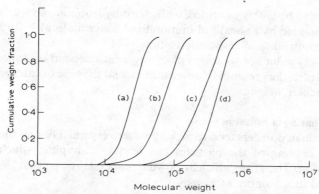

FIG. 4. Cumulative molecular weight distributions of carrageenans. (a) ι, $\bar{M}_n = 14\,300$; (b) ι, $\bar{M}_n = 47\,700$; (c) ι, $\bar{M}_n = 160\,000$; (d) κ, $\bar{M}_n = 266\,000$. (Reproduced with permission from R. L. Davidson (ed.), *Handbook of Water-Soluble Gums*, McGraw-Hill, New York, 1980.)

also vary. Therefore, in order for the carrageenan to perform consistently in a particular application, the carrageenans are adjusted, or standardized, to give a constant value in the following tests:

(a) gel strength in water;
(b) gel strength in milk;
(c) viscosity;
(d) stabilizing efficiency in milk.

Standardization is normally effected by the addition of sugar, or by blending different product batches.

4.d Availability

Carrageenans are normally shipped in 90·8 kg (200 lb) polythene-lined fibre drums. Care should be taken to avoid excess humidity, particularly after the drum has been opened, because the carrageenan will cake and this makes handling difficult.

5 SOLUTION PROPERTIES

Carrageenans typically form highly viscous solutions when dissolved in water. This is due to their linear macromolecular structure and polyelectrolytic nature. The mutual repulsion of the many negatively-charged half-ester sulphate groups along the polymer chain causes the molecule to be highly extended, while their hydrophilic nature causes it to be surrounded by a sheath of immobilized water molecules. Both of these factors contribute to resistance to flow.

Viscosity values of solutions of carrageenans depend on concentration, temperature, the presence of other solutes, and the type of carrageenan and its molecular weight.

5.a Preparing a Solution

Carrageenan powders tend to pick up water very quickly, and the resultant viscous coating of the particles can lead to clumping, which impedes dissolution. Therefore, efficient powder dispersal is essential. This can be achieved in a number of ways.

(a) Always use cold water (or milk) to disperse the carrageenan, then heat to dissolve.
(b) Use a high-shear mixer, which creates a vortex in the solution (without cavitation to entrap air) and add the carrageenan slowly into the vortex.
(c) Add a diluent such as sugar to the carrageenan—at least 3 parts sugar to 1 part of carrageenan.
(d) Use a 'solution-retardant' such as liquid sugar, alcohol or glycerin as an initial dispersant, since carrageenan will not dissolve in them.

All carrageenans are soluble in hot water ($>75°C$). Up to 10% solutions of commercial carrageenan can normally be handled by conventional mixing equipment. Sodium salts of κ and ι are soluble in cold water, while salts of other cations such as Ca^{2+} and K^+ do not dissolve completely but exhibit a varying degree of swelling. λ is fully soluble in cold water.

5.b Measuring Viscosity

In industrial practice, viscosity measurements are commonly made with easily operated rotational viscometers, such as the Brookfield viscometer. Commercial carrageenans are generally available in viscosities ranging from about 5 mPa s to 800 mPa s when measured at 75°C and 1·5% concentration. Carrageenans having viscosities less than 100 mPa s have flow properties very close to Newtonian. The degree of deviation from Newtonian flow increases with concentration and molecular weight of the carrageenan. The solutions exhibit pseudoplastic (shear-thinning) flow properties and are sufficiently shear-dependent that it is necessary to specify the shear rate used for making the viscosity measurement. This is recognized in industry, and the specifying of spindle geometry and rotational speed is well accepted. Where non-Newtonian behaviour is expected, viscosity measurements should be made at a shear rate comparable to that encountered in the application considered.

The pseudoplastic behaviour of carrageenans corresponds closely to the power-law equation:[35]

$$\log \eta_a = \log a + (b-1) \log \dot{\gamma} \qquad (1)$$

where η_a is the apparent viscosity, $\dot{\gamma}$ is the shear rate, a is the consistency index, and b is the flow behaviour index. For carrageenan solutions, b is either unity (Newtonian flow) or less than unity (shear-thinning). For a wide range of shear rates (two or more decades), a better fit to experimental data can be had by adding a quadratic term:

$$\log \eta_a = \log a + (b-1) \log \dot{\gamma} + c \log^2 \dot{\gamma} \qquad (2)$$

The coefficient c in this equation is either zero (strict power-law behaviour) or negative.

5.c Effect of Concentration

Viscosity increases nearly exponentially with concentration. This behaviour is typical of linear polymers carrying charged groups and is a consequence of the increase with concentration of the interaction between polymer chains.

5.d Effect of Salts

Salts lower the viscosity of carrageenan solutions by reducing electrostatic repulsion among the sulphate groups. This behaviour is likewise normal for ionic macromolecules. At high enough salt concentration, however, κ- and ι-carrageenan solutions may gel, with an increase in apparent viscosity. This is particularly true for the strongly gelation-inducing cations, K^+ and Ca^{2+}. At temperatures above the gel melting point, however, Ca^{2+} lowers viscosity to a greater extent than does Na^+ or K^+.

5.e Effect of Temperature

Viscosity decreases with temperature. Again, the change is exponential. It is reversible provided that heating is done at or near the stability optimum at $c.$ pH 9, and is not prolonged to the point where significant thermal degradation occurs (see Section 5.f for more information). Both gelling (κ-, ι-) and non-gelling (λ-) carrageenans behave in this manner at temperatures above the gelling point of the carrageenan. On cooling, however, the gelling types will abruptly increase in apparent viscosity when the gelling point is reached, provided that the counter-ions (notably K^+ or Ca^{2+}) promotive of gelation are present. For κ-carrageenan the transition is sharper with Ca^{2+} than with K^+. The effect for ι-carrageenan has not been reported.

5.f Effect of Molecular Weight

Viscosity increases with molecular weight in accordance with the Mark–Houwink equation:

$$[\eta] = KM^\alpha \tag{3}$$

where $[\eta]$ is intrinsic viscosity, M is an average molecular weight (since carrageenans are polydisperse), and K and α are constants. Intrinsic viscosities correlate well with practical viscosity measurements taken at 75°C and 1·5% concentration.

Loss of molecular weight may occur through depolymerization of carrageenan. Carrageenans are particularly susceptible to depolymerization through acid-catalysed hydrolysis. This is related to 3,6-anhydride content. Cleavage occurs preferentially at the $1 \rightarrow 3$ glycosidic linkages and is promoted by the strained ring system of the anhydride. Sulphation at C-2 of the 1,4-linked units appears to mitigate attack by acid, degradation rates for ι-carrageenan being roughly one-half those for κ-carrageenan under like conditions. Carrageenans in the gel state are more stable to acid than those in the sol state. Depolymerization follows pseudo-zero-order kinetics in either state, but the rate for the gel state is about one-sixth of that for the sol state. The secondary and tertiary structures developed on gelation may act to shield the glycosidic bonds from attack. This effect permits the use of carrageenans in low-pH systems, such as relishes, if enough potassium salt is present to retain the gel state (usually as a broken gel) during the shelf life of the product.

The rate of acid-catalysed depolymerization is proportional to hydrogen-ion activity. Carrageenans are relatively resistant to alkaline degradation, though this does occur through 'peeling' reactions at high pH. As a consequence, carrageenans have maximal stability slightly on the alkaline side, at about pH 9.

The rate of acid-catalysed depolymerization increases with temperature in accordance with the Arrhenius equation, with activation energies in the range of 84–126 kJ/mol. Carrageenans with high 3,6-AG content tend to have activation energies at the low end of this range.

Mathematical relations have been developed that afford good estimates of depolymerization, as measured by viscosity or molecular weight loss, for

TABLE 2
HOLDING TIMES FOR MINIMAL (20–25%) GEL STRENGTH LOSS

Temp. (°C)	Final pH						
	3	3·5	4	4·5	5	5·5	6
120	2 s	6 s	20 s	1 min	3 min	10 min	30 min
110	6 s	20 s	1 min	3 min	10 min	30 min	1·5 h
100	20 s	1 min	3 min	10 min	30 min	1·5 h	5·0 h
90	1 min	3 min	10 min	30 min	1·5 h	5·0 h	15·0 h
80	3 min	10 min	30 min	1·5 h	5·0 h	15·0 h	2·0 days
70	10 min	30 min	1·5 h	5·0 h	15·0 h	2·0 days	6·0 days
60	30 min	1·5 h	5·0 h	15·0 h	2·0 days	6·0 days	20·0 days
50	1·5 h	5·0 h	15·0 h	2·0 days	6·0 days	20·0 days	60·0 days
40	5·0 h	15·0 h	2·0 days	6·0 days	20·0 days	60·0 days	200·0 days

a given set of processing conditions defined by pH, temperature and time. Table 2 shows approximate holding times at various pH values of κ-carrageenan sol, cont[aining] [...] than 20–25% of the orig[inal ...] the gel. In general, each [...] holding time by a factor [...] also decrease holding ti[me ...] dependent on changes i[n ...] other ingredients, such a[s ...]

6.a Preparation of Gels

Gel formation may be t[...] crystallization from solu[tion ...] macromolecule, it may sep[arate ...] entraps the solvent to prod[uce ...] gelling-type carrageenans [...] heat to bring them into so[lution ...] ...utions set to gels. The anionic carrageenans require specific counter-ions, notably potassium and calcium, to be present for gelation to occur.

For example, a 2% aqueous κ-carrageenan gel may be prepared by dispersing 10 g carrageenan and 1 g potassium chloride in 490 ml of distilled water and heating to 80°C in a water bath. Remove from the water bath and adjust the weight to 500 g with distilled water. Allow the solution to cool for at least 1 h until a firm gel has formed.

6.b Measurement of Gel Strength

Gel strength is usually taken as the force required for a plunger of known dimensions to break the gel. Measurement is carried out using any standard gel tester such as the Marine Colloids Gel Tester.

Because many of the applications for carrageenan are milk-based, the gel strength of the carrageenan dissolved in milk is often measured.

6.c Mechanism of Gel Formation

According to Rees,[36] carrageenans that form aqueous gels do so because of association of the molecular chains into double helices. Evidence for double-helix formation has been adduced from X-ray diffraction patterns of fibres of ι-carrageenan.[37] Additional evidence for this model has been obtained from equilibrium studies that demonstrate that segments of carrageenan molecules, too short to form an extended network, reversibly dimerize on cooling of their solutions.[38,39]

At temperatures above the gelling point of the sol, thermal agitation overcomes the tendency of the carrageenan to form helices, and the polymer exists in solution as random coils. A recently advanced model for gelation in these systems postulates that, on cooling, the polymer chains associate into double helices, interrupted by irregularities (1,4-linked galactose 6-sulphate or 2,6-disulphate units), which interlink to form relatively small domains. This occurs regardless of the counter-ions present, and does not directly lead to gelation. For gelation to occur, the

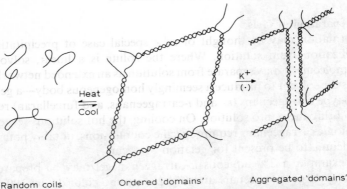

Random coils   Ordered 'domains'   Aggregated 'domains'

FIG. 5. Schematic representation of the domain model of carrageenan gelation.

domains must aggregate to form a three-dimensional network (Fig. 5).[40] Formation of the network in this manner alleviates the topological difficulty associated with the winding-up of helices in an extended network.

In the case of ι- and κ-carrageenans, only when potassium or other gel-promoting cations are present will the domains aggregate into a three-dimensional network. This cation specificity has been studied extensively. Charge-shielding, leading to decreased repulsion between chains, should favour aggregation, but ought not to be specific. The radius of the hydrated cation appears to be a factor; bulky cations, such as tetramethyl-ammonium, do not induce gelation. Bayley,[41] on the basis of X-ray diffraction data, proposed an 'egg-box' structure wherein cations of a certain size lock adjacent chains into an aggregate. However, this assumed a model of carrageenan primary structure that is now known to be incorrect. In the context of the domain model, cations are assumed to lock together helical regions of adjacent domains to build up the network. This can be regarded as a revival of Bayley's concept, but on a tertiary rather than secondary level. Painter,[28] by analogy with the solubilities of monosaccharide sulphates, ascribed the gelation of κ-carrageenan and

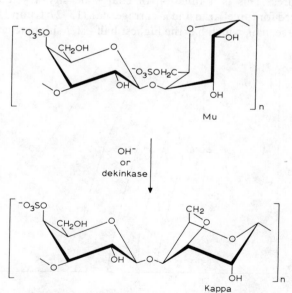

FIG. 6. Conformation of repeating units of μ- and κ-carrageenans. (Reproduced with permission from R. L. Davidson (ed.), *Handbook of Water-Soluble Gums*, McGraw-Hill, New York, 1980.)

furcellaran to reduced solubility in the presence of potassium ions. Despite the appearance of more than 30 publications over the past quarter-century, the nature of this cation specificity is still not completely understood.

An alternative model of carrageenan gelation, which postulates cation-induced aggregation of single helices, has also been proposed.[42]

Regardless of the mechanism, it appears that the occurrence of 6-sulphated 1,4-linked residues in the polymer chain of κ- or ι-carrageenan, or of furcellaran, detracts from the strength of their gels. This is ascribed to kinks, produced by these residues, in the chain. These inhibit the formation of double helices.[43] Alkaline treatment of carrageenan increases the gel strength of the product by removing these kinks through conversion of 6-sulphated residues to the 3,6-anhydride (Fig. 6). Decreased hydrophilicity from the removal of sulphate may also contribute to gelation.

6.d Properties of the Gels
6.d.1 Hysteresis

The sol–gel transition temperatures of carrageenans typically exhibit hysteresis, the melting temperatures being higher than the gelling temperatures. This is a function of charge density. Hysteresis is most pronounced for furcellaran and κ-carrageenan (15–27°C), and least (2–5°C) for ι-carrageenan, which has the highest half-ester sulphate content of the

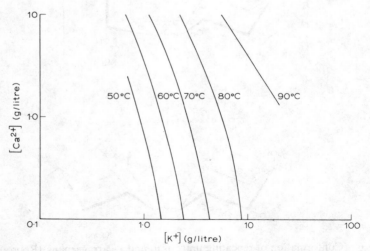

FIG. 7. Gel melting temperature contours for a 1·5% κ-carrageenan gel as a function of K^+ and Ca^{2+} concentrations. K^+ exerts the major effect.

gelling carrageenans. This hysteresis has been studied by optical rotation and can be explained thermodynamically in terms of a free-energy surface with two minima.[44]

6.d.2 Transition Temperatures

The transition temperatures of carrageenans depend primarily on the concentration of gelling cations present, and are relatively insensitive to carrageenan concentration. Although both K^+ and Ca^{2+} gel either κ- or ι-carrageenan, K^+ concentration is the factor controlling the sol–gel transition for κ (Fig. 7), whereas Ca^{2+} concentration controls the transition for ι (Fig. 8).

Transition temperatures have been determined by a variety of rheometric methods, as well as by optical measurements such as optical rotation and light-scattering. Values obtained by different methods vary somewhat, but in general the data are well fitted by functions of the form $\theta_g = a + b\sqrt{c}$, where θ_g is the gelling temperature and c is the concentration of gelling cation. This linear dependence on the square root of ion concentration indicates that Coulombic effects, rather than salting-out effects, control the sol → gel transition. Melting temperatures are similarly related to cation concentration.

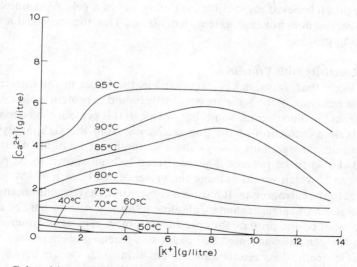

Fig. 8. Gel melting temperature contours for a 1·5% ι-carrageenan gel as a function of K^+ and Ca^{2+} concentrations. Ca^{2+} exerts the major effect.

6.d.3 Syneresis

κ-Carrageenan and furcellaran gels are relatively rigid and subject to syneresis in the presence of calcium ions. If potassium is the sole counter-ion, their gels are compliant, resembling those of ι-carrageenan. As it is difficult to manufacture carrageenans completely free of calcium, and to avoid introduction of this cation in applications, this is not a practical method of achieving the frequently desirable feature of compliancy. Incorporation of certain mannans, notably locust-bean galactomannan and konjac glucomannan, yields a more compliant gel that does not synerese. 'Smooth' regions of the mannan chain (i.e. regions carrying no galactose or glucose side-groups) are believed to bind to the double helices of the κ-carrageenan or furcellaran to reduce their tendency to form tightly-packed aggregates.[36] Guar galactomannan, which has a higher density of galactose side-groups, does not interact in this manner with κ-carrageenan or furcellaran. An alternative model, involving entanglement of mannan chains with the carrageenan gel network, has recently been proposed.[45-47] This model requires no discrete interaction with specific regions of the mannan chains.

ι-Carrageenan by itself yields compliant gels with very little tendency to undergo syneresis. Here the 2-sulphate groups on the 3,6-anhydride residues may act as wedging groups to prevent the tightly-packed aggregation believed responsible for the rigidity of κ gels. Also, unlike κ, ι-carrageenan does not aggregate with mannans. This, too, may be due to the wedging groups.

6.e Reactivity with Proteins

It is likely that agars and carrageenans as they exist in the intercellular matrix materials of the algae are covalently bound to protein moieties. That is, they are proteoglycans. Some, but not all, of this protein can be removed by alkaline extraction of the polysaccharide from the alga; analyses of agars and carrageenans always show a significant amount of nitrogen ($c.$ 0·1–1·0%) to be present. This has recently been shown to be protein nitrogen,[48] which would indicate the presence of as much as 6% protein bound to the carrageenan. It has been conjectured that the enhancement of gel strength of carrageenans by alkaline treatment may in part be due to removal of bound protein, thus producing a more regular polymer, as well as to the generally accepted $S_N 2$ chain-straightening mechanism.

Aside from this, electrostatic interactions can occur between the negatively-charged carrageenans and positively-charged sites on proteins. The commercially important interaction of carrageenans with milk proteins is an example of this. Work has also been done on carrageenans as stabilizers for protein systems other than milk, as precipitants of proteins from industrial waste streams, as enzyme inhibitors, as blood anticoagulants and lipaemia-clearing agents, and as immunologically active substances. With such a diversity of systems, one might expect to find, in detail, a corresponding diversity of mechanisms. However, for the most part, in view of the polyelectrolytic nature of both carrageenans and proteins, these involve electrostatic interactions. These interactions may be either specific or non-specific.

6.e.1 Specific Interactions

The prime example of a specific electrostatic interaction is the reaction of carrageenans with milk proteins above their isoelectric points. This has been investigated extensively (though from different viewpoints) mainly by Poul Hansen and his students at Ohio State University, and by T. A. J. Payens and his associates at NIZO (The Netherlands Institute for Dairy Research). The Ohio State work has been reviewed by Lin[49] and, most recently, by Hansen.[50] The work at NIZO is discussed at length in a thesis by Snoeren[51] and, as part of a general discussion of protein–polysaccharide interactions, by Stainsby.[52]

From this work it appears now to be quite well established that at least one aspect of the 'milk reactivity' of carrageenans is a gelation phenomenon involving a highly specific interaction between carrageenan and κ-casein. In order for it to be manifest, two events must occur:

(a) Carrageenan and κ-casein interact to form a complex.
(b) The resulting complex aggregates into a three-dimensional gel network.

Gelation is overtly manifest in such applications as blancmange-type puddings, while cocoa suspension in seemingly fluid chocolate milk occurs only below a well-defined sol–gel transition temperature.[53] Although carrageenans may interact with other casein fractions (α_{s1} and β) in milk, the binding is much weaker than that with κ-casein, and does not result in gel formation. Presence of the other fractions, as in whole casein, in fact decreases the sharpness of the sol–gel transition.

Specific interaction with κ-casein can occur at pH levels above the isoelectric point of the protein. (Non-specific interaction can, of course, occur below this point.) The former is a most convenient one from a commercial standpoint as it permits the carrageenan to be functional without disturbing the integrity of the casein micelles and of the milk as a

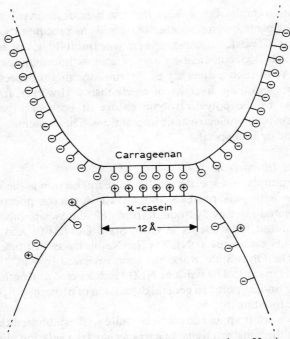

FIG. 9. Ionic interaction of carrageenan and κ-casein; pH $\geq$ isoelectric point. (Reproduced with permission from R. L. Davidson (ed.), *Handbook of Water-Soluble Gums*, McGraw-Hill, New York, 1980.)

whole. This interaction has been ascribed to electrostatic attraction between the negatively-charged sulphate groups of the carrageenan and a predominantly positively-charged region in the peptide chain of κ-casein (Fig. 9).[54] That the interaction is indeed electrostatic has been demonstrated by electrophoresis,[55] light scattering,[53] sedimentation[54] and dye binding.[51]

Interaction occurs despite the fact that the protein above its isoelectric point carries a negative net charge. This can be accounted for by a screened Debye–Hückel potential that, in effect, says that within a positively-charged region of the peptide chain the presence of neighbouring negative charges does not interfere with the electrostatic attraction.

Electron microscopy shows that κ-casein, as isolated from the milk protein system, occurs as independent spherical aggregates (submicelles) about 20 nm in diameter. In the presence of carrageenan these become aligned into thread-like structures likewise about 20 nm in width. These are thicker than the free carrageenan chains, which can also be seen but are only about 2 nm wide.[51,56] It thus appears that the κ-casein particles are attached, presumably by electrostatic attraction, to the carrageenan chain.

Although this association of κ-casein with carrageenan occurs with all carrageenan types, it is primarily the gel-forming types (κ and ι) that exhibit a sol–gel transition with temperature. (λ-Carrageenan of high molecular weight will form a very compliant milk gel, but only at higher concentrations than ordinarily employed for milk gelation with κ or ι.) This is in accord with the practical observation that κ- and ι-carrageenans are effective for the suspension of cocoa in chocolate milk, whereas λ-carrageenan is ineffective. It confirms that gelation is required, and suggests that formation of the gel network may be a function solely of the carrageenan component of the carrageenan–casein complex. Moreover, cocoa suspension does not occur if the carrageenan has a low molecular weight. This parallels the absence of gel structure in solutions of carrageenan of sufficiently low molecular weight, as the chains become too short to support an extended network. Again this supports the notion that the junction zones involve only carrageenan–carrageenan cross-linkages, and not carrageenan–casein or casein–casein linkages.

In milk we are dealing not with κ-casein *per se* but with the intact casein micelle, of which 8–15% is κ-casein. Part of this is exposed on the outer surface of the micelle, and so available to interact.[57] Thus, the interaction between carrageenan and the casein in milk can be explained in the same way as for κ-casein above. Evidence for this can be seen in electron micrographs of κ-carrageenan–milk gels.[58]

The fact that strong milk gels can be prepared at lower carrageenan concentrations (c. 0·2%) than are required for useful water gels led early to the conjecture that milk proteins are somehow involved in the gel cross-linkages. There appears to be no reason why more than one carrageenan chain could not attach to a single casein micelle to form a cross-link. However, the evidence adduced above argues against this interpretation. Furthermore, milk contains the cations (K^+ and Ca^{2+}) required for gelation of carrageenans. If these are replaced by Na^+ through ion exchange, the resulting 'sodium milk' will not gel with carrageenan. To substitute one conjecture for another, one might say that the protein acts solely as a filler to impart strength and rigidity to the gel. Nevertheless, it appears that the effect is specific for κ-casein, as no sol–gel transition occurs with α_{s1} or β-casein under like conditions.[53] The jury is still out on this point.

Also, the fact that carrageenans interact with casein above its isoelectric point early led to the hypothesis that the interaction was mediated by Ca^{2+}, with the divalent cation acting to cross-link negatively-charged sites on the carrageenan with negatively-charged sites on the protein. This may be the case with the α_{s1}- and β-fractions of casein, as these sediment on centrifuging in the presence of carrageenan and Ca^{2+} but not in the absence of Ca^{2+}. The interaction is fairly specific: whey proteins do not sediment under these conditions. κ-Casein, however, sediments with carrageenan both in the presence and absence of Ca^{2+}.[55] Payens[53] noted that whole casein with commercial κ-carrageenan showed a sol–gel transition in a milk salt solution but not in NaCl or KCl at the same ionic strength as the milk salts, nor in milk salts when diluted 1:1. He concluded that Ca^{2+} (from the milk salts) is involved in the phenomenon of milk reactivity. If so, it would appear to be involved in the carrageenan–carrageenan interaction to form junction zones rather than the carrageenan–κ-casein interaction, which should require no cationic mediation. However, this does raise the question of why the transition was not observed with K^+, which should also stabilize the junction zones. The role of calcium (both ionic and non-ionic) in milk systems is complex. Calcium is necessary to maintain the integrity of the casein micelle; however, intact micelles do not seem to be necessary to produce the sol–gel transition. As Payens also noted, a sharper transition is in fact observed with the submicelles of κ-casein. An additional effect of Ca^{2+} may be to neutralize negatively-charged sites on the peptide chain, thereby increasing the probability of interaction of carrageenan with the positively-charged region.

A less-specific type of carrageenan–protein interaction, also somewhat loosely called 'milk reactivity', is the stabilization of milk and other proteins *against* precipitation by Ca^{2+}. (This may seem paradoxical in view of the sedimentation of proteins discussed above. It illustrates the complex role of calcium in milk systems.) This has been extensively studied by Hansen and his students, and reviewed by Lin[49] and Hansen.[50] Although the presence of calcium is necessary to maintain the stability of casein micelles, the balance is a delicate one and an excess of Ca^{2+} can result in destabilization of the milk system with precipitation of the casein. Milks of normal heat stability have Ca^{2+} activities ranging from 2 to 4 mM in the ultrafiltrate and are destabilized if $[Ca^{2+}]$ goes above this upper limit. This can occur naturally ('Utrecht abnormality'[59]) or can be induced by adding Ca^{2+}.

The function of κ-casein in the micelle is as a protective colloid that shields the Ca^{2+}-sensitive caseins (α_{s1}- and β-) from contact with Ca^{2+}. One model of the casein micelle has the charged region of α_{s1}- or β-casein covered with a cap of κ-casein.[60,61] (This is *not* why it is called *kappa-casein*!) This protective action is also shared by carrageenan and furcellaran despite the fact that both α_{s1}- and β-casein are negatively charged above their isoelectric points. There is a direct relationship between the stabilizing efficacy of the carrageenan and its 3,6-AG content.

Electron microscopy of the κ-carrageenan/α_s-casein system shows the latter to exist as discrete aggregates (100–500 nm in diameter) entrapped by interconnecting strands of κ-carrageenan. Little or no protein was seen attached to the polysaccharide strands. The strands presumably are bundles of double helices that form junction zones of low protein reactivity, owing to self-aggregation.[49] Superficially this picture is similar to that of the carrageenan/κ-casein complex in milk gels. However, the mode of interaction appears not to be the same, lacking the highly specific mechanism peculiar to κ-casein. Several other mechanisms have been considered; one is that carrageenan binds Ca^{2+}, thus reducing its activity. However, Lin[62] showed that a large excess (up to 80 mM) of Ca^{2+} reduced stability but did not completely destabilize α_s-casein. This is far more Ca^{2+} than could possibly be bound by the κ-carrageenan (0·2 mM ester sulphate) present in the system studied, so that Ca^{2+} binding evidently does not play an important role.

Free-boundary electrophoresis[63] and sedimentation velocity[62] experiments suggest that no complex formation between α_s-casein and carrageenans occurs in the absence of Ca^{2+}. This has been verified by Skura and Nakai,[64] using a variety of physicochemical techniques. This is suggestive of ionic interaction mediated by Ca^{2+}; however, Lin has discounted this, since the stability of the complex is independent of ionic strength. Lin has therefore postulated that Ca^{2+} acts mainly to neutralize the charged ester sulphate groups of the carrageenan, and thus to promote double-helix formation and aggregation of the carrageenan chains.[49]

This entrapment of discrete casein particles thus seems to be mechanical, somewhat akin to cocoa suspension but on a submicellar scale. Stabilization consists in keeping the particles separated, thus preventing further aggregation that would lead to precipitation of the protein. Stability decreases with increasing α_s-casein concentration, in line with the increased probability of particle collisions leading to aggregation. Likewise, there appears to be a region of carrageenan molecular weight (100 000–300 000 daltons) that is optimal for stabilization. Lin[62] hazards that below 100 000 daltons the helical regions are too short to maintain adequate distance between protein particles, while carrageenans of very high molecular weight may have long helical regions with too few binding sites available to accommodate the protein. However, this concept would bear re-examination in the light of the more recent 'domain' theory of carrageenan aggregation. It is not at all clear how the entrapment zones differ from the rest of the chain.

As might be expected for a network model, the protective action decreases with increasing temperature. However, this does not show the sharp transition (gel melting) characteristic of whole-milk/carrageenan systems such as cocoa suspensions. This shows that the analogy between α_s-casein stabilization and cocoa suspension is not altogether apt, owing to the vast difference in size between casein aggregates and cocoa particles. For the former, destabilization on heating could be due principally to the increased probability of particle collisions from thermal agitation.

The non-gelling λ-carrageenan is relatively ineffective as a stabilizer for α_s-casein. Lin[49] ascribes this to the presence of 6-sulphate on the 1,4-linked galactose units, which prevents aggregation through double-helix formation to form regions to which protein cannot bind. Alkaline modification of λ-carrageenan to form θ-carrageenan enhances its stabilizing power, though not to the point of matching that of κ-carrageenan. θ-Carrageenan, like λ, is non-gelling. Examination of molecular models shows that the presence of 2-sulphate on the 1,3-linked galactose units interferes sterically with the aggregation of double helices.

Carrageenan can also stabilize other proteins against precipitation by Ca^{2+}. These include include the milk proteins, β-casein[65] and para-casein.[66] (Para-casein is casein from which the protective κ fraction has been stripped by enzymic digestion with rennin—see Chapter 4 for more details.) Non-dairy Ca^{2+}-sensitive proteins include soy, peanut, cottonseed and coconut proteins. Chakraborty and Randolph[67] likewise concluded that stabilization was due not so much to carrageenan/protein interaction as it was to conformational features unique to κ-carrageenan.

Other sulphated polysaccharides have also been investigated as stabilizers for α_s-casein.[49] Sulphated cellulose and starch were reported to be ineffective as stabilizers. Sulphated guar was effective only at high concentrations. Sulphated locust-bean gum (SLBG), however, proved effective at much lower concentrations, and indeed was better than κ-carrageenan from *C. crispus*. The system was less turbid than with carrageenan, owing to smaller protein aggregates (10 nm diameter). Lin and Hansen[65] proposed a model wherein protein aggregates were trapped within the 'hairy' regions of the SLBG, while ester sulphate groups on the 'smooth' regions interacted through Ca^{2+} mediation to cross-link the polysaccharide chains in a manner roughly analogous to double-helix cross-linking of carrageenan chains. The notable effectiveness of SLBG vis-à-vis κ-carrageenan or κ-casein for the prevention of the formation of large aggregates of $α_s$-casein was ascribed to the presence of many more protein-reactive centres permitting the entrapment of smaller aggregates. The model, nevertheless, is closely similar to that proposed for stabilization by κ-carrageenan, namely relying largely, if not entirely, on polysaccharide–polysaccharide interactions to form the entrapping network.

The presence of fat in milk products, particularly those that have been concentrated and homogenized such as evaporated milk and infant formulas, may affect the manner whereby carrageenans interact with milk proteins. Badui[68] demonstrated that stratification of carrageenan-stabilized infant formulas on storage and centrifugation resulted in the carrageenan's being found predominantly in the fat-rich layers. This suggests that carrageenan distributes itself according to the available fat globule area. In products of this sort it appears that carrageenans associate with fat-globule surfaces, probably through interaction with adsorbed proteins and phospholipids on the surface. This accounts for their effectiveness in preventing creaming.

There appear to be no specific interactions between carrageenans and native whey proteins, although non-specific, direct electrostatic interactions occur below the isoelectric point of the protein. The principal whey protein, β-lactoglobulin, is denatured on heating at neutral pH and appears to become more casein-like with respect to precipitation on acidification and its tendency to associate with κ-casein. The interaction of denatured β-lactoglobulin with carrageenan has not been investigated in depth, although some study was given to the effect of heat treatment of chocolate milk on its stabilization by carrageenan.[69] Also, statistical analysis has shown that the addition of carrageenan to sterilized milk concentrate, and the interaction between carrageenan and pre-warming, are highly significant in maintaining storage stability and flavour of the product.[70]

6.e.2 Non-specific Interactions

Carrageenans interact with nearly all proteins by direct electrostatic interaction at pH levels below the isoelectric point of the protein. Reaction depends on the protein–carrageenan net charge ratio, and thus is a function of the isoelectric point of the protein, the pH of the system, and the weight ratio of carrageenan to protein.[71] The extent of reaction is greatest when the net charge ratio is equal to unity; usually this results in the precipitation of a protein–carrageenan complex.

Carrageenans interact with β-lactoglobulin at or below pH 4, but not at pH 5 or higher.[72] The use of carrageenan for the recovery of protein from whey has been proposed, but, since the interaction is not specific for carrageenans, cheaper anionic polymers (e.g. carboxymethylcellulose) do the job nearly as well. As might be expected, the interactions of carrageenans with proteins in general, and, in particular, with such a complex colloidal system as milk, can be very complex, and so the mechanisms set forth above must be considered as simplifications (quite likely over-simplifications) of the actual situations. As such they are subject to alternative interpretations, and no doubt much further work could be done to resolve the fine details of the mechanisms. More fundamentally, questions could be directed at the correctness of our present models of the tertiary structures of both carrageenans and proteins.

7 APPLICATIONS

As food additives, carrageenans are more versatile than agars. In addition to being functional as water binders and gellants they also are effective thickening and stabilizing agents. Stabilization is a general term meaning the prevention of phase separation. It may involve water binding, thickening or gelation, or a combination of these functions. Secondary advantages of carrageenan usage include improved palatability and appearance. 'Mouth feel' is a common descriptor related to palatability. It encompasses such subjective characteristics as 'richness', 'smoothness', 'creaminess' and 'body'. Carrageenans can be very effective in imparting these characteristics to a food product.

Food applications of carrageenans are best considered by classifying them according to whether the use is in a milk or a water system. To be sure, milk *is* an aqueous system, but the distinction is a natural one owing to the previously discussed, unique 'milk reactivity' of carrageenans. Thus, carrageenans function differently in milk-based systems and in purely water-based ones.

7.a Milk Applications

Carrageenans may be incorporated in products, commonly dry mixes, that are added to milk by the consumer to produce the desired dish or beverage. In others the carrageenan is incorporated along with milk or milk solids during manufacture of the product. Table 3 gives typical milk (dairy) applications.

Dry mixes include beverage applications (milkshake and instant breakfast powders) wherein carrageenan is used to suspend the ingredients

TABLE 3
TYPICAL MILK (DAIRY) APPLICATIONS OF CARRAGEENAN

Use	Function	Product	Approximate use level (%)
Milk gels			
Cooked flans or custards	Gelation	$\kappa, \kappa + \iota$	0·20–0·30
Cold prepared custards (with added TSPP)	Thickening, gelation	κ, ι, λ	0·20–0·30
Pudding and pie fillings (starch base)	Level starch gelatinization	κ	0·10–0·20
Dry mix cooked with milk		κ	0·10–0·20
Ready-to-eat	Syneresis control, bodying	ι	0·10–0·20
Whipped products			
Whipped cream	Stabilize overrun	λ	0·05–0·15
Aerosol whipped cream	Stabilize overrun, stabilize emulsion	κ	0·02–0·5
Cold-prepared milks			
Instant breakfast	Suspension, bodying	λ	0·10–0·20
Shakes	Suspension, bodying, stabilize overrun	λ	0·10–0·20
Acidified milks			
Yogurt	Bodying, fruit suspension	κ + locust-bean gum	0·20–0·50
Frozen desserts			
Ice-cream, iced milk	Whey prevention, control meltdown	κ	0·010–0·030
Pasteurized milk products			
Chocolate, eggnog, fruit-flavoured	Suspension, bodying	κ	0·025–0·035
Fluid skim milk	Bodying	κ, ι	0·025–0·035
Filled milk	Emulsion stabilization, bodying	κ, ι	0·025–0·035
Creaming mixture for cottage cheese	Cling	κ	0·020–0·035
Sterilized milk products			
Chocolate, etc.	Suspension, bodying	κ	0·010–0·035
Controlled calorie	Suspension, bodying	κ	0·010–0·035
Evaporated	Emulsion stabilization	κ	0·005–0·015
Infant formulations	Fat and protein stabilization	κ	0·020–0·040

(Reproduced with permission from R. L. Davidson (ed.), *Handbook of Water-Soluble Gums*, McGraw-Hill, New York, 1980.)

and impart richness and body to the drink. As these are usually prepared with cold milk, a fine-mesh (<60 μm) λ-carrageenan is used for rapid cold solubility.[35]

Flans and custards using carrageenans are modern versions of the traditional blancmange, and are packaged as dry mixes for addition to milk by the consumer. Both cooked and cold-prepared types are available. For the former, κ-carrageenan is used to produce a delicate, brittle gel. A creamier product results if combinations of κ- and ι-carrageenans are used together with tetrasodium pyrophosphate (TSPP). The phosphate alone induces gelation of the milk, with the carrageenan contributing to firmness. The effect of the phosphate is not well understood, but may be to destabilize the milk system. A wide variety of textures can be produced in this manner. For cold-prepared mixes λ-carrageenan is again used to provide cold solubility, leading to an 'instant' set. Puddings of this type have particularly good flavour release. λ- or ι-carrageenan together with TSPP provide similar functionality to afford syneresis control and texture modification.[35]

A close relative of flan (both texturally and etymologically), *vla* is a Dutch dessert whose popularity has recently begun to spread beyond the confines of Holland. It is a custard-like dessert, but more fluid than flan. Carrageenan in combination with modified waxy maize starch is used to produce a smooth-textured flowable *vla*. It is important that a properly balanced ratio of carrageenan to starch be used to avoid serum separation and gelation during extended storage.[73]

Cooked puddings and pie fillings have traditionally been starch-based. However, κ-carrageenan can be used to provide a more uniform set.[35]

Chocolate milk typically contains 1% cocoa, 6% sugar and from 0·025% to 0·035% carrageenan. Vanillin is added as a flavour modifier. Most commonly these ingredients are sold to the dairy as a blended 'dairy powder' to be incorporated into the milk during pasteurization. On cooling, the interaction between κ-casein and carrageenan produces the delicate gel structure necessary to keep the cocoa suspended and give the drink a rich mouth-feel. κ-Carrageenan is used when the drink is pasteurized; ι and λ are also functional, though less economical.[35] Chocolate syrups are used instead of powders by some dairies, as well as by consumers. Here the syrup is dispersed into cold milk at a ratio of about 1 part syrup to 10–12 parts milk. As in other cold-processed applications, λ-carrageenan is used.[35]

Akin in principle to cocoa suspension in chocolate milk is the suspension of an insoluble calcium salt, or other nutrients, in fortified fluid milk. Tricalcium phosphate at levels of 0·2–1·0% of the milk is of sufficiently low solubility that the Ca^{2+} activity of the milk does not rise high enough to destabilize the system. Carrageenan at a level of 0·025–0·040% will suspend the calcium phosphate and inhibit whey-off without imparting any palpable gel structure or flavour to the milk.[74]

Ice-cream and sherbet have been defined as partly frozen foams comprising a gas (air) dispersed as small cells in a partially frozen continuous aqueous phase. This phase contains fat dispersed as the inner phase in an oil-in-water emulsion, non-fat milk solids and stabilizers in colloidal solution, and sugar and salt in true solution.[75–77] Both a primary and a secondary stabilizer are commonly used in the formulation. The primary stabilizer serves to prevent the separation or uneven distribution of fat and other solids, to prevent growth of large crystals of ice and/or lactose, and to impart a desirable texture to the product.

Carrageenans *per se* are unsuitable for primary stabilization, as they increase the viscosity of the mixture to such an extent that it is difficult or impossible to incorporate enough gum to stabilize it adequately. Locust-bean gum, guar and carboxymethylcellulose, separately or in combination, are the preferred primary stabilizers. However, they have the common defect of sometimes causing whey separation in the ice-cream mix. Employed at a low level as a secondary stabilizer, κ-carrageenan effectively prevents whey-off, both during storage of the unfrozen mix and thereafter under freeze–thaw conditions.

Filled and skim milks benefit from the addition of low levels of ι- or κ-carrageenan to stabilize emulsified fat and to simulate the appearance and mouth feel of whole milk.[35]

Another class of fluid milk products comprises articles that have been sterilized, rather than pasteurized, using high-temperature short-time (HTST) or ultrahigh-temperature (UHT) processing. These are packaged or canned so as to maintain sterility until consumed. κ-Carrageenan is used in them to provide suspension and bodying, and to prevent fat and protein separation.[78]

As little as 0·005% of κ-carrageenan in evaporated milk prevents fat and protein separation. Prior to this use of carrageenan it was necessary to turn over the cases of milk in storage to effect remixing of the separated phases. Carrageenan is now included in the US FDA Standards of Identity for evaporated milk.[35,78]

Infant formulations are sold either as concentrates to be diluted with water prior to use or in ready-to-use single-strength form. Most of these products comprise milk solids or caseinates and/or soy protein plus vegetable oil or butterfat. Partially alkali-modified κ-carrageenan is used to prevent fat and protein separation, manifest in such defects as 'fat capping'.

ι-Carrageenan may be used to replace part of the starch in canned, ready-to-eat milk puddings. This confers advantages during processing (reduced viscosity resulting in better heat transfer) and to the finished product (controlled degree of set, minimal syneresis and improved flavour release).[35]

Addition of carrageenan to natural cream confers improved stability to the whip in whipping cream. λ- or κ-carrageenan is used depending on whether addition is to the cold cream or during pasteurization.[35]

κ-Carrageenan, alone or together with locust-bean gum, is used to stabilize both natural and artificial aerosol-propelled cream toppings without development of excessive viscosity in the container. κ-Carrageenan at a level of about 0·03% suffices for natural cream, but a higher level (0·05–0·10%) is needed for the artificial type.[35]

Combinations of κ- and λ-carrageenans are used in frozen whipped toppings to improve body and reduce syneresis under freeze–thaw conditions.[35]

κ-Carrageenan can be used to stabilize yoghurt to which fruit is added. Dissolution of the gum is achieved by adding it prior to pasteurization and inoculation.[35]

Imitation milk products use sodium caseinate and/or soy protein in place of milk solids, while butterfat is replaced by vegetable fat. ι- and λ-carrageenan are added to stabilize the fat emulsion and provide body to the product.[35]

Carrageenan in combination with locust-bean gum is used in cottage-cheese production to prevent fat and whey separation and to provide 'cling' to the curd. It also induces small curd formation under low-pH conditions. κ-Carrageenan is used at levels of 0·01–0·05%, usually together with locust-bean gum, to induce curd formation, impart shape retention, and prevent syneresis in cottage and creamed cheese products.[35,79]

Carrageenan has been used in the production of hard cheeses to improve yield. Here it acts to effect precipitation of whey proteins along with the casein.

Imitation cheese is designed to duplicate the textural properties and functions of natural cheese but at a substantially lower cost. The principal ingredients are water, hydrogenated vegetable fat and sodium caseinate. κ-Carrageenan at a level of about 2·5% is used as a gellant to bind the ingredients together into a block. It also aids in texture modification, improves sliceability and shredability, and controls meltdown. The artificial product can be used effectively as an extender for natural cheese at ratios of up to 50% natural cheese.[80]

7.b Water Applications

Table 4 lists typical water applications. As with milk applications, the final product may be a dry mix to be reconstituted by the consumer or a manufactured ready-to-use product. Dry mixes may be formulated either for hot-water preparation or for cold-water, instant preparation.

Gelatine-based, hot-water-prepared dessert gels have long been deservedly popular for their sparkling clarity, smooth texture and palatability. Gelatine gels melt at mouth temperature, resulting in very attractive organoleptic characteristics, including rapid flavour release and smooth texture (see Chapter 7 for more details). These are most evident if the gel is consumed within a few hours after preparation. On prolonged storage, however, gelatine gels toughen, making them difficult to spoon and detracting markedly from their tender texture. ι-Carrageenan affords compliant gels that are strikingly similar to gelatine gels but suffer the disadvantages of having a high melting temperature, which detracts from

TABLE 4
TYPICAL WATER APPLICATIONS OF CARRAGEENAN

Use	Function	Carrageenan type	Approximate use level (%)
Dessert gels	Gelation	$\kappa + \iota$	0·5–1·0
		$\kappa + \iota +$ locust-bean gum	
Low-calorie jellies	Gelation	$\kappa + \iota$	0·5–1·0
Pet foods (canned)	Fat stabilization, thickening, suspending, gelation	$\kappa +$ locust-bean gum	0·5–1·0
Fish gels	Gelation	$\kappa +$ locust-bean gum, $\kappa + \iota$	0·5–1·0
Syrups	Suspension, bodying	$\kappa + \lambda$	0·3–0·5
Fruit drink powders and frozen concentrates	Bodying	Sodium κ	0·1–0·2
	Pulping effects	λ, potassium/calcium κ	0·1–0·2
Relishes, pizza, barbecue sauces	Bodying	κ	0·2–0·5
Imitation milk	Bodying, fat stabilization	ι, λ	0·5
Imitation coffee creams	Emulsion stabilization	λ	0·1–0·2
Whipped toppings (artificial)	Stabilize emulsion, overrun	κ, ι	0·1–0·3
Puddings (non-dairy)	Emulsion stabilization	κ	0·1–0·3

(Reproduced with permission from R. L. Davidson (ed.), *Handbook of Water-Soluble Gums*, McGraw-Hill, New York, 1980.)

the mouth feel. However, the fact that carrageenan gels do not melt at room temperature and do not require refrigeration to set the gel is a distinct advantage in the tropical market or where refrigeration is not available. A further advantage is that, unlike gelatine gels, they do not become tough on ageing. This is important for ready-to-eat desserts, items that are popular in Europe. They also are stable under freeze–thaw conditions.[81]

κ-Carrageenan yields rigid, fracturable gels quite unlike the compliant gelatine and ι-carrageenan gels. In the 1940s it was discovered that incorporation of locust-bean gum or agar into a κ-carrageenan gel improved compliance to yield gels of acceptable texture.[82–84] Commercial locust-bean gum, however, yields hazy dispersions, so that gels prepared using it are not sparkling clear. Methods for clarifying locust-bean gum have been developed, and some of the major carrageenan manufacturers also produce the clarified gum for use with their carrageenans.

ι-Carrageenan may also be used in combination with κ-carrageenan and locust-bean gum for further control of gel texture.[78]

Since the gelling carrageenans are not soluble in cold water, they cannot normally be used for the cold preparation of dessert gels. The sodium salts of κ- and ι-carrageenans, however, are cold-soluble, and combinations of these salts together with gel-inducing potassium or calcium salts have been patented.[85]

As with milk gels, carrageenan-based water dessert gels are also packaged as ready-to-eat, single-serving items. These do not require refrigeration and have excellent shelf life. They are usually processed using a high-temperature short-time regime. Since the final product has a pH between 3·5 and 4·0, it is necessary to process rapidly or to add the acidulant to the containers near the end of the production cycle, in order to minimize hydrolysis of the carrageenan. Table 2 provides a guide to processing conditions. Mixtures of κ- and ι-carrageenans, with or without clarified locust-bean gum, are suitable gellants for these products.[35]

Conventional fruit jellies consist of 65% sugar (sucrose or glucose) plus high-methoxyl pectin and an acidulant. Since the sugar is required to induce gelation with high-methoxyl pectin (see Chapter 10 for more details), it cannot be replaced with a non-caloric sweetener to produce a low-calorie jelly if pectin is to be used. Mixtures of κ- and ι-carrageenans or κ-carrageenan plus clarified locust-bean gum can be used with non-caloric sweeteners to provide a very acceptable low-calorie product. As with canned dessert gels, addition of acid should be done late in the process so as to avoid excessive hydrolysis of the carrageenan.[86]

Fruit drink mixes that can be reconstituted in cold water consist of sugars (or a non-caloric sweetener), acid and flavouring. λ-Carrageenan or the sodium salt of κ-carrageenan at levels of 0·1–0·2% based on the prepared drink may be used to provide body and a pleasing mouth feel.[35]

Sorbet is a frozen, non-dairy dessert comprising fruit purée, a sweetener and a stabilizer system. High solids (40–50%), low overrun (less than 30%) and an appropriate stabilizer produce a creamy alternative to ice-cream, but without the fat. As with ice-cream, the limiting factor for the stabilizer is the mix viscosity. A 'fine tuned' ratio of a gelling-type carrageenan together with a supplementary stabilizer, which may be guar gum, locust-bean gum or pectin, is essential to prepare a smooth-textured sorbet. A 50/50 ratio of carrageenan to supplement is recommended as a starting point, the total being at a level of 0·4% in the sorbet. Too high a ratio of carrageenan will begin to mask the fruit flavour, while too low a ratio may tend to produce an icy, grainy texture.[87]

κ-Carrageenan has been used in combination with guar gum in pet foods to prevent fat separation during processing and impart body to the gravy.[88] In this application, extracted carrageenan has now been almost entirely displaced by seaweed flour in which the κ-carrageenan has been alkali-modified without extraction from the seaweed matrix.[89] Heat processing of the pet food extracts the carrageenan so that it becomes functional. This product is highly cost-effective for pet foods and some non-food applications, but has yet been approved for human consumption.

κ- or ι-carrageenan can be used at concentrations up to 0·5% to provide texture, sheen and improved adhesiveness to relishes, and to pizza and barbecue sauces.[35] In low-oil or no-oil salad dressings they are functional in simulating the mouth feel of a high-oil system and in suspending the condiments.[90]

Carrageenan in combination with xanthan gum serves to stabilize the oil-in-water emulsion and provide body and cling in a reduced-oil imitation mayonnaise. Varying the gum concentration allows a wide range of rheological properties while maintaining uniformity and cling.[91]

κ-Carrageenan functions as a gellant in pumpkin pie fillings to produce firm but tender slices. As the pie is cooled, the stabilizing action progresses rapidly to produce a custard-like texture that resists shrinkage and subsequently prevents surface cracking and skin formation.[92]

7.c Meat Applications

Combinations of κ- and ι-carrageenans, or of κ plus locust-bean gum, gel the broth and preserve the flavour of fish packed in cans or jars (e.g. gefilte fish). Seasonings visible in the broth remain suspended.[35,78]

A solution of κ-carrageenan, locust-bean gum and potassium chloride produces a gelled film coating on frozen fish. This protects the fish from freezer burn and mechanical disintegration during processing. The frozen fish is passed through a 0·4% solution of the mixture and then returned to freezer storage.[35]

The fat content (normally 30%) of frankfurters can be halved by the use of κ-carrageenan at a level of 0·3% of the formulation. Product quality was judged by a test panel to be as good as that of a standard frankfurter containing 30% fat.[93]

In ham pumping and tumbling, κ-carrageenan binds free water and interacts with the protein to maintain the moisture and soluble protein content of the ham. The pumping solution contains salts, and it is essential to the pumping process that this brine has a low viscosity. The gums are therefore dispersed in the brine after dissolution of the salts. This inhibits swelling that would lead to excessive viscosity development. During subsequent cooking, the gums dissolve and so become functional.[94]

The number of processed poultry products continues to increase as a result of consumer demand for ease of preparation, price stability and low-fat, high-protein foods (dietary considerations). Processors face inherent problems when poultry meats are subjected to thermal/mechanical processing. Among these are protein denaturation, which results in textural changes and moisture loss ('purge'); fat oxidation, leading to off-flavours and spoilage; and insufficient freeze–thaw stability, resulting in ice crystal growth that causes breakdown of meat texture. The poultry industry has made efforts to control these problems through the use of phosphates, salts, starches, proteins and, most recently, carrageenan. As with ham processing, these ingredients are incorporated into a brine that is introduced into the meat either by injection or by tumbling.

κ-Carrageenan products specially developed for poultry processing improve yield and have beneficial effects on moisture retention, oxidation/flavour, texture, sliceability and freeze–thaw stability. As these carrageenans disperse easily with minimal swelling in cold brine solutions, the resulting low-viscosity brines improve the flow rate and uniformity of distribution within the poultry in both pumping and tumbling processes.[95]

7.d Furcellaran

Furcellaran generally finds applications similar to those for κ-carrageenan. Historically, furcellaran has dominated two major European application fields: tart or cake glaze powders and flan powders. Today the special properties of excellent gel texture and flavour release make furcellaran a preferred product for use in milk-pudding powders.[96,97]

7.e Regulatory Aspects†

Carrageenan is used in virtually every country of the world and is recognized by all regulatory agencies as a safe and functional ingredient for foods. Some countries have specific regulations pertaining to use levels and types of foods carrageenan may be used in. These represent local geographic considerations. In the United States, carrageenan ('chrondrus extract') is Generally Recognized As Safe (GRAS: 21 CFR 182.7255), and carrageenan is an approved food additive (21 CFR 172.620) in accordance with Food and Drug Administration (FDA) regulations.[98] Monographs for carrageenan in the *Food Chemicals Codex*[99] and *United States Pharmacopeia/National Formulary*[100] are recognized by the FDA for food and pharmaceutical products, respectively.

In the European Economic Community (EEC), carrageenan is similarly recognized for its safety and utility, and is referenced by the designation E407.[101] The Japanese view carrageenan as a 'natural product' and it is therefore not subject to food additive regulations other than Good Manufacturing Practice (GMP).

In 1984 the Joint Expert Committee on Food Additives (JECFA), a working group of the Codex Alimentarius Commission of the World Health Organization (FAO/WHO), completed an intensive toxicological review of carrageenan.[102] The committee, which included expert toxicologists from around the world, confirmed that carrageenan could safely be used in foods and that it was unnecessary to specify an acceptable daily intake (ADI).

At times the safety of carrageenan has been questioned owing to confusion with degraded carrageenan, which is derived by extensively depolymerizing carrageenan in the presence of acid. Degraded carrageenan is now known as poligeenan as a result of the carrageenan industry working with the United States Adopted Names Council (USAN), a division of the American Medical Association. In 1988 USAN published a monograph on poligeenan that characterizes the chemistry of this product.[103] Poligeenan has been used in the past to treat peptic ulcers and is used as a dispersing aid and deflocculent for barium sulphate suspensions used in diagnostic X-ray testing.

The utility and safety of carrageenan has been clearly established by many members of the scientific community and is fully recognized by international regulatory authorities. Carrageenan continues to be an important functional ingredient in foods.

† Section 7.e has been contributed by David W. Manning, Marine Colloids Division, FMC Corporation.

REFERENCES

1. Towle, G. A., in *Industrial Gums*, ed. R. L. Whistler & J. N. BeMiller. Academic Press, New York, 1973, pp. 83–114.
2. Hirase, S. & Watanabe, K., The presence of pyruvate residues in λ-carrageenan and a similar polysaccharide. *Bull. Inst. Chem. Res. Kyoto Univ.*, **50** (1972) 332–6.
3. DiNinno, V. L., McCandless, E. L. & Bell, R. A., Pyruvic acid derivative of a carrageenan from a marine red alga (*Petrocelis* species). *Carbohydrate Res.*, **71** (1979) C1–C4.
4. McCandless, E. L. & Gretz, M. R., Biochemical and immunochemical analysis of carrageenans of the Gigartinaceae and Phyllophoraceae. *Hydrobiologia*, **116/117** (1984) 175–8.
5. Hirase, S., Araki, C. & Watanabe, K., Component sugars of the polysaccharide of the red seaweed *Grateloupia elliptica*. *Bull. Chem. Soc. Japan*, **40** (1967) 1445–8.
6. Nunn, J. R. & Parolis, H., A polysaccharide from *Aeodes orbitosa*. *Carbohydrate Res.*, **6** (1968) 1–11.
7. Allsobrook, A. J. R., Nunn, J. R. & Parolis, H., Sulphated polysaccharides of the Grateloupiaceae family. Part V. A polysaccharide from *Aeodes ulvoidea*. *Carbohydrate Res.*, **16** (1971) 71–8.
8. Parolis, H., The polysaccharides of *Phyllymenia hieroglyphica* (*P. belangeri*) and *Pachymenia hymantophora*. *Carbohydrate Res.*, **93** (1981) 261–7.
9. Smith, D. B. & Cook, W. H., Fractionation of carrageenin. *Arch. Biochem. Biophys.*, **45** (1953) 232–3.
10. Smith, D. B., O'Neill, A. N. & Perlin, A. S., Studies on the heterogeneity of carrageenin. *Can. J. Chem.*, **32** (1955) 1352–60.
11. Rees, D. A., The carrageenan system of polysaccharides. Part I. The relation between the κ- and λ-components. *J. Chem. Soc.* (1963) 1821–32.
12. Dolan, T. C. S. & Rees, D. A., The carrageenans. Part II. The positions of the glycosidic linkages and sulphate esters in λ-carrageenan. *J. Chem. Soc.* (1965) 3534–9.
13. Anderson, N. S., Dolan, T. C. S. & Rees, D. A., Carrageenans. Part III. Oxidative hydrolysis of methylated κ-carrageenan and evidence for a masked repeating structure. *J. Chem. Soc.* (C) (1968) 596–601.
14. Anderson, N. S., Dolan, T. C. S., Penman, A., Rees, D. A., Mueller, G. P., Stancioff, D. J. & Stanley, N. F., Carrageenans. Part IV. Variations in the structure and gel properties of κ-carrageenan, and the characterisation of sulphate esters by infrared spectroscopy. *J. Chem. Soc.* (C) (1968) 602–6.
15. Anderson, N. C., Dolan, T. C. S., Lawson, C. J., Penman, A. & Rees, D. A., Carrageenans. Part V. The masked repeating structures of λ- and μ-carrageenans. *Carbohydrate Res.*, **7** (1968) 468–73.
16. Lawson, C. J. & Rees, D. A., Carrageenans. Part VI. Reinvestigation of acetolysis products of λ-carrageenan. Revision of the structure of α-1,3-galactotriose, and a further example of the reverse specificities of glycoside hydrolysis and acetolysis. *J. Chem. Soc.* (C) (1968) 1301–4.
17. Anderson, N. S., Dolan, T. C. S. & Rees, D. A., Carrageenans. Part VII.

Polysaccharides from *Eucheuma spinosum* and *Eucheuma cottonii*. The covalent structure of ι-carrageenan. *J. Chem. Soc. Perkin I* (1973) 2173–6.
18. Lawson, C. J., Rees, D. A., Stancioff, D. J. & Stanley, N. F., Carrageenans. Part VIII. Repeating structures of galactan sulphates from *Furcellaria fastigiata, Gigartina canaliculata, Gigartina chamissoi, Gigartina atropurpurea, Ahnfeltia durvillaei, Gymnogongrus furcellatus, Eucheuma cottonii, Eucheuma spinosum, Eucheuma isiforme, Eucheuma uncinatum, Agardhiella tenera, Pachymenia hymantophora* and *Gloiopeltis cervicornis*. *J. Chem. Soc. Perkin I* (1973) 2177–82.
19. Penman, A. & Rees, D. A., Carrageenans. Part IX. Methylation analysis of galactan sulphates from *Furcellaria fastigiata, Gigartina canaliculata, Gigartina chamissoi, Gigartina atropurpurea, Ahnfeltia durvillaei, Gymnogongrus furcellatus, Eucheuma isiforme, Eucheuma uncinatum, Agardhiella tenera, Pachymenia hymantophora* and *Gloiopeltis cervicornis*. Structure of ξ-carrageenan. *J. Chem. Soc. Perkin I* (1973) 2182–7.
20. Penman, A. & Rees, D. A., Carrageenans. Part X. Synthesis of 3,6-di-*O*-methyl-D-galactose, a new sugar from the methylation analysis of polysaccharides related to ξ-carrageenan. *J. Chem. Soc. Perkin I* (1973) 2188–91.
21. Penman, A. & Rees, D. A., Carrageenans. Part XI. Mild oxidative hydrolysis of κ- and λ-carrageenans and the characterisation of oligosaccharide sulphates. *J. Chem. Soc. Perkin I* (1973) 2191–6.
22. Welti, D., Carrageenans. Part XII. The 300 MHz proton magnetic resonance spectra of methyl β-D-galactopyranoside, agarose, *kappa*-carrageenan, and segments of *iota*-carrageenan and agarose sulphate. *J. Chem. Res. (S)* (1977) 312–13.
23. Pernas, A. J., Smidsrod, O., Larsen, B. & Haug, A., Chemical heterogeneity of carrageenans as shown by fractional precipitation with potassium chloride. *Acta Chem. Scand.*, **21** (1967) 98–110.
24. Stancioff, D. J. & Stanley, N. F., Infrared and chemical studies on algal polysaccharides. In *Proc. XIth Int. Seaweed Symp.*, ed. R. Margalef. Subsecretaria de la Marina Mercante, Madrid, 1969, pp. 595–609.
25. Lawson, C. J. & Rees, D. A., An enzyme for the metabolic control of polysaccharide conformation and function. *Nature*, **227** (1970) 390–3.
26. Stanley, N. F., Process for treating a polysaccharide of seaweeds of the Gigartinaceae and Solieriaceae families. US Patent 3 094 517 (1963).
27. Greer, C. W. & Yaphe, W., Characterization of hybrid (beta–kappa–gamma) carrageenan from *Eucheuma gelatinae* J. Agardh (Rhodophyta, Solieriaceae) using carrageenases, infrared and ^{13}C-nuclear magnetic resonance spectroscopy. *Bot. Marina*, **27** (1984) 473–8.
28. Painter, T. J., The location of the sulphate half-ester groups in furcellaran and κ-carrageenan. In *Proc. 5th Int. Seaweed Symp.*, ed. E. G. Young & J. L. McLachlan. Pergamon Press, London, 1966, pp. 305–13.
29. Sand, R. E. & Glicksman, M., Seaweed extracts of potential economic importance. In *Industrial Gums*, ed. R. L. Whistler & J. N. BeMiller. Academic Press, New York, 1973, pp. 147–94.
30. Furneaux, R. H. & Miller, I. J., Isolation and ^{13}C-NMR spectral study of the water soluble polysaccharides from four South African red algae. *Bot. Marina*, **29** (1986) 3–10.

31. Laserna, E. C., Veroy, A. H., Luistro, A. H. & Cajipe, G. J. B., Extracts from some red and brown seaweeds of the Philippines. In *Proc. 10th Int. Seaweed Symp.*, ed. T. Levring. Walter de Gruyter, Berlin, 1981, pp. 443–8.
32. Greer, C. W., Shomer, I., Goldstein, M. E. & Yaphe, W., Analysis of carageenan from *Hypnea musciformis* by using κ- and ι-carrageenases and ^{13}C-NMR spectroscopy. *Carbohydrate Res.*, **129** (1984) 189–96.
33. Stanley, N. F., The properties of carrageenans as related to structure. In *Proc. CIC Conference on the Marine Sciences*, ed. J. Rigney. University of Prince Edward Island, Charlottetown, 1970, Paper No. 13.
34. Deslandes, E., Floc'h, J. Y., Bodeau-Bellion, C., Brault, D. & Braud, J. P., Evidence for λ-carrageenan in *Soliera chordalis* (Solieriaceae) and *Callibepharis jubata*, *Callibepharis ciliata*, *Cystoclonium purpureum* (Rhodophyllidaceae). *Bot. Marina*, **28** (1985) 317–18.
35. Guiseley, K. B., Stanley, N. F. & Whitehouse, P. A., Carrageenan. In *Handbook of Water-Soluble Gums and Resins*, ed. R. L. Davidson. McGraw-Hill, New York, 1980, pp. 5-1–5-30.
36. Rees, D. A., Mechanism of gelation in polysaccharide systems. In *Gelation and Gelling Agents, British Food Manufacturing Industries Research Association*, *Symp. Proc.* No. 13, London, 1972, pp. 7–12.
37. Arnott, S., Scott, W. E., Rees, D. A. & McNab, C. G. A., ι-Carrageenan: molecular structure and packing of polysaccharide double helices in oriented fibres of divalent cation salts. *J. Mol. Biol.*, **90** (1974) 253–67.
38. Bryce, T. A., Clark, A. H., Rees, D. A. & Reid, D. S., Concentration dependence of the order–disorder transition of carrageenans. Further confirmatory evidence for the double helix in solution. *Eur. J. Biochem.*, **122** (1982) 63–9.
39. Bryce, T. A., McKinnon, A., Morris, E. R., Rees, D. A. & Thom, D., Chain conformations in the sol–gel transitions, and their characterisation by spectroscopic methods. *J. Chem. Soc., Faraday Disc.*, **57** (1974) 221–9.
40. Morris, E. R., Rees, D. A. & Robinson, G., Cation-specific aggregation of carrageenan helices: Domain model of polymer gel structure. *J. Mol. Biol.*, **138** (1980) 349–62.
41. Bayley, S. T., X-ray and infrared studies of *kappa*-carrageenin. *Biochim. Biophys. Acta*, **17** (1955) 194–205.
42. Paoletti, S., Smidsrod, O. & Grasdalen, H., Thermodynamic stability of the ordered conformation of carageenan polyelectrolytes. *Biopolymers*, **23** (1984) 1771–94.
43. Mueller, G. P. & Rees, D. A., Current structural views of red seaweed polysaccharides. In *Drugs from the Sea*, ed. H. D. Freudenthal. Marine Technology Society, Washington DC, 1968, pp. 241–55.
44. Rees, D. A., Steele, I. W. & Williamson, F. B., Conformational analysis of polysaccharides. III. The relation between stereochemistry and properties of some natural polysaccharides. *J. Polymer Sci., Part C*, **28** (1969) 261–76.
45. Carroll, V., Morris, V. J. & Miles, M. J., X-ray diffraction studies of *kappa* carrageenan-tara gum mixed gels. *Macromolecules*, **17** (1984) 2443.
46. Cairns, P., Miles, M. J. & Morris, V. J. *Int. J. Biol. Macromol.*, **8** (1986) 124.
47. Cairns, P., Miles, M. J. & Morris, V. J., X-ray diffraction studies on konjac mannan-*kappa* carrageenan mixed gels. *Carbohydrate Polymers*, **8** (1988) 99–104.

48. King, G. M. & Lauterbach, G. E., Characterization of carrageenan nitrogen content and its susceptibility to enzymatic hydrolysis. *Bot. Marina*, **30** (1987) 33–9.
49. Lin, C-F., Interaction of sulfated polysaccharides with proteins. In *Food Colloids*, ed. H. D. Graham. Avi Publishing Co., Westport, Connecticut, 1977, pp. 320–46.
50. Hansen, P. M. T., Hydrocolloid–protein interactions: Relationship to stabilization of fluid milk products. A review. In *Gums and Stabilisers for the Food Industry*, ed. G. O. Phillips, D. J. Wedlock & P. A. Williams. Pergamon Press, Oxford, 1982, pp. 127–38.
51. Snoeren, Th. H. M., *Kappa*-carrageenan. A study on its physicochemical properties, sol–gel transition and interaction with milk proteins. Thesis, Nederlands Instituut voor Zuivelonderzoek, Ede, The Netherlands, 1976.
52. Stainsby, G., Proteinaceous gelling systems and their complexes with polysaccharides. *Food Chem.*, **6** (1980) 3–14.
53. Payens, T. A. J., Light scattering of protein reactivity of polysaccharides, especially of carrageenans. *J. Dairy Sci.*, **55** (1972) 141–50.
54. Snoeren, Th. H. M., Payens, T. A. J., Jeunink, J. & Both, P., Electrostatic interaction between κ-carrageenan and κ-casein. *Milchwissenschaft*, **30** (1975) 393–6.
55. Grindrod, J. & Nickerson, T. A., Effect of various gums on skim milk and purified milk proteins. *J. Dairy Sci.*, **51** (1968) 834–41.
56. Snoeren, Th. H. M., Both, P. & Schmidt, D. G., An electron-microscopic study of carrageenan and its interaction with κ-casein. *Neth. Milk Dairy J.*, **30** (1976) 132–41.
57. Schmidt, D. G. & Payens, T. A. J., Micellar aspects of casein. *Surf. Colloid Sci.*, **9** (1976) 162–229.
58. Hood, L. F. & Allen, J. E., Ultrastructure of carrageenan-milk sols and gels. *J. Food Sci.*, **42** (1977) 1062–5.
59. Boorgaerdt, J., Instability of milk due to a high content of calcium ions. *Nature (London)*, **174** (1954) 884.
60. Rose, D., A proposed model of micelle structure in bovine milk. *Dairy Sci. Abstr.*, **31** (1969) 171.
61. Waugh, D. F., Creamer, L. K., Slattery, C. W. & Dresdner, G. W., Core Polymers of casein micelles. *Biochemistry*, **9** (1970) 786.
62. Lin, C-F., The casein stabilizing function of sulfated polysaccharides. PhD dissertation, Ohio State University, Columbus, 1971.
63. Hansen, P. M. T., Stabilization of α_s-casein by carrageenan. *J. Dairy Sci.*, **51** (1968) 192–5.
64. Skura, B. J. & Nakai, S., Physicochemical verification of non-existence of α_{s1}-casein–κ-carrageenan interaction in calcium-free systems. *J. Food Sci.*, **45** (1980) 582–91.
65. Lin, C-F. & Hansen, P. M. J., Stabilization of casein micelles by carrageenan. *Macromolecules*, **3** (1970) 269–74.
66. O'Loughlin, K. & Hansen, P. M. T., Stabilization of rennet-treated milk protein by carrageenan. *J. Food Sci.*, **37** (1972) 719–21.
67. Chakraborty, B. K. & Randolph, H. E., Stabilization of calcium sensitive plant proteins by κ-carrageenan. *J. Food Sci.*, **37** (1972) 719–21.

68. Badui, S., Fate of carrageenan in processed milk. PhD dissertation, Ohio State University, Columbus, 1977.
69. Stanley, N. F., unpublished work.
70. Swanson, A. M., Maxwell, G. E. & Roehrig, P. C., Sterilized milk concentrate properties as affected by certain processing treatments and additives. (Abstract) *J. Dairy Sci.*, **51** (1968) 92.
71. MacMullan, E. A. & Eirich, F. K., The precipitation reaction of carrageenan with gelatin. *J. Colloid Sci.*, **18** (1963) 526–37.
72. Hidalgo, J. & Hansen, P. M. T., Interactions between food stabilizers and β-lactoglobulin. *Agric. Food Chem.*, **17** (1969) 1089–92.
73. Anon., Dutch vla/pourable custard. *Application Bulletin D-54*, FMC Corporation, Marine Colloids Div., 1988.
74. Anon., Calcium fortified fluid milk. *Application Bulletin E-53*, FMC Corporation, Marine Colloids Div., 1988.
75. Nielsen, B. J., Function and evaluation of emulsifiers in ice cream and whippable emulsions. *Gordian*, **76** (1976) 200–25.
76. Keeney, P. G. & Kroger, M., Frozen dairy products. In *Fundamentals of Dairy Chemistry*, 2nd edn, ed. B. H. Webb, A. H. Johnson & J. A. Alford. Avi Publishing Co., Westport, Connecticut, 1974, pp. 873–913.
77. Arbuckle, W. S., *Ice Cream*, 2nd edn. Avi Publishing Co., Westport, Connecticut, 1972.
78. Glicksman, M., Red seaweed extracts (agar, carrageenans, furcellaran). In *Food Hydrocolloids*, Vol. 2, ed. M. Glicksman. CRC Press, Boca Raton, Florida, 1983, pp. 73–113.
79. Anon., Cottage cheese dressing. *Application Bulletin E-27*, FMC Corporation, Marine Colloids Div., 1988.
80. Anon., Imitation cheese-block. *Application Bulletin B-11*, FMC Corporation, Marine Colloids Div., 1988.
81. Glicksman, M., Frozen gels of *Eucheuma*. US Patent 3 250 621 (1966).
82. Baker, G. L., Edible gelling composition containing Irish moss extract, locust bean gum and an edible salt. US Patent 2 466 146 (1949).
83. Baker, G. L., Gelling compositions. US Patent 2 669 519 (1954).
84. Baker, G. L., Carrow, G. W. & Woodmansee, C. W., Three-element colloid makes better low-solid gels. *Food Ind.*, **21** (1949) 617–19, 711, 712.
85. Glicksman, M., Farkas, E. & Klose, R. E., Cold water soluble *Eucheuma* gel mixtures. US Patent 3 502 483 (1970).
86. Anon., Reduced calorie jam and jelly (imitation). *Application Bulletin B-19*, FMC Corporation, Marine Colloids Div., 1988.
87. Anon., Sorbet. *Application Bulletin B-49*, FMC Corporation, Marine Colloids Div., 1987.
88. Anon., Canned retorted pet food. *Application Bulletin B-30*, FMC Corporation, Marine Colloids Div., 1987.
89. Lewis, J. G., Stanley, N. F. & Guist, G. G., Commercial production and applications of algal hydrocolloids. In *Algae and Human Affairs*, ed. C. A. Lembi & J. R. Waaland. Cambridge University Press, Cambridge, 1988, pp. 205–36.
90. Anon., No oil/low oil pourable Italian type dressing. *Application Bulletin C-5*, FMC Corporation, Marine Colloids Div., 1988.

91. Anon., Imitation mayonnaise. *Application Bulletin C-61*, FMC Corporation, Marine Colloids Div., 1988.
92. Anon., Pumpkin pie filling. *Application Bulletin D-17*, FMC Corporation, Marine Colloids Div., 1988.
93. Anon., Low-fat frankfurters. *Application Bulletin B-60*, FMC Corporation, Marine Colloids Div., 1988.
94. Anon., Meat application–ham pumping/tumbling. *Application Bulletin B-41*, FMC Corporation, Marine Colloids Div., 1987.
95. Anon., Processed poultry products. *Application Bulletin B-51*, FMC Corporation, Marine Colloids Div., 1987.
96. Bjerre-Petersen, E., Christensen, J. & Hemmingsen, P., Furcellaran. In *Industrial Gums*, 2nd edn, ed. R. L. Whistler & J. N. BeMiller. Academic Press, New York, 1973, pp. 123–36.
97. Stanley, N. F., Production, properties and use of carrageenan. In *Production and Utilization of Products from Commercial Seaweeds*, ed. D. J. McHugh. FAOUN, Rome, 1987, pp. 97–147.
98. United States Code of Federal Regulations (Title 21). Government Printing Office, Washington, DC, 1985.
99. *Food Chemicals Codex*, 3rd edn. National Academy of Sciences, Washington, DC, 1981, pp. 74–5.
100. *United States Pharmacopeia/National Formulary*. United States Pharmacopeial Convention, Rockville, Maryland.
101. *Official Journal of the European Communities*. Specifications for standards of purity. No. 1 232/12 14/8/78.
102. *Evaluation of Certain Food Additives and Contaminants* (E407—Carrageenan). Twenty-eighth report of the Joint FAO/WHO Expert Committee on Food Additives. World Health Organization Technical Report Series 710, Geneva, 1984.
103. Anon., List No. 297 USAN Council New Names Column. *Clinical Pharmacology and Therapeutics*, **44** (1988) 246–8.

Chapter 4

CASEIN

P. F. Fox and D. M. Mulvihill
Department of Dairy and Food Chemistry, University College, Cork, Republic of Ireland

1 INTRODUCTION

Milk, a fluid produced by female mammals for the nutrition of their young, is intended to be consumed as a liquid, but man has exploited its rather unique gelling properties for at least 8000 years in the production of a wide variety of fermented foods that fall into two general categories: cheeses and fluid fermented products. Taken together, these products probably represent the longest-established and most widely produced food protein-based gels: $c.\ 10^7$ t of cheese are produced world-wide annually from $c.\ 10^8$ t (10^{11} litres) of gelled milk. Milk proteins are used in many other food gel systems but, although these are quite significant in their own right, they are, quantitatively, trivial in comparison with cheese and fermented milks. However, gelation of milk proteins is not always desirable, e.g. age gelation of sterilized milks and coagulation/gelation of milk, especially concentrated milk, during sterilization.

This chapter will be restricted primarily to a review of rennet- and acid-produced casein gels. Although the whey proteins are incorporated into these gels, actively or inactively, and influence their characteristics, sometimes very significantly, the active gelling agent is casein. Caseins may be present in other food gel systems, e.g. meat products, but they are not the primary gelling agent; hence, the definition of 'casein gels' is to some extent subjective.

2 STRUCTURE

2.a Molecular and Physicochemical Properties of the Caseins

The majority of milk-based gels depend on either of two characteristics of the casein system: (1) insolubility in the region of their isoelectric points ($c.$

pH 4·6), which is exploited in the production of fermented milks and some fresh cheeses, and (2) instability in the presence of Ca^{2+} following specific, very limited, proteolysis by selected proteinases (rennets), which is exploited in the manufacture of most cheeses. Compared to most other proteins, the caseins are very heat stable at pH > 6·5 and, therefore, do not normally form thermally-induced gels, although they may do so under certain conditions, which may cause problems in the manufacture of sterilized, concentrated products.

It appears appropriate to commence this chapter with a brief review of the principal molecular and physicochemical properties of the caseins.

2.a.1 Heterogeneity of the Milk Protein System

Bovine milk contains c. 3·5% protein, which, like the milks of apparently all other mammals, falls into two principal groups: the caseins, i.e. phosphoproteins insoluble at pH 4·6 (the isoelectric pH) at 20°C, and the whey (or non-casein) proteins, which are soluble under these conditions. The ease with which isoelectric casein, which represents c. 80% of the total nitrogen of bovine milk, can be prepared, made it the subject of study by the pioneer protein chemists, e.g. Hammersten (1880–1890). Isoelectric (acid) casein was considered to be homogeneous until the pioneering work of Linderstrom-Lang (1925–1930), Pedersen (1936) and Mellander (1939) showed that it contained at least three proteins, designated α, β and γ in order of decreasing electrophoretic mobility. These proteins were isolated to apparent homogeneity by Hipp et al. (1952) and α-casein was fractionated by Waugh and von Hippel (1956) into Ca-sensitive (α_s) and Ca-insensitive (κ) fractions.

Although free-boundary electrophoresis permitted considerable progress in the study of casein heterogeneity and fractionation, the full extent of its complexity did not become apparent until the application of gel electrophoresis by Wake and Baldwin (1959), who showed that isoelectric casein contains c. 20 individual proteins. It appears inappropriate to review in detail the developments in casein purification over the past 30 years; the interested reader is referred to the periodic reports of the Milk Protein Nomenclature Committee of the American Dairy Science Association (see Ref. 1) and the numerous reviews on the subject, e.g. Refs 2–6.

Suffice it to state that isoelectric casein is now known to consist of four primary proteins (gene products), α_{s1}, α_{s2}, β and κ, in the approximate ratio 40:10:35:12, and several minor proteins, most of which originate via posttranscriptional proteolysis of the primary caseins by the action of indigenous alkaline milk proteinase, plasmin. These polypeptides include γ-

casein (γ^1, γ^2, γ^3) and proteose peptones 5- and 8-fast, all derived from β-casein, λ-casein derived from α_{s1}-casein and at least 30 as yet unidentified peptides, most of which are probably derived from the principal caseins and are included in the classical proteose peptone fraction. (The proteose peptones have, traditionally, been classified as whey proteins because they are soluble at pH 4·6; however, since many or most of them are derived from the caseins, it is now recommended[1] that they be classified as caseins.)

The four primary caseins exhibit 'microheterogeneity' due to variations in the degree of phosphorylation or glycosylation, disulphide-linked polymerization and genetically-controlled amino-acid substitution (genetic polymorphism).

2.a.2 Phosphorylation and Disulphide Groups

All the caseins are phosphorylated, but to variable extents; the phosphate is esterified to the polypeptides as monoesters of serine (or rarely of threonine). α_{s1}-Casein usually contains 8 moles P per mole protein, but occasionally an additional serine is phosphorylated, producing what was previously called α_{s0}-casein; these proteins are now designated α_{s1}-casein-8P and α_{s1}-casein-9P, respectively.[1] α_{s2}-Casein consists of four proteins with a common polypeptide but with 10–13 moles P per mole protein; these proteins were previously called α_{s2}-, α_{s3}-, α_{s4}- and α_{s6}-caseins but have been renamed α_{s2}-CN-10P, α_{s2}-CN-11P, α_{s2}-CN-12P and α_{s2}-CN-13P, respectively.

β-Casein usually contains 5 moles P per mole protein but some genetic variants contain only four phosphate residues. Most molecules of κ-casein contain only 1 mole P per mole protein, but recent evidence[7] suggests that some may contain two phosphate groups.

α_{s2}-Casein contains two half-cystine residues per mole, which apparently exist naturally as an intermolecular disulphide bond; disulphide-linked α_{s3}- and α_{s4}-caseins give α_{s5}-casein but presumably any two of the four α_{s2}-caseins may exist as a disulphide-linked dimer. κ-Casein also contains two half-cystine residues per mole; these normally form homogeneous intermolecular disulphide bonds so that κ-casein normally exists as a highly cross-linked polymer, probably predominantly on the casein micelle surface.[8,9] In raw milk, κ-casein does not appear to form disulphide-linked polymers with α_{s2}-casein or with the whey proteins but following heat denaturation, β-lactoglobulin—which contains a buried, unreactive sulphydryl in the native state—forms a disulphide-linked complex with κ-casein[10] with major effects on many technologically important properties of milk, e.g. rennet coagulation, heat stability and age gelation of sterilized milks. κ-Casein may also undergo sulphydryl–disulphide interchange reactions with α-lactalbumin[11] (which contains four intramolecular disulphides but no sulphydryl group). κ-Casein may also form disulphide-linked polymers with α_{s2}-caseins during heating.[12] α_{s1}-Caseins and β-caseins contain no cysteine or cystine.

κ-Casein is the only glycosylated protein in the casein group. The carbohydrate moiety is composed of N-acetylneuraminic acid (NANA), galactose (Gal) and N-acetylgalactosamine (NAcGal), which exists as a trisaccharide or tetrasaccharide:[13]

$$\text{Thr} \xrightarrow{\beta\text{-}1} \text{NacGal} \xrightarrow{\beta\text{-}1,3} \text{Gal} \xrightarrow{\alpha\text{-}2,3} \text{NANA}$$

$$\text{Thr} \xrightarrow{\beta\text{-}1} \text{NacGal} \xrightarrow{\beta\text{-}1,3} \text{Gal} \xrightarrow{\alpha\text{-}2,3} \text{NANA}$$
$$\downarrow \alpha\text{-}2,6$$
$$\text{NANA}$$

The full extent of κ-casein heterogeneity is still not definitely established.

All the principal milk proteins exhibit genetic polymorphism, first demonstrated in β-lactoglobulin by Aschaffenburg and Drewry (1957), usually due to substitution of one or two amino acids; in two rare variants, α_{s1}-A and α_{s2}-D, deletion of a sequence of 13 residues occurs. To date, four variants of α_{s1}-casein, four of α_{s2}-casein, eight of β-casein and two of κ-casein have been identified. Since genetic polymorphism is normally detected by gel electrophoresis, only substitutions leading to changes in protein charge are detectable; it is almost certain that substitutions that do not cause a change in charge also occur but would be detected only by very precise amino-acid analysis.

The frequency of occurrence of genetic variants is species- and breed-dependent,[14] and has been of some value in tracing the phylogenetic development of the genus *Bos* and of breeds of cattle. A number of recent studies have shown that genetic polymorphism has technological significance, especially in relation to cheese manufacture.[15]

2.a.3 Post-secretion Proteolysis in Milk

The occurrence of an indigenous proteinase in milk was suggested by the work of Babcock in the 1890s, but until recently it was felt that the proteolytic activity in milk, which is very low, was probably due to microbial enzymes. However, it is now well established that milk contains blood plasmin (an alkaline, serine proteinase), a cathepsin-like acid proteinase, and possibly thrombin and aminopeptidases (for reviews see Refs 16–18).

Although the occurrence of γ-casein was demonstrated by Mellander (1939) and of 'proteose peptone' by Rowland (1938), the interrelationship between these two fractions and their formation from β-casein via proteolysis by plasmin did not become apparent until recently.[18] Plasmin is highly specific for peptide bonds to which lysine or arginine supply the carboxyl group. β-Casein contains 11 or 12 lysyl and 4 or 5 arginyl residues per mole (depending on variant) but only Lys–Lys (28–29), Lys–His (105–106) and Lys–Glu (107–108) are hydrolysed at significant rates, giving rise to six polypeptides, i.e. γ^1-casein (residues 29–209), γ^2 (106–209), γ^3 (108–209), PP 8-fast (1–28), PP 5 (1–105/107) and β-CN (f29–105/107). γ-Caseins normally represent c. 3% of whole casein but may be as high as 10% in late-lactation milks, probably owing to higher plasmin activity.

α_{s2}-Casein, which contains 24 lysyl and 6 arginyl residues, is rapidly hydrolysed by plasmin, but the cleavage sites have not been identified; it has three Lys–Lys, one Lys–Arg, one Lys–His and two Lys–Glu bonds, which, by analogy with β-casein, may be the cleavage sites. Isolated α_{s1}-casein is quite susceptible to plasmin but it undergoes little proteolysis in milk; there is some evidence that λ-casein is produced from α_{s1}-casein by plasmin action. κ-Casein is very resistant to plasmin.

2.a.4 Molecular Characteristics of the Caseins

Since the caseins have been available in homogeneous form for many years, they have been characterized in considerable detail, including primary structures and estimates of secondary and tertiary structures.[19] All the caseins are relatively small molecules, c. 20 000–24 000 daltons, and all are insoluble around their isoelectric points, pH 4·5–4·9, which fact has a major influence on their functional properties, including gelation. As previously indicated, α_{s1}- and β-caseins contain no cysteine or cystine, while α_{s2}- and κ-caseins each contain two half-cystine residues that normally exist as homogeneous intermolecular disulphide bonds. The small number of cystine/cysteine residues in the caseins precludes extensive polymerization and gelation via disulphide bonds, as in gluten.

The caseins are strongly hydrophobic in the order $\beta > \alpha_{s1} > \kappa > \alpha_{s2}$. However, primary structures indicate that the hydrophobic and polar or charged residues are not uniformly distributed throughout the sequences (see Ref. 19); clustering is particularly apparent for the phosphoseryl residues and this has a marked influence on the metal-binding properties of the caseins. All the caseins contain high levels of proline, which are fairly uniformly distributed throughout the sequences. Prolyl residues interrupt α-helical and β-sheet structures but give the caseins a type of stable poly-L-proline helical structure. The caseins are, therefore, relatively small, amphipathic randomly-coiled, unstructured, open molecules; these features markedly influence their functional properties.

α_{s1}-Casein has a highly charged (net charge at pH 6·6 is $-20·6$), highly solvated region between residues 41 and 80 that contains the eight phosphoseryl residues, and three strongly hydrophobic regions: 1–40, 90–113 and 132–199. α_{s2}-Casein, which contains 10–13 phosphoseryl residues, is the most hydrophilic of the caseins and has an average hydrophobicity in the range of most globular proteins. The molecule can be divided clearly into hydrophobic and hydrophilic domains: the N-terminal half includes three charged segments (8–16, 56–61, 129–133), which contain the phosphoseryl groups, while the C-terminal segment, from residues 160 to 207, is strongly hydrophobic; the sequence 90–120 is also hydrophobic. The sequence of α_{s2}-casein from Lys 158 to the C-terminal contains no anionic residues and has a net charge at pH 6·6 of $+9·5$, while the N-terminal 68 residues have a net charge of -21 at pH 6·6; electrostatic interactions should therefore be important in the structure of this protein. The C-terminal half of α_{s2}-casein probably exists in a globular conformation while the N-terminal probably forms a randomly structured hydrophilic tail.

The N-terminal section (residues 1–21) of β-casein, the most strongly hydrophobic of the caseins, is highly charged (net charge at pH 6·6 is -12) and contains four of the five phosphoseryl residues; the remainder of the molecule has no net charge and has a high content of hydrophobic residues. κ-Casein is also amphipathic; its C-terminal region is strongly anionic (net charge at pH 6·6 is -10 and there is no cationic residue among the 53 C-terminal amino acids), contains all the strongly hydrophilic carbohydrate moieties, no aromatic and few hydrophobic residues; this is the (glyco)macropeptide produced by chymosin during the rennet coagulation of milk, release of which is responsible for the gelation of casein. The N-terminal two-thirds of κ-casein is strongly hydrophobic and represents the para-κ-casein produced on renneting.

Studies using circular dichroism and optical rotary dispersion suggest that the caseins have little secondary structures. However, theoretical considerations[19,20] suggest that α_{s1}-casein probably possesses some secondary structures. Up to 10% of the β-casein molecule may exist as an α-helix, 13% as β-sheet and 77% as unordered structures;[21] Swaisgood[19] predicts a fairly high content of β-turns in view of the high content of proline. He also predicts that α-helices are most likely to occur in the sequences 57–103 and 138–146 with β-sheets at 52–60, 77–87 and 187–195. Creamer et al.[20] propose some additional structural regions.

κ-Casein appears to be the most highly structured of the caseins with 23% of the sequence as α-helices, 31% as β-sheets and 24% as β-turns.[22] It is predicted that the region around the chymosin-susceptible bond, Phe_{105}–Met_{106}, is likely to adopt an α-helical structure but it also has a strong tendency to form β-sheets, which would favour hydrogen bonding to the active site of chymosin. Moreover, the sequence 102–109 is located between two predicted β-turns, 98–101 and 109–112, with another β-turn at 113–116, which should cause the chymosin-susceptible bond to protrude from the surface of the κ-casein molecule, enabling it to fit into the active site cleft of the acid proteinases (rennets).

The hydrophobicity and the open, amphipathic structure of the caseins are of considerable technological significance:

(1) The caseins associate strongly via hydrophobic bonding, necessitating the use of strong dissociating agents in their isolation by chemical or chromatographic means. It is probably also significant in such functional properties as viscosity, gelation, swelling and water sorption.

(2) Some protein, especially the very hydrophobic β-casein, dissociates from the casein micelles on cooling, possibly rendering them more susceptible to proteolysis by plasmin,[17] although this could not be confirmed,[15,18,23] and extracellular proteinases from psychrotrophs.

The open structure of the caseins renders them much more susceptible to proteolysis than the whey proteins; while this characteristic may be desirable nutritionally, it renders the caseins more susceptible to proteolytic spoilage by indigenous milk proteinase and psychrotroph proteinases than whey proteins.

(3) Enzymatic hydrolysates of casein have a marked propensity to bitterness owing to the high content of hydrophobic peptides, which tend to be bitter.

(4) The caseins have good surface activity owing to their small size, which allows them to migrate quickly to air–water or oil–water interfaces; their open conformations allow them to spread readily at interfaces and their amphipathic structures facilitate orientation of the hydrophobic residues into the air or oil phases with the hydrophilic residues in the aqueous phase, thus optimizing their surfactant properties.[5]

2.a.5 Casein Associations

Because of their high content of hydrophobic residues, the caseins undergo strong temperature-dependent association; the presence of highly charged segments enables electrostatic interactions. Since these associations influence gelation, a brief description is warranted (see Refs 24 and 25 for comprehensive reviews).

α_{s1}-Casein associates in a series of consecutive steps,

$$\alpha i + 1\alpha \rightleftharpoons \alpha i + 1 \qquad i = 1, 2, 3, \ldots$$

that are strongly dependent on pH and ionic strength but essentially independent of temperature up to 30°C. Association is governed by electrostatic repulsion and attraction owing to hydrophobic and hydrogen bonding (about 13 hydrophobic and 1–2 hydrogen bonds are formed). α_{s2}-Casein associates via a series of consecutive steps similar to α_{s1}-casein; its association is strongly dependent on ionic strength, being maximal at $\mu = 0.2$.

The association of β-casein is also dependent on ionic strength and pH, but in contrast to α_{s1}- and α_{s2}-caseins its association is strongly temperature-dependent; β-casein is monomeric at 4°C but considerable association is apparent at 8.5°C. The association may be described as

$$i\beta \rightleftharpoons \beta i \qquad i \gg 1$$

which is characteristic of the association (micelle formation) of ionic detergents, undoubtedly a reflection of the strongly amphipathic nature of β-casein. At low ionic strength (0.05) and pH ~ 7, the value of i is 13 or 14, while at $\mu = 0.2$ values of $i = 23, 49$ or 58 have been reported, indicating that while the formation of β-casein micelles at high temperatures is generally accepted, the size of the micelles is a matter of some controversy. Formation of β-casein micelles depends on a critical protein concentration, the value of which depends on ionic strength and temperature.

κ-Casein forms trimers of molecular weight $c.$ 60 000 via intermolecular disulphide bonds; the trimers aggregate further by hydrophobic bonding. Reduced κ-casein associates in a manner similar to β-casein with a value of $i \sim 30$, i.e. molecular weight $\sim 600\,000$.

Association between casein species, especially α_{s1}- and κ-caseins, is important in micelle formation, and has been extensively investigated.[15,24]

2.a.6 Ion Binding of the Caseins

Because of the high content of phosphoseryl residues, α_{s1}-, α_{s2}- and β-caseins bind polyvalent cations strongly, principally Ca^{2+} but also Zn^{2+}, which is nutritionally important, leading to charge neutralization, aggregation and eventually precipitation. Many of the technologically important properties of milk, e.g. heat stability, alcohol stability, rennet coagulability and the strength and syneresis properties of rennet gels, are strongly influenced by [Ca^{2+}]; in the manufacture of certain dairy products, e.g. sterilized concentrated milk, processed cheeses, cream liqueurs, it is necessary to reduce [Ca^{2+}] by addition of chelators, usually ortho- or polyphosphates or citrates, to prevent gelation.

Calcium binding by caseins, which is influenced by temperature, pH and ionic strength, has been studied extensively.[26-29] At low [Ca^{2+}], as in milk, Ca^{2+} are probably bound exclusively to phosphoseryl residues, but at higher [Ca^{2+}] binding may also occur to aspartyl or glutamyl residues. Binding of Ca^{2+} reduces protein charge, permitting protein association; in the case of α_{s1}-casein, octamers are formed initially but these aggregate and precipitate at higher [Ca^{2+}]. Calcium-mediated association of β-casein, which is highly hydrophobic, proceeds in a similar manner to that of α_{s1}-casein, although the precipitating particles are larger. Since the aggregation of β-casein depends on hydrophobic interactions, and hence is temperature-dependent, the precipitation of β-casein by Ca^{2+} is more strongly dependent on temperature than is that of α_{s1}-casein and β-casein is soluble in the presence of $Ca^{2+} < 20°C$.[29]

κ-Casein, which contains only one phosphoseryl residue, does not bind Ca^{2+} strongly and is soluble in the presence of high [Ca^{2+}]. Further, κ-casein associates with α_{s1}- and/or β-casein, and in the presence of Ca^{2+} κ-casein stabilizes these Ca-sensitive caseins against precipitation by Ca^{2+}, with the formation of stable colloidal particles that are generally similar to, but less stable than, native casein micelles.[25,30,31]

2.a.7 The Casein Micelle
In normal milk, c. 95% of the casein exists as coarse colloidal particles, micelles, with molecular weights of c. 10^8 and mean diameters of c. 100 nm (range 50–300 nm). On a dry-weight basis, the micelles consist of c. 94% protein and c. 6% small ions, principally calcium, phosphate, magnesium and citrate, referred to collectively as colloidal calcium phosphate (CCP). In milk, the micelles are highly hydrated, typically c. 2 g H_2O/g protein.

The technological properties of milk are very strongly influenced by the stability of the casein micelles, the structure of which has been the subject of considerable research (see Refs 6, 24, 25 and 30–36 for reviews). While the detailed structure of the casein micelle is still not known, it is widely, but not universally, accepted that the micelles are composed of spherical submicelles, 10–15 nm in diameter, and have porous, open structures. The stoichiometry of the four principal caseins is believed to vary between individual submicelles and it appears that κ-casein, the principal micelle stabilizing factor, is located predominantly at the surface, although some of the other caseins also occur on the surface.

The submicelles are linked together in the micelles by CCP and hydrophobic bonds, with contributions from other secondary forces. Removal of CCP by acidification/dialysis, EDTA, citrate or dialysis against calcium-free phosphate buffers causes dissociation of the micelles to particles of molecular weight $c. 10^6$ but it is not known whether these represent the original submicelles. Casein micelles are also dispersed by urea (>4M), sodium dodecyl sulphate (>2%) or pH ~9, presumably without dissolution of CCP; the size and nature of the resultant submicelles have not been investigated but it is apparent that hydrophobic, hydrogen and electrostatic bonds play significant roles in micelle stabilization.

Pyne and McGann[37] suggest that CCP has a citrate apatite-type structure: $3Ca_3(PO_4)_2 \cdot CaH$ Citr. However, Schmidt[25] suggests $Ca_9(PO_4)_6$ as the most likely structure. Further possibilities are discussed by Holt and collaborators,[38-41] who suggest that the structure of CCP is $CaHPO_4 \cdot 2H_2O$. The nature of the attachment of CCP to casein is also unresolved. McGann and Pyne[42] concluded that attachment via organic seryl phosphates is most likely; Schmidt[25] essentially concurs with this view but suggests that attachment is mediated by Ca^{2+}, which adsorbs onto the CCP particles giving them a positive charge. Schmidt[25] has calculated that, per micelle (molecular weight $= 10^8$), there are 70 600 calcium atoms and 30 100 inorganic phosphate groups from which 5000 $Ca_9(PO_4)_6$ clusters might be formed, leaving 25 300 calcium ions. There are about 25 000 organic phosphate residues per micelle, about 40% of which can be linked in pairs via $Ca_9(PO_4)_6$, with Ca^{2+} acting as counter-ions for most of the remainder.

Numerous models of casein micelle structure have been proposed, most of which suffer limitations to a greater or lesser degree (see Refs 25, 34 and 35). The latest model[25,43] proposes a subunit structure in which the subunits contain variable levels of surface κ-casein; a proportion of each of the other caseins is also located on the surfaces of the submicelles. The κ-casein-deficient submicelles are considered to be located within the micelle with the κ-casein-rich submicelles concentrated at the surface, giving the intact micelles a surface rich in κ-casein, although some of each of the other caseins are also at the micelle surface. The submicelles are linked together principally by CCP. Heth and Swaisgood[44] provide direct experimental evidence for this type of structure. It is suggested[43] that the hydrophilic C-terminal regions of κ-casein protrude from the micelles, giving them a hairy appearance, and contribute to micelle stabilization by steric hindrance. The micelles have zeta potentials of $c. -20$ mV under the ionic environment in milk and this is also a major factor in micelle stability.

2.a.8 Micellar Destabilization

Considering that $c.$ 85% of isoelectric casein is very insoluble at $[Ca^{2+}] > 6$ mM, natural casein micelles are remarkably stable. However, they are destabilized by any of several treatments or circumstances, e.g. acid, ethanol or other organic solvents, heat, limited proteolysis, Ca^{2+} or combinations of these. Many of these destabilizations result in gelation and represent the principles for the production of a range of dairy products; the more important of these are discussed below. Unfortunately, micellar destabilization also creates serious problems during the processing and storage of certain dairy products, e.g. coagulation (gelation) during the sterilization of concentrated milk and age gelation of sterilized concentrates, especially UHT-treated milks.

3 SOURCE AND PRODUCTION

Although several mammalian species are used for milk production, commercial dairying is essentially confined to cattle, buffalo, sheep and goats, from which approximately 460×10^6, 33×10^6, 8.6×10^6 and 7.8×10^6 t of milk, respectively, were produced in 1985.[45] Approximately 13×10^6 t of cheese (all types) are produced annually from $c. 10^8$ t of milk,[44] i.e. $c.$ 20% of total milk production (a value of 31% of total milk for cheese manufacture is cited in Ref. 46). World-wide data for the production of yoghurt and other fermented milks are not available, but total consumption of these products in the principal dairying countries in 1984 was $c. 9.5 \times 10^6$ t.[47] Although world production of milk is impressive, production is very unevenly distributed throughout the world, e.g. production in Africa, North America, Latin America, Asia, Europe, Oceania and USSR was 11·7, 73·1, 35·1, 42·4, 183·8, 14·2 and 97·8 million tonnes, in 1985, respectively.[45]

4 COMMERCIAL AVAILABILITY

The production data in the previous section indicate that milk is readily available in Europe, North America, Oceania and USSR; it is probably reasonable to conclude that in these regions rennet and some acid cheeses are made mainly, if not entirely, from fresh raw, thermized or pasteurized milks. For reasons to be discussed later, fermented milks are usually produced mainly from heat-treated milk, frequently fortified with skim milk powder or other milk solids. In regions of the world where the local milk supply is inadequate to meet demands, reconstituted/recombined low-heat milk powders may be used for the manufacture of rennet and/or acid cheeses but, as discussed below, excessive heating destroys the cheese-making properties of milk. Milk powders are produced extensively in the more intensive dairying areas of the world.

5 SOLUTION PROPERTIES OF CASEIN

Caseins are soluble below about pH 3 and above about pH 5·5 but are insoluble in their isoelectric region, about pH 4·6, at temperatures above about 5°C. At both acidic and neutral pH values, caseins form highly viscous solutions, the viscosity of which depends strongly on protein concentration, temperature, pH and calcium concentration.[5] However, caseinates are not normally used to control the viscosity of food products, although they are widely used for fat emulsification and water binding.[5]

6 GEL PROPERTIES

6.a Renneted Milk Gels

It is now well known that the rennet coagulation of milk is a two-stage process, the first involving enzymatic production of para-casein and peptides, the second involving the precipitation of para-casein by Ca^{2+} at temperatures above 20°C. Both stages, especially the primary phase, are fairly clearly understood.

6.a.1 Rennets
The rennets traditionally used in cheese manufacture are crude preparations of gastric proteinases from calves, lambs or kid goats. Chymosin accounts for >90% of the milk-clotting activity in rennets from very young animals, but, as the animal ages, the proportion of chymosin decreases while that of pepsin increases. The molecular characteristics of chymosins and pepsins have been reviewed by Foltmann.[48,49] Rennets from very young animals are considered to give best results in cheese manufacture but the supply has been inadequate for many years and has led to a search for

suitable 'rennet substitutes'. Most proteinases can coagulate milk under suitable conditions but most are too proteolytic relative to their milk-clotting activity; consequently, they hydrolyse the coagulum too quickly, causing weak gels, reduced cheese yields and/or defective cheese with a propensity to bitterness. Only six rennet substitutes have been found to be more or less acceptable for some or all cheese varieties: bovine, porcine and chicken pepsins, and the acid proteinases from *Mucor miehei*, *M. pusillus* and *Endothia parasitica*. Chicken pepsin is the least suitable of these and is used widely only in Israel and Czechoslovakia. Bovine pepsin is probably the most satisfactory and many commercial 'calf rennets' contain 40–60% of bovine pepsin. Microbial rennets, especially those from *M. miehei*, are widely used for cheese in the United States and for rennet casein elsewhere. Calf rennet is still used predominantly in Europe, Australia and New Zealand; in fact, rennet substitutes are prohibited in some European countries. For reviews on rennet substitutes, see Refs 50–53.

Although there has been an interest in immobilized rennets for many years, doubts have been expressed[54,55] as to whether an immobilized rennet can coagulate milk; none is used commercially. There is also interest in the production of 'calf' chymosin by genetically engineered micro-organisms; the technical feasibility of doing this has been demonstrated and such rennets are now commercially available.

6.a.2 Primary Phase of Rennet Action
With the isolation of κ-casein, and the demonstration that it is responsible for micelle stability and that its micelle-stabilizing properties are lost on renneting,[56] it became possible to define the primary phase of rennet action precisely. Only κ-casein is hydrolysed during rennet coagulation,[57] and only its Phe_{105}–Met_{106} bond is cleaved.[58] This particular bond is many times more susceptible to hydrolysis by acid proteinases than any other in the milk protein system. The peptide containing residues 1–105 of κ-casein is called para-κ-casein, while the sequence 106–169, which varies with respect to carbohydrate content, is the macropeptide.

The unique sensitivity of the Phe–Met bond has aroused interest. The dipeptide, H-Phe.Met-OH, is not hydrolysed, nor are tri- and tetrapeptides containing a Phe–Met bond. However, this bond is hydrolysed in the pentapeptide, H.Ser-Leu-Phe-Met-Ala-OMe, and reversing the positions of serine and leucine in this pentapeptide, to give the correct sequence of κ-casein, increases the susceptibility of the Phe–Met bond to hydrolysis by chymosin. Both the length and the sequence around the sectile bond are important determinants of enzyme–substrate interaction. $Serine_{104}$

appears to be particularly important and its replacement by Ala in the above pentapeptide renders the Phe–Met bond resistant to hydrolysis by chymosin but not by pepsin.

Extension of the above pentapeptide from the N- and/or C-terminal to reproduce the sequence of κ-casein around the chymosin-susceptible bond increases the efficiency with which the Phe–Met bond is hydrolysed by chymosin[60] and the peptide representing the sequence His_{98}–Lys_{111} of κ-casein is hydrolysed 66 000 times faster than the original peptapeptide and at a similar rate to intact κ-casein. κ-Casein is readily hydrolysed at pH 6·6 (and at pH 4·7) but small peptides are not hydrolysed at a significant rate at pH 6·6 until the sequence is extended to His_{98}–Lys_{111}, which is readily hydrolysed at pH 4·7 and 6·6. Thus, this sequence appears to contain the necessary determinants for rapid cleavage of the Phe–Met bond by chymosin and presumably by other acid proteinases.

The Phe_{105}–Met_{106} bond of κ-casein is the only such bond in milk proteins; however, these two residues are not intrinsically essential for hydrolysis and there are several Phe and many Met residues in all milk proteins. In the above pentapeptide, Phe may be replaced by $Phe(NO_2)$ and Met by Nle with relatively small changes (c. 3-fold) in k_{cat}/K_m. Neither porcine or human κ-caseins possess a Phe–Met bond (they have a Phe–Ile bond at this position), yet both are readily hydrolysed by calf chymosin, although more slowly than bovine κ-casein; in contrast, porcine milk is coagulated more efficiently than bovine milk by porcine chymosin, indicating that unidentified subtle structural changes influence chymosin action. Thus, the sequence around the Phe–Met bond, rather than the bond itself, contains the important determinants for hydrolysis. The particularly important residues are Ser_{104}, at least one of the three histidines (98, 100 or 102), some or all of the four prolines (99, 101, 109, 110) and Lys_{111}.

The importance of the proline residues may reside in their significance in secondary structures. Loucheux-Lefebvre et al.[22] propose that the chymosin-susceptible bond of κ-casein is situated between two β-turns, which may cause this sequence to protrude from the κ-casein molecule; this conformation should enable the relevant sequence to fit into the active-site cleft of acid proteinases. Theoretically, the peptide around the Phe–Met bond should be capable of forming α-helical or β-sheet structures; the latter should facilitate hydrogen bonding with the active site of the enzyme. For references to the original work in this area, see Refs 29, 55, 59 and 60.

The primary phase of rennet action has a $Q_{10°C}$ of c. 2 and is optimal at 45°C and pH 5·1–5·5. However, in cheese manufacture, milk is coagulated at c. 31°C and pH $\sim$ 6·6.

While the hydrolysis of κ-casein by acid proteinases other than calf chymosin has not been thoroughly studied, it is generally assumed that all commercially used rennets hydrolyse the Phe–Met bond. However, as discussed below, the aggregation characteristics of casein micelles altered by rennet substitutes differ from those of chymosin-treated micelles, suggesting differences in the extent and/or specificity of κ-casein hydrolysis. Furthermore, the common rennets differ in their action on synthetic peptides and this may be used to quantify the individual enzymes in a mixed rennet or as the basis for defining a rennet activity unit.[59]

6.a.3 Secondary Phase of Coagulation

The macropeptides released from the casein micelles by rennet are highly charged and hydrophilic; hence hydrolysis of κ-casein during the primary phase of rennet action reduces the zeta potential of the casein micelles from $-10/-20$ mV to $-5/-7$ mV and removes the protruding peptides from their surfaces, thereby reducing the intermicellar repulsive forces (electrostatic and steric) and destroying the colloidal stability of the caseinate system. When $c.$ 85% of the κ-casein has been hydrolysed, the casein micelles in milk begin to aggregate into a gel but an individual micelle cannot participate in gelation until $c.$ 97% of its κ-casein has been hydrolysed. Reducing the pH or increasing the temperature from the normal values ($c.$ 6·6 and $c.$ 31°C, respectively) permits coagulation at a lower degree of κ-casein hydrolysis.

Coagulation of rennet-altered micelles is dependent on a certain concentration of Ca^{2+}, which may act by cross-linking casein micelles, e.g. via serine phosphate residues, or by charge neutralization. Colloidal calcium phosphate (CCP) also plays an essential role: milk containing < 80% of the indigenous level of CCP fails to coagulate on renneting unless the $[Ca^{2+}]$ is increased. Various cationic species predispose casein micelles to coagulation and may even coagulate unrenneted micelles. Chemical modification of histidine, lysine or arginine residues inhibits coagulation, suggesting that electrostatic interactions are important in coagulation.[55,59] Dephosphorylation, especially of β-casein, also increases the rennet coagulation time (RCT).[91]

Pyne[61] claimed that pH has essentially no effect on the coagulation of rennet-altered micelles, but Kowalchyk and Olson,[62] using milk that had been treated with immobilized rennet, showed that there is a direct relationship between pH and the duration of the non-enzymatic phase of coagulation. The coagulation of renneted micelles is very temperature-dependent ($Q_{10°C} \sim 16$) and normal bovine milk does not coagulate below

c. 18°C unless the $[Ca^{2+}]$ is increased. The marked difference between the temperature dependence of the enzymatic and non-enzymatic phases of rennet coagulation has been exploited in the study of the effects of environmental factors on the rennet coagulation of milk, in attempts to develop a system for the continuous coagulation of milk for cheese or casein manufacture and in the application of immobilized rennets.

Para-κ-casein flocculates in the absence of Ca^{2+} but the rate of flocculation depends on the rennet used,[63] presumably reflecting differences in proteolytic specificity that have not been elucidated. The rate of firming of rennet gels is also influenced by the type of rennet, presumably for similar reasons.

RCT of milk is weakly correlated with casein concentration within the range normally encountered. Concentration of milk by ultrafiltration causes a slight increase in RCT at a constant level of rennet, although the rate of firming of the gel increases markedly with increasing concentration; diluting milk with ultrafiltrate or synthetic milk serum causes a marked increase in RCT.[59]

6.a.4 Effect of Pre-Renneting Treatment of Milk on Rennet Coagulation

The total RCT of milk increases markedly with increasing pH, especially above pH ~ 6·4, to an extent characteristic of the type of rennet; porcine pepsin is the most pH-dependent of the commonly used rennets. Addition of 1·5–2% starter decreases the pH of milk by c. 0·15 units and cognizance must be taken of this when neutralized or concentrated starters are used.

RCT decreases during the first 20% of lactation but increases thereafter, slowly at first but quite rapidly toward the end of lactation. The reasons for this phenomenon are not clear: increasing pH is obviously a factor, but adjusting the pH does not fully restore rennetability; the concentrations of protein, calcium and CCP in late lactation milk should be favourable. The sodium content, which increases with advancing lactation, may be a contributory factor,[18] as may the action of indigenous milk proteinase, plasmin,[64] although a recent study[65] showed that very extensive proteolysis by plasmin did not affect rennet coagulation or syneresis of the coagulum.

It is fairly common practice to add $CaCl_2$ (c. 0·04%) to cheese milk. This practice causes three changes in milk, all of which reduce RCT: increased $[Ca^{2+}]$, increased [CCP] and reduced pH. These three factors also increase gel strength, which may improve cheese yield. Less rennet is required for a standard RCT if $CaCl_2$ is added.

Milk for most cheese varieties is renneted at c. 31°C, which is far below the optimum for rennet coagulation. While RCT or the amount of rennet required may be reduced by increasing the renneting temperature, only a small increase is permissible owing to inhibition of mesophilic starters and the adverse effect of higher temperature on gel development.

Both the primary and secondary phases of rennet action are retarded if the cheese milk is heated at temperatures that cause denaturation of whey proteins; coagulation is completely inhibited following severe heat treatment. This change is due to disulphide-linked interaction between heat-denatured β-lactoglobulin (and possibly α-lactalbumin) and κ-casein that makes the rennet-susceptible bond inaccessible and also renders renneted micelles incapable of gelation; consequently, the strength of renneted milk gels is weakened. The adverse effects of heat treatment on the rennet coagulation of milk, both coagulation time and gel strength, may be counteracted by acidification, with or without reneutralization, apparently as a result of increased $[Ca^{2+}]$ (Ref. 60). Cooling and cold storage of milk following severe heat treatment exacerbates the adverse effects of heat treatment, an effect known as rennet hysteresis, owing to re-establishment of the calcium phosphate equilibrium by solution of some indigenous CCP.

Cold storage of raw milk also adversely affects rennet coagulability owing to one or more of the following:

(1) solution of CCP, which is readily reversible by HTST pasteurization or by heating at 40°C × 10 min;
(2) dissociation of some micellar casein, especially β-casein, which is also reversible;
(3) proteolysis by indigenous milk proteinase, plasmin;
(4) proteolysis by psychrotroph proteinases, which become important when these organisms exceed 10^6–10^7 cfu/ml.

Various aspects of the secondary phase of rennet coagulation and post-coagulation phenomena have been reviewed by Green and Grandison.[66]

6.a.5 Gel Assembly

The viscosity of renneted milk remains constant or decreases slightly during a period equivalent to $c.$ 60% of the visually observed RCT,[67,68] probably owing to a decrease in the voluminosity of the casein micelles following release of the protruding macropeptide segments of κ-casein; quasi-elastic light scattering showed that the hydrodynamic diameter of casein micelles decreases from $c.$ 139 nm to $c.$ 128 nm during the primary phase of rennet action.[69] Following the initial lag period, the viscosity increases exponentially up to the onset of visual coagulation or gelation.

During the first 60% of the visually-observed RCT, the micelles retain their individuality;[68] no changes are observed in the size, surface structure or general appearance of the micelles during this period. Approximately 85% of the κ-casein has been hydrolysed at 60% of the RCT, and at between 60% and 80% of the RCT the rennet-altered micelles begin to aggregate progressively with no sudden changes in the type or extent of aggregation.[68] Small, chain-like aggregates form initially and at visual coagulation most of the micelles have aggregated into short chains, which then begin to aggregate with the formation of a network. Initially, most micelles are linked by bridges (65 nm long and 40 nm across) and do not touch; the nature of the bridging material is unknown but a large proportion of the surface of the participating micelles is involved in the bridging.

Aggregation of rennet-altered micelles can be described by the Smoluchowski theory for the diffusion-controlled aggregation of hydrophobic colloids when allowance is made for the need to produce, enzymatically, a sufficient concentration of particles capable of aggregation, i.e. casein micelles in which $>97\%$ of the κ-casein has been hydrolysed.[70-75]

The overall rennet coagulation of milk can be described by combining three factors:[73] (a) proteolysis of κ-casein, which may be described by Michaelis–Menten kinetics; (b) the requirement that $c.$ 97% of the κ-casein on a micelle must be hydrolysed before it can aggregate; (c) the aggregation of para-casein micelles by the Smoluchowski process. The relationships have been described by eqns (1)[76] or (2):[77]

$$t_c = t_{prot} + t_{agg} \sim \frac{K_m}{V_{max}} \ln\left(\frac{1}{1-\alpha_c}\right) + \frac{c}{V_{max}} S_0 + \frac{1}{2k_s C_0}\left(\frac{M_{crit}}{M_0} - 1\right) \quad (1)$$

where K_m and V_{max} are the Michaelis–Menten parameters, α_c is the extent of κ-casein hydrolysis, S_0 is the initial concentration of κ-casein, k_s is the rate constant for aggregation, C_0 is the concentration of aggregating material, M_{crit} is the weight average molecular weight at t_α (~ 10 micellar units) and M_0 is the weight average molecular weight at $t=0$. Alternatively,

$$t_c = \frac{1}{V}\left[S_0 + \frac{1}{C_m}\exp(-C_m S_0) - 1\right] + \frac{W_0 \exp(-C_m S_0)}{k_s}\left[\frac{1}{n_c} - \frac{1}{n_0}\right] \quad (2)$$

where V is the velocity of enzymatic hydrolysis of κ-casein, S_0 is the initial concentration of κ-casein, C_m is a constant relating the stability of the casein micelle to κ-casein concentration, W_0 is the initial stability factor for casein micelles, n_0 is the initial concentration of casein micelles and n_c is the concentration of casein aggregates at the observed clotting time.

About 90% of the micelles are incorporated into the curd at the visual RCT.[78,79] The micelles that are 'free' at or after the RCT may react differently from those already coagulated—prior to coagulation all micelles are free in solution and can aggregate randomly but, once a matrix has started to form, free micelles may react either with the gel matrix or with each other. Therefore, gel assembly may be regarded as a two-stage process and the properties of the final curd may be affected by the amount of casein 'free' at the RCT.

The rate of firming of rennet gels as a function of time shows two maxima:[80] firming was first observed $c.$ 2·5 min after the visual clotting and the firming rate increased to a maximum 10 min later; the rate decreased over the next 10–15 min to $c.$ 80% of the maximum value, after which it either remained constant or increased slightly for a further 15 min and thereafter decreased steadily. It was suggested that the second maximum may be due to the incorporation of free micelles into the gel network. The two phases responded differently to assay conditions:[81] the rate and magnitude of the first phase increased with decreasing pH (6·6 to 6·0), while pH had relatively little effect on the second phase. In contrast, temperature (24–40°C) had little effect on the first phase (slight maximum at 35°C), while the second phase was very temperature-dependent, being very marked at 25°C but absent at 35 and 40°C. Rennet concentration had little effect on either phase, while the rates of both phases increased with increasing concentrations of calcium and casein.

The viscometric data of Tuszynski[82] also suggest that gel assembly is a two-stage process: flocculation and gelation. Additional evidence for a two-stage gel assembly process is presented by Surkov et al.,[83] although the terminology is not clear. They suggest the following mechanism for rennet coagulation:

$$E + S \underset{k_{-1}}{\overset{k_1}{\rightleftharpoons}} ES \overset{k_2}{\longrightarrow} E + P \overset{k_c}{\longrightarrow} P^* \overset{k_s}{\longrightarrow} P_n$$

The first two steps are the Michaelis–Menten model for the primary, enzymatic phase and are essentially as proposed by Payens et al.:[70]

$$E + S \underset{k_{-1}}{\overset{k_1}{\rightleftharpoons}} ES \longrightarrow E + P_1 + M$$

where P_1 = para-κ-casein and M = macropeptide. Payens et al.[70] suggest that the second non-enzymatic phase may be represented by

$$iP_1 \overset{k_s}{\longrightarrow} P_i$$

Surkov et al.[83] suggest that the enzyme-altered micelles (para-casein micelles, P) undergo a cooperative transition in quaternary structure to yield clot-forming particles (P*) with a rate constant $= k_c$; $E_a = 191$ kJ/mol and $Q_{10°C} = 12$. The nature of this transition is not indicated but the altered particles, P*, then undergo gelation according to Smoluchowski's diffusion-controlled kinetics, the rate constant for which is k_s; $E_a = 34$ kJ/mol and $Q_{10°C} = 1·6$. These values are very similar to those obtained by Tuszynski.[82]

A four-stage kinetic model for gelation has been proposed by Johnston[84] on the basis of polymer cross-linking theory. In the first stage, κ-casein is enzymatically converted to para-κ-casein; when sufficient conversion has occurred, the micelles are able to coagulate or cross-link, i.e. the second stage. When the cross-linking index, γ_1, has reached a value of 1, a weak gel forms. In the third stage, the proportion of gelled material ($W_g = 1 - (\gamma^1/\gamma)$, γ^1 = cross-linking index of material in the sol phase) increases from zero to its limiting value, which is set by the number of reacted sites per initial unit; the number of cross-links within the gelled material also increases during this phase. The fourth phase commences when all the particles in the sol have been cross-linked, i.e. when the ultimate value of the cross-linking index, γ_∞, has been reached. During this phase, gel strength increases only by the formation of further cross-links within the gel. Johnston[84] provides experimental support for the validity of the rubber network theory as a model for the interpretation of milk gel formation and properties.

The nature of the sites involved in the aggregation process is not known. Reduction of the micellar zeta potential by proteolysis of κ-casein facilitates linkage of particles, possibly via calcium bridges or hydrophobic interactions (which the marked temperature dependence of the second phase would indicate). The hydrophobic amino-terminal segment (residues 14–24) of α_{s1}-casein appears to be important in the formation of a protein network in rennet curd;[85] the softening of cheese texture during ripening is considered to be due to the break-up of the network on hydrolysis of α_{s1}-casein to α_{s1}-I-casein.[85,86] The fact that modification of histidyl, lysyl or arginyl residues inhibits the secondary phase of rennet coagulation[87–89] suggests that a positively-charged cluster on para-κ-casein interacts electrostatically with unidentified negative sites on casein micelles. Hill[89] suggests that in native micelles this positive site is masked by the macropeptide segment of κ-casein but becomes exposed and reactive when this peptide is released. Culioli and Sherman[90] suggest that the conversion of —SH groups to —S—S— bonds is favoured somehow by the hydrolysis of κ-casein and that the formation of a rennet gel is analogous to the sulphydryl/disulphide reactions that occur in flour dough development. It is difficult to envisage how this operates in casein, which has very few —SH groups. Pearse et al.[91] concluded that the phosphate groups of casein, particularly those of β-casein, are directly involved in the micelle–micelle interactions that occur during rennet coagulation and syneresis.

After the visual coagulation, the network appearance becomes gradually more pronounced;[68] the strands of the network are c. 5 micelles thick and c. 10 micelle diameters apart. The bridges between the micelles contract slowly, forcing the micelles into contact and to fuse partly. The fate of the bridging material on micelle fusion is unknown; it may be that its disappearance is responsible for the second maximum in the firming rate–time curve[80] and for the reported[82–84] flocculation and gelation stages in the gel-assembly process.

6.a.6 Relationship between Rennet Concentration and Coagulation Time

As early as 1870, Storch and Segaliki showed that the rennet coagulation time (RCT) is inversely related to enzyme concentration: $Ct = k$, where C is the enzyme concentration and $t = $ RCT (Ref. 92). This equation, which assumes that visually observed coagulation is dependent only on the enzymatic process, was modified by Holter (see Ref. 92) to take account of the duration of the secondary, non-enzymatic phase:

$$C(t - x) = k$$

where x is the time required for the coagulation of the enzymatically-altered casein micelles and $(t - x)$ is the time required for the enzymatic stage. Foltmann,[92] who rearranged the equation to a more convenient form, $t_c = k(1/C) + x$, showed that this equation is valid within a certain range of rennet concentrations and under certain conditions of temperature and pH.

A very good linear relationship has been demonstrated between clotting time and the reciprocal of pepsin concentration using a turbidimetric milk clotting assay.[93] Using the Formagraph to measure RCT, McMahon and Brown[94,95] found a very good linear relationship between t_c and $1/E$ (E is the enzyme concentration) or $\log t_c$ and $\log(1/E)$ over a wide range of rennet concentrations (t_c ranged from 0·7 min to 66 min. Under their test conditions (skim milk powder dissolved in 0·01M $CaCl_2$, held overnight at 2°C, equilibrated at the assay temperature, 35°C, for 30 min, pH unspecified), $t_c = 0·685 + 0·075(4/E)$.

The coagulation equations developed by Dalgleish[76] and Darling and von Hooydonk[77] might be regarded as refinements of these simpler equations and reduce to them on first approximation. Payens and collaborators[70-72] expressed clotting time, t_c, in an alternative form:

$$t_c \left(\frac{k_s V_{max}}{2} \right)^{1/2} = C$$

where k_s, the diffusion-controlled flocculation rate constant according to von Smoluchowski's theory (non-enzymatic phase), is proportional to the concentration of reactive (coagulable) particles (proteolysed micelles) and hence to enzyme concentration; V_{max} is the maximum velocity in Michaelis–Menten kinetics (enzymatic phase) and is proportional to enzyme concentration; C = constant depends on the experimental conditions.

6.a.7 Curd Tension (Gel Strength)

The strength of rennet-coagulated milk gels (curd tension) is particularly important from the viewpoint of cheese yield. For most cheese varieties a period roughly equal to the RCT is allowed after visual coagulation for the gel to become sufficiently firm before cutting. If the gel is too soft when cut, fat and casein losses in the whey will be high (see Ref. 96 for a recent study of the influence of curd firmness on cheese yield and for references thereon). Much of the current interest in curd tension stems from the desire to standardize cheese manufacture, which is particularly important with mechanized, semi-automated cheese-manufacturing systems.

A wide range of techniques, using penetrometers, suspended bodies, torsion viscometers, manometers, light absorption and ultrasonics, have been used to measure curd tension. These have been reviewed by Thomasow and Voss.[97] Modifications of the thrombelastograph, e.g. the Formagraph and lactodymograph, are now available and have been used in several studies on the coagulation properties of milk.[15,94,98] A method based on oscillatory deformation of the curd (a pressure transmission system) was introduced by Vanderheiden[99] and assessed with satisfactory results.[100-102] The Instron Universal Testing Machine has also been applied satisfactorily.[80,81,103,104] The various penetrometers, suspended bodies and some viscometers permit measurement at only a single point in time, which is a serious limitation in kinetic studies, and they also require meticulous test conditions since CT increases with time after renneting. Some viscometers, e.g. thrombelastograph-type instruments (see Ref. 97 for references), the Brookfield LVT-Helipath viscometer,[105] vibrating-reed-type viscometers,[102] the Instron, oscillatory deformation and manometric methods,[106] permit continuous measurement.

As a quality control measure in cheese manufacture, the measurement apparatus should be usable in the cheese vat and be easily operated and robust. A speed-compensated 'torsiometer' was developed by Burnett and Scott Blair[107] for in-vat measurement of gel rigidity. Vibrating reed-type viscometers and the oscillatory deformation system appear to be very suitable for the continuous measurement of gel strength in the cheese vat. Marshall et al.[102] conclude that while the former is probably more suitable for laboratory investigations, the latter may be more useful for indicating optimum cutting time during cheesemaking. The 'vatimeter', a suspended body-type apparatus, was also specifically designed for in-vat use.[108]

Rennet gels behave as a 'Burger's body', i.e. a system with a viscous element (η_1) in series with an immediate elastic element (G_1), followed in series by a unit consisting of a viscous element (η_2) in parallel with a second elastic element (G_2).[106] η_1/η_2 is the relaxation time (t_{rel}); η_2/G_2 is the retardation time (t_{ret}); t_{rel}/t_{ret} is an indication of springiness. These indices are influenced independently and the effects of various factors thereon have been described.[109,110]

In general, there is an inverse relationship between CT and RCT.[111,112] CT is directly related to casein and colloidal calcium phosphate (CCP) concentrations[112] and increases markedly with decreasing pH[112] to a maximum at pH 5·9–6·0;[102,111] the decrease in CT at lower pH values may be due to solubilization of CCP as the pH is decreased. Addition of calcium increases CT[102,111–113] up to a maximum at c. 10 mM; the decrease at higher [Ca^{2+}] may be due in part to the concomitant decrease in pH, but a similar effect was observed when the pH was maintained constant.[111] Addition of calcium chelators reduces CT,[111] while addition of NaCl or KCl increase CT up to 100 mM but markedly decrease it at higher concentrations, possibly by displacing micellar calcium.[111]

The curd tension of gels from mastitic milk is lower than that of gels from normal milk,[114,115] probably owing to a combination of factors, e.g. high pH, low casein. Heat treatment of milk to temperatures that denature whey proteins reduces CT.[102,113,116,117] CT increases with increasing rennet concentration and increasing temperature but the magnitude of the response depends on the method of measurement.[102] There are numerous reports (e.g. Refs 105 and 118) that CT is strongly influenced by the type of rennet used: calf chymosin gives a more rapid increase in CT than microbial rennets, although the substrate on which the rennets were standardized for clotting activity is of some significance. The fact that rennets standardized to equal clotting activities cause different rates of curd firming that respond differently to compositional factors, e.g. [Ca^{2+}], suggests possible differences in the extent and/or specificity of proteolysis during the primary, enzymatic phase of rennet coagulation. As discussed previously, the primary phase of coagulation by the principal coagulants involves cleavage of the Phe_{105}–Met_{106} bond; however, additional bonds may be hydrolysed by microbial rennets.

6.a.8 Curd (Gel Syneresis)

The rennet coagulation process is common to all rennet cheese varieties; the gel assembly process is presumably also essentially common to all varieties, although the firmness of the gel at cutting varies somewhat with variety. However, the subsequent treatments of the rennet gel vary markedly among varieties. The properties of the gel (curd, cheese) are strongly influenced by these treatments and hence the individuality of cheese varieties. Not only is the texture of the curd altered by post-gelation treatments but through their effect on cheese composition, particularly moisture content, they also strongly influence the biological, biochemical and chemical reactions that occur in the cheese during ripening.

Many of the post-gelation treatments of rennet gels can be classified as dehydration: cheese manufacture essentially involves concentrating the fat and casein about 10-fold, with the removal of lactose and soluble salts in the whey. Although there are certain common features, e.g. acidification, the method of dehydration is characteristic of the type of cheese, e.g. for Cheddar and Swiss varieties dehydration is accomplished mainly in the cheese vat by finely cutting the gel and by extensive 'cooking' of the curd–whey mixture with vigorous agitation. For soft (high-moisture) varieties, the gel may be scooped directly into moulds without cutting or cooking and dehydration occurs mainly in the moulds. Curd for some varieties, e.g. Cheddar and Swiss, is subjected to considerable pressure in the moulds, thus aiding whey removal, while soft cheeses are pressed only under their own weight. All cheeses are salted during manufacture and this contributes to whey removal. However, the point during manufacture at which salt is applied and the method of application, e.g. dry mixing (Cheddar type), submersion in brine (Dutch and Swiss types) or surface application of dry salt (Blue), vary and add a further variable in the dehydration process.

Dehydration is made possible by the marked tendency of rennet gels to contract (synerese) with the expulsion of whey. Most of the published studies on syneresis have been concerned mainly with the factors that affect it during the early stages of dehydration, i.e. mainly during cooking, but it is assumed that basically the same mechanism operates throughout.

Despite its accepted importance in the control of cheese moisture, and consequently on cheese ripening, the syneresis mechanism for rennet gels is not well understood, although there is considerable information on factors that influence syneresis.

Poor methodology for studying syneresis is mainly responsible for the lack of information; the number of principles exploited in methods used to measure syneresis attests to their unsuitability. These include measurement of the volume of whey expressed from cut curd under controlled conditions; the moisture content, volume or density of the curd; use of tracers or markers to measure whey volume; or changes in the electrical resistance of the curd.[119,120] Some authors have attempted to simulate cheese manufacture (e.g. use of starter, stirring and a cooking profile) but the accuracy and precision of many of the methods are poor. However, using these methods, as well as cheese-making experiments, the influence of several factors on syneresis is well established in general terms. The subject has been reviewed recently[119,120] and only a summary is included here.

6.a.8.1 INFLUENCE OF COMPOSITIONAL FACTORS ON SYNERESIS

The syneresis properties of renneted milk gels are influenced by milk composition as affected by feed, stage of lactation and animal health. Fat tends to reduce and protein to improve syneresis; increasing the fat content of milk increases cheese yield by $c.$ 1·2 times the weight of additional fat. Concentration suppresses syneresis (this is possibly related to gel strength), although the rigidity modulus at the time of cutting appears to have little effect on syneresis. Syneresis is inversely related to pH, and the rate of syneresis is optimal at the isoelectric point (i.e. pH 4·6–4·7). Addition of $CaCl_2$ to milk improves syneresis but the effect is less than might be expected and may be negative at certain pH values and at high calcium concentrations, especially if the gel is held for a long period before cutting; Ca^{2+} presumably reacts with protein —COO^{2-} groups, leading to an increased net positive charge, swelling of the protein and suppression of syneresis. It is likely that a more firm gel, such as would be obtained on longer holding, would also be more resistant to syneresis. The influence of colloidal calcium phosphate does not appear to have been investigated. Low levels of NaCl increase the rate of syneresis but higher levels retard it.

6.a.8.2 INFLUENCE OF PROCESSING VARIABLES ON SYNERESIS

The extent of syneresis, and hence the moisture level in cheese, is influenced by the size of cut, agitation, rate of acid development, removal of some whey, extent of dry stirring, size of curd blocks during cheddaring and the amount of salt added to the curd. Syneresis is promoted by increasing temperature, especially with unstirred curd, but in the temperature range normally used in cheese-making (35·5–38°C) the effect is slight and may be negative above 38–40°C owing to the inhibitory effect of higher temperatures on starter growth and acid production. Heating milk under conditions that cause whey protein denaturation reduces the tendency of rennet gels made from it to synerese. Homogenization of whole milk has a similar effect.

6.a.8.3 KINETICS AND MECHANISM OF SYNERESIS

Syneresis is initially a first-order reaction because the pressure depends on the amount of whey in the curd; holding curd in whey retards syneresis owing to back pressure of the surrounding whey, while removing whey promotes syneresis. When the curd is reduced to 70% of its initial volume, syneresis becomes dependent on factors other than whey volume.

Syneresis is due to protein–protein interactions and may be regarded as a continuation of the gel-assembly process in rennet coagulation.[121] The inhibitory effects of high salt ($CaCl_2$, NaCl, KCl) concentrations on syneresis suggest ionic attractions, and the effectiveness of pH in promoting syneresis is probably due to a reduction of overall charge as the isoelectric point is approached. Marshall[121] considers that hydrophobic interactions within the casein network are most likely responsible for the advanced stages of syneresis, which is in accord with the promotion of syneresis by reduced pH and low levels of $CaCl_2$, which reduce micellar charge and increase hydrophobicity, and by increased temperatures, which increase hydrophobic interactions.

The foregoing discussion on syneresis pertains especially to that occurring in the cheese vat, i.e. mainly to hard and semi-hard varieties. Syneresis continues after moulding and represents the major portion of syneresis in soft varieties. Presumably, the mechanism of syneresis in moulds is the same as in the cheese vat, although the range of treatments that can be applied at this stage is rather restricted. External pressure is applied to many cheese varieties after moulding and makes a significant contribution to whey removal; in general, the drier the cheese at hooping, the higher the pressure applied, which reflects the greater difficulty in ensuring fusion of dry curds.

Salting contributes significantly to syneresis, even in 'dry' varieties like Cheddar. Brine-salting probably causes greater syneresis than dry salting (blending or surface application). During salting, NaCl diffuses into the curd chips or cheese with a concomitant outward movement of water according to the equation

$$-\Delta W_x \sim p\,\Delta S_x$$

where ΔW and ΔS are the number of grams of H_2O and NaCl, respectively, per 100 g solids-not-salt and x is the distance from cheese surface; p decreases from $c.$ 3·7 at the interface to <1 at a zone in the middle of the cheese, i.e. the outflow of water from the surface exceeds the inflow of salt. Initially, a positive moisture gradient and a negative salt gradient exist from the surface to the centre of the cheese but these more or less disappear during storage (see Ref. 122 for a review of the various aspects of the salting of cheese).

6.a.9 Microstructure of Cheese

From the organoleptic viewpoint, the flavour and body (texture) are more or less of equal importance and, in addition, the texture affects certain physicochemical properties, e.g. meltability of processed and other cooked cheeses, elasticity of Swiss cheese (gas retention). The body of cheese is closely related to its (micro)structure,[123] which may be regarded as gel-like and which is modified by proteolysis and other biochemical changes during ripening.

The structure of the renneted milk gels is reflected to some extent in that of the finished cheese, but certain cheese-making operations, e.g. cheddaring or stretching (pasta filata cheeses), result in a definite fibrous structure that is readily observed using light microscopy.[124] The boundaries of curd grains are readily apparent in Cheddar, Emmental and Edam cheese (and presumably in other varieties) under the light microscope;[124,125] these grain boundaries are almost fat-free. Light microscopy clearly shows that cheese consists of a protein network in which are embedded irregularly-shaped fat particles of variable size (probably non-globular) and bacteria that appear to be concentrated in crevices.[125,126] As ripening proceeds, the fine protein structures tend to disappear.[127]

An electron microscopy study[127] of Cheddar cheese showed that cooking caused the casein network formed during gelation to form strands, some of which were linked by casein bridges: whey drains through the channels between the protein strands. During whey drainage the spaces between the strands are reduced and the protein matrix becomes more compact, pressing and distorting the fat globules, which still retain their individuality although some coalescence occurs. The overall structure can be visualized as a casein sponge in which fat globules and bacteria are trapped; the casein matrix, formed by progressive fusion of micelles and adjacent aggregates, becomes extensively folded as manufacture progresses and individual micelles lose their identity. Pressure and curd flow during cheddaring cause further deformation, with progressive elimination of interstitial spaces. Fat globules become very deformed and although they still appear to be essentially discrete, the natural membrane appears to be replaced by casein in most cases. The structural details become less clear during maturation; some fat globules are still apparent in mature cheese but in many cases the boundaries between them have disappeared and many are surrounded by 'debris', probably hydrolysed protein.

Scanning electron microscopy (SEM) shows the sponge-like structure of cheese very clearly.[128-134] The protein matrix is readily apparent and fat globules, or rather the holes left by them following solution in fixing solvents, and bacteria can be seen clearly. A comparison[128] of the ultrastructure of Cheddar, Cheshire and Gouda showed that the size of protein aggregates decreases with decreasing cheese pH: in Gouda, which has a high pH at the end of manufacture, the protein aggregates (particles) are more or less like those in milk, whereas in Cheshire, the most acid of the cheeses studied, the particles are smallest, with Cheddar occupying an intermediate position; presumably, acidification disperses casein micelles by dissolving CCP. Fat globules are more aggregated and coalesced in Cheddar than in Cheshire and Gouda, presumably because of the higher cooking temperature and the pressure and deformation applied during cheddaring. The manufacturing procedures for Gouda, Cheddar and Cheshire cheese are not markedly dissimilar, at least in principle, yet the texture, and flavour, differ considerably. Obviously, relatively small differences in manufacture, especially the rate of acidification, influence the microstructure of cheese curd, which in turn is reflected in the body (and texture) of the finished cheese, e.g. Gouda has a resilient, springy texture, Cheshire has a hard, brittle, crumbly texture, while Cheddar occupies an intermediate position with a hard, compact texture. Further research is required to elucidate the subtle changes in gel assembly and microstructure that lead to these differences in body and texture.[135]

The fibrous nature of Cheddar, Provolone and Mozzarella is very clear in SEM, whereas it is absent in unstretched varieties, e.g. Edam, Gouda, Brick, in which the protein particles appear granular.[131] Cheddar made using calf rennet has a more compact and organized structure than that made using bovine or porcine pepsins.

Kalab and Emmons[136] proposed a mechanism for the development of the texture of Cheddar cheese: bundles of casein micelle chains (formed during gel assembly) partly fuse during scalding and become uniformly oriented during cheddaring. Spaces between protein fibres are filled with clusters of fat globules that fuse into spherical masses when the stress of cheddaring is removed. During salting and pressing, the elongated bundles of relatively separate fibres fuse to form a solid homogeneous mass; thus, while an overall fibrous structure is retained, the typical 'chicken-breast' texture of freshly cheddared curd is lost.

The coarseness of the protein network increases in proportion to the degree of concentration of milk concentrated by ultrafiltration; the protein is packed in larger, more compact areas while the fat is more segregated and the fat–protein interfacial area is smaller in concentrated milks;[134] these structural features influence cheese texture and flavour development. These authors confirmed earlier observations[127] on the structure of cheese curd made from unconcentrated milk and showed that, during ripening, the microstructure became more open and less well defined, presumably reflecting degradation of the protein network by proteolysis. Loss of structural detail during ripening is also reported by other investigators (see, e.g., Refs 127, 132 and 133). The application of electron microscopy to dairy foods, including several cheese varieties, has been reviewed by Kalab.[137,138]

6.a.10 Use of Milk Concentrated by Ultrafiltration for Cheese Manufacture

Only about 75% of the milk protein is recovered in cheese manufactured by conventional methods. Stimulated by the possibilities of improving protein recovery and developing a continuous method for cheese manufacture, attention has been focused during the past 15 years on the use of ultrafiltration to pre-concentrate the colloidal phase of milk to that characteristic of the cheese variety being produced and to convert this 'pre-cheese' to curd by added starter and rennet but without whey drainage.

Ultrafiltration (UF) has been particularly successful, and is widely used commercially, for the manufacture of Feta and Quarg; it has also been successfully used for several soft cheeses, e.g. Blue, St Paulin, Camembert, Coulommier. However, numerous problems, especially in relation to texture, have been encountered with hard and semi-hard varieties but some success has been achieved.[139-143]

Some features of the rennet coagulation and gel-assembly process in UF-concentrated milks differ considerably from those for standard milk and merit comment. At a fixed rennet concentration, the rennet coagulation time (RCT) of milk remains constant or increases very slightly on concentration by UF up to 3-fold;[144] thus, on a casein basis, the amount of rennet required to give a constant RCT is reduced to $c.$ 20% of that normally required.[139,141] However, this appears to be true only at high enzyme levels.[90] Although RCT remains essentially constant with increasing casein concentration at fixed rennet levels, a decreasing percentage of the casein micelles are incorporated into the gel at the point of visual coagulation:[79] while c. 90% of the micelles in normal milk are incorporated into the curd at the RCT, only c. 50% are incorporated at the corresponding stage in a 4-fold concentrate. As discussed previously, micelles that are free at the RCT will be incorporated into the gel differently from those incorporated before RCT and may lead to an altered gel structure. The degree of casein incorporation into the gel may be maintained essentially constant by increasing the amount of rennet used pro rata with the degree of concentration, but this markedly reduces RCT: e.g. for a 4-fold concentrate, the RCT would be c. 33% of that of a normal milk. An alternative approach is to allow a longer time for gel assembly. Incomplete incorporation of micelles into the gel matrix may be of little consequence in soft cheese in which the gel is not cut but may be important in hard varieties and may partly explain the textural problems encountered with these varieties.

The effects of protein concentration on the rennet coagulation of milk are influenced by reaction pH and type of enzyme (chymosin or bovine pepsin).[145] The initial velocity of the enzymatic phase of rennet action increases with increasing casein concentration but the magnitude of the increase depends on pH and enzyme. Non-protein nitrogen (NPN) at RCT as a percentage of total N decreases with increasing protein concentration, which is another way of expressing the observation[79] that an increasing proportion of casein micelles in concentrated milk remains free at RCT. In contrast to the results of Dalgleish,[144] who showed that the degree of concentration had little effect on RCT, Garnot and Corre[145] showed that at pH 6·7 RCT decreased with increasing degree of concentration, especially for bovine pepsin, but at pH 6·2 RCT increased markedly with increasing concentration; the results obtained at pH 6·2 were somewhat like those of Culioli and Sherman[90] at low enzyme concentrations. The reported differences in the relationship between RCT and degree of concentration may be due to differences in the method of sample preparation but these are not apparent from the details provided; further work on the influence of concentration on RCT under a range of experimental conditions appears warranted.

The rigidity modulus, G, of rennet gels increases markedly with increasing concentration.[90] The time required for a gel to reach maximum G decreases as the amount of added rennet is increased and is independent of protein concentration; hence, the rate of firming, dG/dt, increases with degree of concentration. The kinetics of gel formation in milk, irrespective of casein concentration, may be described by the equation

$$G = G_\infty e^{-\lambda/t}$$

developed by Scott Blair and Burnett,[146] where G is the rigidity modulus at any time, t, after visual coagulation, $G_\infty = G$ at infinite time and λ is the time required for G to become equal to G_∞. This kinetic equation is also applicable to UF-concentrated milks, independently of concentration.[147] Plots of ln G against ln protein concentration (C_p) were linear[90] and of the form

$$G = d(C_p)^b$$

with $d = 19.5$ and $b = 2.58$. Regardless of the degree of concentration, the creep compliance–time behaviour of the gels can be represented by six viscoelastic parameters,[90] which suggests that their internal structures were essentially similar and that they differed only in degree. This conclusion is at variance with that of Dalgleish.[79,144]

The influence of protein concentration, C_p, on the rate of firming depends on the rigidity of the gel when measurements are made.[147] Thus, plots of the time interval required to reach three degrees of rigidity, $G = 0.15$ cm ($\Delta t_{0.15}$), $G = 2.7$ cm ($\Delta t_{2.7}$) and $G = 15$ cm (Δt_{15}), as a function of C_p showed that $1/\Delta t_{0.15}$ increased up to $C_p = 90$ g/kg but decreased at higher values of C_p; $1/\Delta t_{2.7}$ also increased up to $C_p = 90$ g/kg but plateaued at higher values of C_p, while $1/\Delta t_{15}$ increased almost linearly up to $C_p = 120$ g/kg. The initial rate of firming ($\Delta t_{0.15}$) is favoured by C_p up to 90 g/kg owing to the increased rates of κ-casein hydrolysis and protein aggregation at higher values of C_p; the decrease above $C_p = 90$ g/kg is considered to be due to the high proportion of unhydrolysed micelles, which are unable to participate in gelation, in the very highly concentrated samples. As the time elapsed after rennetting is extended, i.e. $\Delta t_{2.7}$ and especially Δt_{15}, the extent of κ-casein hydrolysis becomes progressively more complete, allowing greater participation of micelles in gel formation, and hence the rate of firming becomes increasingly highly correlated with C_p with increasing time after RCT. These authors also showed that G_∞, λ and RCT were not significantly influenced by the fat content of the milk but that small differences in C_p had a major effect on G_∞.

Both Culioli and Sherman[90] and Garnot *et al.*[147] emphasize the difficulty of visually determining the coagulation point in concentrated milks and recommend rheological methods.

The rate of firming of rennet gels, measured by a vibrating reed viscometer or oscillating diaphragm instrument, also increased markedly with concentration factor, even when the RCT was increased by reducing the amount of rennet used.[148] When concentrated milks were made into Cheddar cheese, the proportion of total milk protein retained in the curd increased slightly with concentration factor but the proportion of fat retained decreased markedly above 2-fold concentration, reducing cheese yield and giving a curd of abnormal composition. Other investigators do not appear to have experienced this problem and certainly not to the same extent. The pH of the cheese increased directly with degree of concentration, owing to increased buffering capacity. Sutherland and Jameson[140] overcame this problem by acidifying the milk before concentration. Lipolysis in maturing cheese was not significantly affected by concentration, but proteolysis and flavour development were retarded, probably because of the reduced concentration of rennet retained in curd from concentrated milk since the proportion of rennet retained was essentially the same for all curds, i.e. $2 \cdot 7 \pm 0 \cdot 6\%$ of that added.

Electron microscopy has shown that as more concentrated milk is used, the protein masses at and after maximum scald become larger and the protein network coarser with more segregation of fat and protein, leading to a reduction in fat/protein interfacial area.[134] The firmness, cohesiveness, force to cause fracture, crumbliness, granularity and dryness of the cheese all increased as the concentration factor of the milk increased but adhesiveness decreased and there was little change in elasticity. The textural changes were related to the lower fat content, lower level of proteolysis, stronger links in the protein matrix, and reduced capacity of the fat and protein phases to move relative to each other in the cheese made from the concentrated milks.

6.b Acid Coagulation of Milk

When milk is acidified, the colloidal calcium phosphate in the casein micelles progressively solubilizes[149] and aggregation of the casein occurs as the isoelectric point (pH 4·6) is approached. If the pH of agitated milk at temperatures $>20°C$ falls rapidly to $<5·0$, the casein aggregates to such an extent that syneresis occurs; the aggregates precipitate from solution and may be separated from other milk components by filtration, which is the basis of acid casein production. When the pH of milk at temperatures $>20°C$ falls slowly under quiescent conditions, e.g. as a result of fermentation of lactose to lactic acid or as a result of acid formation from an added acidogen, e.g. glucono-δ-lactone (GDL), or when milk is acidified at temperatures $<4°C$ and subsequently heated without agitation, there is less coalescence and a gel network with good water-holding capacity is formed; this is the principle for the production of fermented milk products and some fresh cheeses, including yoghurt, quarg and cottage cheese.

6.b.1 Mechanism of Acid Gelation

In milk of normal pH, the casein micelles are stabilized by hydration and steric repulsion due to their net negative charge. On acidification, the micelles become unstable and in an unstirred solution coagulation occurs with the formation of a gel network in which casein is the main component.[150] Early views on the mechanism of gelation[137,151-154] suggested that on acidification casein micelles aggregate as a result of charge neutralization, leading to the formation of chains and clusters that are somehow linked together to give a three-dimensional network. Glaser et al.[154] observed an increase in mean micelle or particle size from 87·6 nm in milk to 180 nm at gelation in curd made using GDL as acidulant. They noted that the early stages of gelation involved aggregation of micelles to form chains in which the micelles retained their identity. The chains developed into multiple strands in which chain junctions appeared as clusters of micelles.

Davis et al.[153] suggested that, during yoghurt gel development, micelles tend to group together and partly lose their integrity as coalescence occurs. Kalab et al.[152] noted that in unheated skim milk inoculated with a commercial yoghurt culture and incubated at 44°C the first signs of gelation became evident at pH 5·14 and gelation was complete at pH 4·92. About 30 min before gelation, micelles began to expand and undergo surface changes. As gelation proceeded, non-micellar protein accumulated on the micellar surfaces as ray-like projections. About 13 min before gelation, micelles began to fuse, initially into groups of two or three and later into larger units. A three-dimensional network of chains and clusters developed in which the outlines of the large micellar aggregates became sharp and smooth. The authors concluded that the large aggregates were tightly fused clusters of unaltered micelles.

The view that the gel network in acid casein gels is formed as a result of interactions between casein micelles or aggregates of micelles was supported by most researchers in the field until very recently. However, as pointed out by Heertje et al.,[155] this view does not envisage any significant role for colloidal calcium phosphate despite its importance in maintaining the integrity of the casein micelles and its solubilization and removal from the micelles on acidification. Thus, Heertje et al.[155] queried whether acid-induced gelation of milk should be described as an aggregation of the original individual micelles, as they occur in milk, or whether partial disintegration of the micelles precedes aggregation. This question was extensively considered by Heertje et al.,[155] Roefs et al.[156] and Roefs.[150]

Heertje et al.[155] showed that on acidification of skim milk with GDL at 30°C the colloidal calcium phosphate was completely solubilized at pH ~ 5·0. Electron microscopy, using a freeze-fracture technique, showed that during acidification from pH 6·6 to 5·5 the casein micelles retained their integrity, shape and dimensions. At pH 5·5, some small micellar particles were evident, suggesting the onset of micelle disintegration. At pH 5·2, the micelles lost their integrity and an aggregated network structure was evident. As the pH was reduced to 4·8, contraction and rearrangement occurred, leading to the formation of a particulate network. Similar changes in micellar structure were observed in heated and unheated milks, although the pH values at which the changes occurred differed.

Roefs et al.[156] and Roefs[150] found little change in the hydrodynamic diameter and light-scattering intensities of a reconstituted skim milk on decreasing the pH from 6·8 to 4·6 with HCl at 8°C, although a slight minimum in both parameters was observed at pH 5·2. These results suggest that little change in micelle size occurs on acidification. However, a marked change in casein solubility was observed over the same pH range. At 5–8°C, soluble casein (non-sedimentable at 60 000g for 2 h) increased from c. 13% at pH 6·6 to c. 60% at pH 5·4 but decreased thereafter to 0% at pH 4·6. The levels of non-sedimentable α_s-, β- and κ-caseins increased as the pH was decreased to 5·4, with α_s-casein being least solubilized. Particle diameter remained almost constant despite considerable solubilization of casein as the pH was decreased, resulting in maximum voluminosity at pH 5·4.[156] Electron microscopy confirmed that disaggregation of casein micelles occurred as the pH was lowered to 5·4; at this pH, casein particles, with a range of diameters similar to that in milk at pH 6·7, were still evident, although these particles had a less compact structure than native micelles and a large amount of finely dispersed casein was also present. As the pH was decreased to 4·6, the dispersed casein reaggregated.[150] Heertje et al.[155] suggested that as the pH drops to 5·2, weakly bound β- and κ-caseins are released from the micelles, while a size-determining framework of α_s-casein remains intact, resulting in little decrease in particle size and an increase in micelle voluminosity. After partial disintegration at pH 5·5–5·2, small aggregated structures become evident, indicating interaction of free caseins. As pH drops to 4·9, the aggregated structures contract to form discrete particles that are larger than the original micelles. These particles then rearrange and aggregate to form the final casein network. Heertje et al.[155] explained the changes that occur in terms of the zeta potential of casein micelles and the behaviour of β-casein. Normally, on lowering the pH, the zeta potential of colloidal proteins decreases continuously and becomes zero at the isoelectric point; however, the pH dependence of the zeta potential of casein micelle shows an anomalous behaviour.[157] A first minimum in the zeta potential occurs at pH 5·2, which is the isoelectric point of β-casein; it coincides with the pH of maximum voluminosity and the start of aggregation of the casein solubilized as a result of CCP dissolution. As pH is lowered further, positively charged β-casein is re-adsorbed onto the framework of negatively charged α_s-casein, causing a decrease in voluminosity and an increase in potential; the latter causes contraction, leading to the separation (formation) of particles that are distinctly different from the original micelles in milk. The potential begins to decrease again at pH 4·9–5·0 owing to the decreasing charge on α_s-casein and the newly formed particles aggregate into chains and clusters, forming the final network.

Different stages of aggregation during the gelation of unheated skim milk that was acidified in the cold and subsequently heated to induce gelation were distinguished by Roefs.[150] On acidification, the casein molecules within micelles show two tendencies: (1) to aggregate as the pH approaches the isoelectric point, and (2) to dissociate owing to dissolution of CCP. On acidification to pH 5·5, dissociation predominates but the residual particles are similar in size to the native micelles owing to swelling. As the pH is decreased further, soluble casein associates with existing casein particles and also self-associates to form new aggregates that are distinctly different from the native micelles in milk at pH > 6·0. On heating, the casein particles coagulate into small strands and aggregates, from which larger conglomerates are formed. Subsequently, particles of variable size fuse and are inhomogeneously distributed throughout the available space to form the gel network. Rearrangement of the casein molecules within particles and of some complete particles in the network occurs on ageing, leading to an increase in gel strength.

In the production of fermented products such as quarg and yoghurt, where syneresis is not desirable, it is normal industrial practice to pre-heat milk at temperatures in excess of the denaturation temperature of whey protein; this contrasts with the use of HTST pasteurized milk, without further heating, for the production of acid casein and fresh acid cheeses, e.g. cottage cheese, where syneresis is necessary. In severely heated milks, acid gelation proceeds more rapidly than in unheated milk, although the pH decline is similar in both. Therefore, heated milk gels at a higher pH than unheated milk and the gelation mechanism is somewhat different from that in unheated milk.[152,155] On heating milk of normal pH at high temperatures (>85°C for longer than 5 min), the whey proteins are denatured and β-lactoglobulin interacts with κ-casein on the micelle surface, leading to the appearance of filamentous appendages projecting irregularly from the micelles or to micelles with ragged and fuzzy surfaces.[152]

On acid development by lactic acid fermentation, the casein particles in heated milk coalesce less extensively than those in unheated milk.[152,153,158] Similarly, on acidification with GDL, the gel network from heated milk consists of casein particles that are considerably smaller and more uniformly distributed than those in the network formed from unheated milk.[159,160] It is envisaged that the presence of denatured β-lactoglobulin on the micelle surfaces inhibits micelle contact and fusion;[152,153,155,158,159] the casein particles in yoghurt made from heated milk have a mean diameter of 230 nm compared to 460 nm for the particles in yoghurt from unheated milk.[152] Normal pasteurized milk behaves similarly to unheated milk; the changes associated with heating are evident only when milk is heated to above 85°C.

Heertje et al.[155] propose two possible mechanisms for the onset of gelation at a higher pH in heated milk: (1) interaction of whey proteins with casein micelles on heating milk at its natural pH may increase the hydrophobicity of the micellar surface and decrease the hydration barrier against aggregation, thus allowing aggregation and gelation to occur at a higher pH value than in unheated milk. They suggest that disintegration of the casein micelles might be incomplete owing to incomplete dissolution of CCP and the gel might, therefore, be regarded as a network of micelles that have retained much of their integrity. Alternatively, on heating milk there is an increase in the concentration of serum β-casein and a decrease in serum α_s-casein, leading to a micellar framework that is more calcium-sensitive and thus aggregates at a higher pH; in this case micellar integrity is lost and the gel network is composed of newly formed casein particles and not of native micelles. The authors favour the second mechanism, based on electron micrographs of structure formation during the gelation of pre-heated milk. Also, it has been reported recently[161] that heating milk at pH 6·9 results in dissociation of κ-casein from micelles, and this would further sensitize the α_s-casein framework to calcium-induced aggregation.

Skim milk (10% total solids) pre-heated to 90°C and acidified or heated to 90°C after acidification to pH 5·5 in the cold with GDL, HCl, citric acid or oxalic acid all display an unusual microstructure:[159,160] the casein particles consist of a solid protein core, over 300 nm in diameter, surrounded by a distinct outer layer (lining) 30–50 nm thick, and separated from the core by a space 50–80 nm wide. The core and outer layer appear to be composed of micelle aggregates.[159] When oxalic acid is used as acidulant, the core is partly disintegrated, probably as a result of chelation of calcium; however, citric acid, which is also a sequestering agent, does not cause disintegration of the core.

When gels were formed by adding GDL to skim milk at 70, 80 or 90°C to reduce the pH to 5·5, the core-and-lining effect was only just visible at 70°C but was clearly evident at 80 and 90°C, showing that the phenomenon is temperature dependent.[160] A mechanism for the formation of the core–lining structure was proposed by Harwalkar and Kalab:[160] at temperatures above 70°C, β-lactoglobulin and κ-casein interact, with the appearance of filamentous appendages protruding from the surfaces of the casein particles. These appendages interact with each other, culminating in the formation of the lining. However, Heertje et al.[155] propose that the effect is associated with the release of β-casein from the micelles on acidification: at 20°C and pH 5·5, considerable β-casein is released from the micelles and micellar voluminosity is high; at higher temperatures, β-casein redeposits on the surfaces. They further suggest that the lining may be due to differences in contraction of the various proteinaceous components of the reconstituted particles as the temperature is increased above 70°C. It is possible that a combination of the two proposed mechanisms may be involved: serum β-casein, which redeposits on the micelle surfaces as the temperature is increased, may be trapped by and interact with appendages of complexed β-lactoglobulin and κ-casein formed at high temperatures that protrude from the micelle surface. The absence of the lining effect at pH 4·6[159,160] may be due to β-casein, having passed through its isoelectric point (pH 5·2), being positively charged and interactive with the negatively charged, mainly α_s-casein, core rather than with the β-lactoglobulin–κ-casein appendages.

6.b.2 Strength and Syneresis Properties of Acid Casein Gels

The most important physical characteristics of acid casein gels are the strength of the gel once formed and its tendency to expel whey, i.e. synerese. In the case of acid cheeses, e.g. cottage cheese, the yield and properties of the finished product are influenced by the physical properties of the curd,[154,162,163] while the consistency of fermented products such as quarg or yoghurt is important in determining consumer acceptance. Syneresis due to gel contraction is desirable during the manufacture of acid cheeses but is undesirable in products like yoghurt and quarg. Both these physical characteristics are influenced by compositional and environmental factors during gelation.

6.b.2.1 INFLUENCE OF CASEIN CONCENTRATION

A casein gel network is assembled as a result of interaction between casein particles and, as pointed out by Roefs,[150] it is unlikely that all particles will contribute equally to the overall gel properties since some particles may occur in large non-contributing aggregates. Therefore, the physical properties of the gel depend on the spatial distribution of particles throughout the gel network and on the number and strength of bonds between particles.[150]

On cold acidification of a reconstituted skim milk or sodium caseinate solution followed by subsequent heating to induce gelation, Roefs[150] reports that a minimum casein concentration of $c.\,1{\cdot}0\%$ is needed for gel formation. He also reports that the strength of gels made by similar methods from reconstituted skim milk or a 2·8% sodium caseinate solution develops at similar rates, indicating that acid-induced skim milk gels are essentially casein gels.

In industrial practice, casein concentrations far in excess of 1·0% are used for the preparation of acid casein gels. Gels prepared from whole or skim milk contain $c.\,2{\cdot}5{-}3{\cdot}2\%$ casein and it is normal industrial practice for yoghurt manufacturers to increase the casein concentration in the mix prior to fermentation by adding milk solids, e.g. milk powder (full cream, skim or buttermilk) or caseinate, or by concentration of the milk by evaporation, ultrafiltration or reverse osmosis.[164,165] This increase in milk solids is carried out to improve the viscosity/consistency of the end-product, as the firmness of acid casein gels increases with casein concentration.[150,159,166,167] Supplementation of skim milk, containing 3·5% protein, with sodium caseinate, milk protein concentrate or skim milk powder (SMP) to bring the protein content to 5%, or with 0·5% gelatine, yielded yoghurts with firmness values of 117·9, 77·1, 79·7 and 52·4 g, respectively.[167] The microstructures of the yoghurts fortified with SMP or milk protein concentrate were similar (both contained casein micelles held together in chains by short, thin links), while the microstructure of that fortified with sodium caseinate consisted of larger casein micelles that were fused more extensively.[168]

The firmness and microstructure of yoghurts made from milks fortified by addition of SMP or sodium caseinate or by evaporation, ultrafiltration or reverse osmosis to protein contents ranging from 5·09% to 5·49% and total solids from 11·79% to 16·01% were studied by Tamime et al.[164]

Firmness decreased in the order: Na caseinate > reverse osmosis > ultrafiltration > evaporation > SMP. Microstructural studies supported the conclusions of Modler and Kalab[168] regarding addition of caseinate or skim milk powder. Yoghurts from milks fortified by reverse osmosis or evaporation were similar to that from milk fortified with SMP, while those fortified by ultrafiltration were similar to caseinate-fortified samples but the chain linkages were less robust. Thus, increasing the casein content increases the firmness and changes the microstructure of the gel.

Harwalkar and Kalab[159] reported an increase in the firmness of acid skim milk gels prepared using GDL on increasing total solids (TS) from 10% to 30%. Micrographs of gels containing 10%, 20% or 30% TS and acidified to pH 4·6 differed only slightly; at 10% and 20% TS, large clusters of closely-fused casein particles as well as long chains of particles were evident; at 30% TS, smaller clusters of casein particles were more densely distributed throughout the gel network and there were fewer chains. Differences in microstructure due to differences in TS content were more obvious in gels acidified to pH 5·5; in the 10% TS gel, casein particles in the form of a solid core surrounded by a lining were arranged mostly in clusters with few chains; at 20% TS, the lining was closer to the core and the network was mainly in the form of long chains of casein particles up to c. 500 nm in diameter; at 30% TS, casein particles were smaller (c. 300 nm in diameter), were surrounded by tendrils or spikes protruding from the particles and were arranged in chains.

Roefs[150] prepared gels, by cold acidification with HCl followed by heating to 30°C, from SMP reconstituted in distilled water, skim milk concentrated by ultrafiltration, skim milk diluted with UF permeate or sodium caseinate dissolved in 0·12M NaCl, all containing casein concentrations in the range from c. 1·0% to 6·0%. He found that gel strength (storage modulus) increased exponentially with casein concentration and concluded that the number of stress-carrying strands was not proportional to the volume fraction of casein. He proposed that the casein gel network is heterogeneous and, based on electron microscopic analysis, concluded that the network is composed of coagulated particles that have only partly fused and are not homogeneously distributed throughout the available space but are grouped in dense areas with large free spaces corresponding to the gel meshes. From permeability measurements he concluded that the gel can be represented as a porous medium consisting of spherical aggregates of casein particles with diameters in the order of micrometres, and since the effective porosity is c. 0·5 the diameter of the pores is approximately equal to the diameter of the aggregates. The permeability of the gels decreases rapidly as the casein concentration is increased, suggesting that the heterogeneity of the gel network depends on casein concentration. Gels prepared by acidification of a sodium caseinate solution are less porous, owing to smaller casein particles and a more homogeneous gel network. It is widely accepted that increasing the casein concentration reduces the susceptibility of acid casein gels to syneresis.[165,167-170]

6.b.2.2 INFLUENCE OF HEAT TREATMENT

As previously discussed, it is normal industrial practice to pre-heat milk for the production of many fermented products. As well as influencing the mechanism of gelation, pre-heat treatment of the milk and the gel ageing temperature also influence the rheological properties of the resulting gels. Pre-heating milk to temperatures greater than the denaturation temperature of the whey proteins (>85°C) increases gel firmness and reduces gel syneresis;[152,153,166,169] however, ultra-high-temperature processing (149°C × 3·3 s) results in a lower yoghurt gel strength than in-vat heating at 63°C for 30 min.[171] The increase in firmness and decrease in susceptibility to syneresis of gels from pre-heated milk is attributed to the finer microstructure of the casein network from heated milk. It is envisaged that the presence of the whey protein–casein complex reduces coalescence and fusion of casein particles, leading to a more even distribution of particles throughout the gel network and a better immobilization of the liquid phase. Accumulation of casein in large particles in unheated milk results in fewer chains, fewer junctions and a more open structure, leading to softness and syneresis.[152,159] Also, heating milk increases the hydrophilic properties of the protein.[152,169,170,172] Walstra et al.[170] suggested that this is due to the increased voluminosity of the whey proteins on heating and that their association with the casein results in a higher effective volume fraction of protein in the network, thus reducing the tendency for syneresis.

In contrast to yoghurt, quarg, etc., where syneresis is undesirable, milk for the production of cottage cheese should be heated as little as possible, since in this case syneresis is desirable and the curd is cut and cooked to promote whey expulsion.[137,162] In cottage cheese manufacture, increasing the cooking temperature (between 42°C and 64°C) and increasing the heating time (between 20 and 142 min) results in increased whey expulsion.[162] The initial gel has an open, porous network structure but, during cooking, the porosity of the network decreases, possibly because the chains and strands of casein particles break at weak points and shrink or fold into one another and cross-link to form clusters. Cooking promotes shrinkage of the curd microstructure and causes it to appear more homogeneous; the mean size of casein aggregates increases from 88 nm in milk to 180 nm at gelling and to >200 nm during cooking.[154]

During the preparation of acid gels by GDL, the temperature of the milk when the acidulant is added has a large influence on gel strength and susceptibility to syneresis. Increasing the temperature at addition from 40°C to 90°C increases gel firmness and decreases susceptibility to syneresis, irrespective of whether the milk is pre-heated.[159,160,166]

With milks pre-acidified to pH 4·6 < 4°C and subsequently heated, Roefs[150] reports that the minimum temperature for gel formation is 8–10°C and between 10°C and 20°C a very strong temperature effect on the rate of gelation is evident. Gel strength (storage modulus) increases from 13 N m^{-2} after ageing at 10°C for 16 h to 250 N m^{-2} after ageing at 20°C for the same period. On ageing at temperatures up to 40°C, gel strength increased linearly with the logarithm of time and continued to increase without reaching a plateau value even after 7 days, suggesting that bonds between casein particles continue to form owing to slow protein conformational rearrangements. Stronger gels formed at higher ageing temperatures are also more permeable. Therefore, the strongest gels are the most inhomogeneous at the network level. However, it must be remembered that gel strength depends on the number of effective bonds between particles as well as on the distribution of particles, and it appears that, although the particles are not as homogeneously distributed, the number of effective bonds between particles increases at higher ageing temperatures.[150] Also, when a gel prepared by cold acidification and subsequent heating at 30°C for 18 h is then cooled to 4°C, the gel state does not reverse, and in fact gel strength increases very markedly, suggesting an increase in the number or strength of bonds between casein particles. This latter increase in gel strength on cooling can be reversed by re-heating. A rapid increase in the temperature of cold-acidified milks results in weak, highly permeable gels, possibly because of irregular coagulation; also, the temperature to which the acidified milk is heated and the temperature at which the subsequent gel is aged influence the distribution of casein particles and thus gel permeability.[150]

Gel temperature also influences the rheological character of acid casein gels, which become more fluid at low temperatures and more elastic or solid-like at higher temperatures.[150]

6.b.2.3 INFLUENCE OF pH AND ACIDIFICATION METHOD

The pH at the onset of gelation depends on the pre-heat treatment of the milk, its temperature and the acidulant used. In acid casein gels produced by fermentation at 44°C, the first signs of gelation of unheated milk occur at pH 5·14 and at pH 4·92 gelation is visually complete. In milk pre-heated at 90°C, the first signs of gelation are apparent at pH 5·36 and gelation is visually complete at pH 5·17.[152] On cold acidification of milk with HCl followed by heating to 30°C or on addition of GDL at 30°C, gel formation occurs in the pH range 4·3–4·9.[150] On addition of GDL to milk at 90°C, gels can be formed at pH values from 3·95 (and possibly lower) to 6·0.[166] Gel strength increases as pH is decreased from that at the onset of gelation to 4·0.[150,159,160,164,166] As the pH decreases, the extent of fusion of casein particles into chains and clusters increases, giving larger casein aggregates,[159] the gel network becomes more inhomogeneous and permeability increases,[150] but an increase in the number of effective bonds between particles results in increased gel strength.

The susceptibility to syneresis of acid casein gels made by fermentation does not vary significantly in the pH range 3·8–4·5.[169,170] The method of acidification influences both the strength of gels and their susceptibility to syneresis. Harwalkar and Kalab[160] found that skim milk gels made by acidification to pH 4·6 or pH 5·5 in the cold with HCl, citric or oxalic acid or GDL differed significantly in both of these properties. Gels were not formed under any conditions on acidification with oxalic acid and gels were not formed on acidification with any acid to pH 5·5 and subsequent heating to 40°C. Gels formed on acidification to pH 4·6 with HCl, GDL or citric acid and subsequently heated to 40°C did not differ significantly in firmness. On acidification to either pH 5·5 or pH 4·6 and subsequent heating to 90°C, the firmness of citric acid-induced and HCl-induced gels were similar while those made with GDL were much stronger. GDL-induced gels were also stronger than gels of a similar pH formed by fermentation.[166] However, Roefs[150] reported that acid gels of similar pH made by acidification of sodium caseinate or skim milk with GDL at 30°C were much weaker than gels made by cold acidification with HCl followed by heating to 30°C.

Acid milk gels formed by cold acidification and subsequent quiescent heating show little or no syneresis when kept undisturbed for some hours at 30°C, even if they are then cut or wetted; however, acid gels formed by fermentation do exhibit syneresis. Walstra et al.[170] suggest that syneresis of gels prepared by fermentation may be due to slow proteolysis of the caseins, i.e. the casein may become like para-casein and therefore more insoluble, resulting in syneresis. There is a similar tendency for syneresis in gels formed by heating milk to 40°C following cold acidification to pH 4·6 with HCl or citric acid, but gels made in a similar manner with GDL are less susceptible to syneresis.[160] When GDL is added at 90°C, syneresis is reduced still further. Microstructural studies suggest that the casein particles in GDL-induced gels are larger than those in gels prepared in a similar manner with HCl or citric acid,[160] or in gels of a similar pH formed by fermentation. GDL-induced gels are also more permeable,[150] suggesting that the gel is more inhomogeneous at the network level.

6.b.2.4 INFLUENCE OF IONIC STRENGTH
A minimum salt concentration of 0·10 mol NaCl per kg dispersion is required for the formation of an acid casein gel by cold acidification and subsequent heating of a sodium caseinate solution. At lower salt concentrations, casein precipitation occurs at pH 4·6 even at 0°C, while above 0·24M NaCl prevents gel formation.[150] In the salt range 0·11–0·20M NaCl, gel strength decreases with increasing salt concentration.[150] Addition of up to 50 mmol NaCl per kg of reconstituted skim milk, with an initial estimated ionic strength of 0·13–0·14M at pH 4·6, reduced slightly the strength of gels made by cold acidification and subsequent heating. Addition of between 50 and 150 mmol NaCl/kg decreased gel strength progressively and with more than 150 mmol gelation was inhibited.[150] The permeability of acid casein gels also decreased with increasing NaCl concentration, suggesting that added NaCl decreases the number of effective interparticle bonds in the gel network and increases the homogeneity of distribution of the casein particles.[150] Addition of $CaCl_2$ at concentrations of 10–20 mmol per kg reconstituted sodium caseinate while maintaining ionic strength constant results in the formation of large casein particles at pH 6·7. The strength of acid gels formed from these solutions by cold acidification and subsequent heating decreases with increasing $CaCl_2$ concentration. It was envisaged that the large casein particles formed on addition of $CaCl_2$ are only partly disrupted during acidification, since the final gel network is also very coarse and has high permeability.[150]

6.b.2.5 INFLUENCE OF STABILIZERS AND NON-CASEIN PROTEINS
On fortification of a yoghurt mix by addition of milk powders or by ultrafiltration, reverse osmosis or evaporation, the ratio of casein to non-casein protein remains approximately constant and the properties of the yoghurts are influenced by the overall protein content rather than by the type of protein. On fortification by addition of either casein or whey protein, the properties of the acid gel are influenced by the altered ratio of casein to non-casein proteins as well as by the overall protein content.

Yoghurts prepared by addition of whey protein concentrate to skim milk containing 3·5% protein to increase the protein content by 0·5%, 1·0% and 1·5% were less firm and more susceptible to syneresis than when casein was added at the same levels.[167] The microstructures of the whey protein-fortified yoghurts were also different, as the casein particles were not fixed but were present as individual particles with the interparticle spaces spanned by what was presumed to be finely flocculated denatured whey protein.[167] This type of microstructure is not capable of holding water as well as the more fused particle microstructure obtained on addition of casein. In general, the recommended level of addition of whey powder to a yoghurt mix is 1–2%, since higher levels impart an undesirable 'whey' flavour.[165]

Addition of hydrocolloids or stabilizers to regulate the strength and syneresis properties of the gel is also widely practised. Several stabilisers are available and may be added singly or as blends.[165] Their use is governed by legislation in most countries and their mode of action is two-fold: firstly, by binding water, and secondly, by increasing gel firmness. Among the most widely used stabilizers are gelatine, alginates and modified starches.[137] At the level normally used in yoghurt manufacture, gelatine does not appear to influence the microstructure of the acid casein gel network whereas carrageenan and starch alter the microstructure considerably.[173] At 0·4%, carrageenan results in the formation of large clusters of casein particles that are connected by long, thin fibres of carrageenan, while addition of pre-gelatinized starch results in the formation of flat fibres and sheets between clusters of casein particles.

6.b.2.6 INFLUENCE OF HOMOGENIZATION OF THE MIX AND MECHANICAL HANDLING OF THE COAGULUM

During homogenization of a yoghurt mix, new fat globules are formed and the membranes surrounding these contain primarily casein and serum proteins. On acid gelation of the mix, these fat globules may become part of the overall casein matrix, since they are capable of interacting with other casein particles owing to the surface coverage of casein.[174] Also, on homogenization some denaturation of serum proteins may occur, and as a result of this casein–whey protein interactions may occur. These changes increase the viscosity of acid gels made from a homogenized mix and also increase water-holding capacity or reduce syneresis.[165,175]

The properties of acid casein gels produced by fermentation may also be influenced by mechanical disturbance of the gel after manufacture. When acid production by fermentation is carried out in the retail container without stirring, the yoghurt is referred to as a 'set' yoghurt product while 'stirred' yoghurt is produced when fermentation in a large incubation tank is followed by mechanical disturbance of the coagulum by agitation or pumping. Stirring of yoghurt decreases the gel strength and reduces the viscosity of the coagulum; stirring also leads to the production of a network with fewer chains and more clusters of casein particles joined together by thin fibres compared to the 'set' product.[137,173] However, the overall microstructure is preserved and the decrease in viscosity is attributed to only occasional ruptures of the microstructure.[137] Separation of whey is significantly increased when yoghurts are stirred and the gel matrix is disturbed.[169,170]

6.b.3 Type of Interaction Forces Involved in Acid Gelation of Casein

On the basis of observations on the mechanism of acid gelation of casein and on the effects of environmental factors on the rheological properties of gels once formed, Roefs[150] discusses the forces involved in gelation and distinguishes different stages of aggregation. He divides the overall aggregation process into four stages: (1) aggregation of individual casein molecules leading to the formation of small aggregates, (2) further aggregation of these to larger metastable particles, (3) the aggregation of these particles into strands and conglomerates, and (4) the ultimate aggregation of the latter to form a continuous gel network. He suggests that the formation of interparticle bonds depends on the structure and state of aggregation of individual casein molecules within particles and this in turn is influenced by intramolecular bonding influencing the secondary and tertiary structures of the casein molecules and intermolecular bonding influencing the association of these casein molecules to form particles. He suggests that hydrophobic bonding, electrostatic interactions, van der Waals attractions and probably hydrogen bonding contribute to the conformation of the casein molecules, while these forces together with Ca^{2+} bridging and steric interactions will contribute to the spatial structure of the casein particles.

Considering the effects of pH and NaCl on the rheological properties of gels Roefs suggests that electrostatic bonds and van der Waals attractions are important interparticle forces. Since gel strength decreases with increasing measurement temperature, he expresses doubt about the contribution of hydrophobic bonding directly to interparticle interactions but explains the effect in terms of an increase in temperature increasing the propensity for hydrophobic bonds within casein particles, thus decreasing particle size and thereby decreasing the number of interparticle bonds. He discounts hydrogen bonding between casein particles because it needs special spatial orientations of carboxylic groups on different casein particles, while he suggests that steric factors still play an important part in interparticle interactions at pH 4·6.

7 OTHER CASEIN GELS

While rennet- and acid-produced casein gels are by far the most commercially important, casein may be gelled by other agents/treatments that merit mention; such gels are generally considered as defects, although some of them could be considered as potential dairy products.

7.a Ethanol Coagulation

It has long been known that milk is coagulated on addition of ethanol, usually to about 40% (i.e. on mixing equal volumes of milk and 80% ethanol). The ability of milk to withstand treatment with ethanol without coagulation (ethanol stability) has long been used as a selective test for milk for the manufacture of sterilized concentrated milk. The ethanol stability of milk is strongly influenced by $[Ca^{2+}]$ and pH.[176,177]

A method for the preparation of micellar casein by ethanol coagulation has been described.[178] Coagulation of cream liqueurs during storage is a common problem, apparently related, at least in part, to ethanol coagulation of casein. It appears tenable that a gelled product, e.g. a dessert-type product, could be produced by ethanol coagulation of a concentrated milk.

7.b Heat-induced Coagulation

Casein is an extremely heat-stable protein: at its normal pH (c. 6·7) milk may be heated at 140°C for 20 min before coagulation occurs but is less stable (c. 7 min) at about pH 6·9; serum protein-free casein micelles in milk salts buffer are less stable than milk at pH 6·4–6·8 but somewhat more stable at pH > 6·9. The heat coagulation time–pH profile of milk is reproduced by adding β-lactoglobulin or α-lactalbumin to a serum protein-free system. Sodium caseinate dissolved in milk salts buffer is more heat stable than milk, and sodium caseinate in water is more stable still (HCT > 60 min at 140°C) but both are destabilized at pH ∼ 6·9 by addition of β-lactoglobulin. The literature on the heat stability of milk and caseinate systems has been reviewed by Fox and Morrissey[179] and Fox.[180]

Normally, milk with a high heat stability is desirable and that of milk of normal concentration is usually adequate to withstand all processes to which milk is normally subjected. The heat stability of concentrated milk is marginal for the manufacture of sterilized evaporated milk but can usually be made adequate by process manipulations or through the use of additives, e.g. orthophosphates.

Below pH 6·4, the heat stability of milk decreases abruptly, e.g. at pH 6·1 milk will coagulate on heating to about 90°C and a concentrated milk would coagulate at a lower temperature. Thus, it would be possible to produce thermoset milk gels at moderate temperatures. We are not aware of any research on the properties of such gels, which might offer an interesting line of new dairy products.

Thermoset milk gels may also be prepared from milk containing $c.$ 1·2% β-lactoglobulin on heating at 90°C for 10 min.[161] This study was made using purified β-lactoglobulin but presumably a similar effect could be achieved using whey protein isolate or concentrate. Gelation is not due to β-lactoglobulin *per se*; rather, β-lactoglobulin promotes the dissociation of micellar κ-casein on heating above 90°C and the κ-casein-depleted micelles are more sensitive to Ca^{2+} (and to ethanol, heat or rennet) than normal micelles. The properties of these gels have not been investigated but they may also offer the basis for a new range of dairy products.

7.c Age-gelation of Sterilized Milks

In-container sterilized evaporated milk occasionally gels during storage. However, UHT-sterilized milk, especially concentrated milk, is very prone to gelation during storage (age-gelation), limiting the usefulness of such products. The cause(s) and mechanism of age-gelation have not been definitely established (see Ref. 181) but there are probably two separate causes. Extracellular proteinases secreted by psychrotrophic bacteria are major causative factors in the gelation of UHT milk produced from raw milk of poor microbiological quality. Although good-quality raw milk yields a product with a much longer shelf-life, such UHT milks do eventually gel, apparently as a result of physicochemical changes in the casein micelles. It has been recognized for many years that whey proteins dissociate from the casein micelles in UHT milk during storage; it is possible that the dissociating whey proteins promote the dissociation of micellar κ-casein and that the κ-casein-depleted micelles are rendered calcium-sensitive and gel via a mechanism similar to those of the gels described in the previous section. However, age-gelation of UHT sterilized milk is not prevented by prior treatment with formaldehyde to 5 mM (Fox, unpublished), which prevents the dissociation of micellar κ-casein.[182]

7.d κ-Carrageenan-induced Casein Gels

Carrageenans, especially κ-carrageenan, are frequently added to dairy products, e.g. ice-cream, custards, chocolate milk, to improve viscosity. Moderately high concentrations of κ-carrageenan cause the casein to gel as a result of electrostatic interactions between the caseins, especially κ-casein and κ-carrageenan. These gels will not be considered further here and the interested reader is referred to Lin[183] and Chapter 3.

REFERENCES

1. Eigel, W. N., Bulter, J. E., Ernstrom, C. A., Farrell, H. M. Jr, Harwalkar, V. R., Jenness, R. & Whitney, R. McL., *J. Dairy Sci.*, **67** (1984) 1599–631.
2. Fox, P. F. (ed.), *Developments in Dairy Chemistry—1—Proteins*. Applied Science Publishers, London, 1982.
3. Fox, P. F. & Mulvihill, D. M., *J. Dairy Res.*, **49** (1982) 679.
4. Fox, P. F., *Proc. 6th Int. Congr. Food Sci. Technol.*, Bootle Press, Dublin, **5** (1983) 177.
5. Fox, P. F. & Mulvihill, D. M., *Proc. IDF Symposium, Physico-chemical Aspects of Dehydrated Protein-rich Milk Products*, Hensingor, Denmark, 1983, pp. 188–259.
6. McMahon, D. J. & Brown, R. J., *J. Dairy Sci.*, **67** (1984) 499–512.
7. Doi, H., Ibuki, F. & Kanamori, M., *J. Dairy Sci.*, **62** (1979) 195.
8. Pepper, L. & Farrell, H. M. Jr, *J. Dairy Sci.*, **65** (1982) 2259.
9. Carroll, R. J. & Farrell, H. M. Jr, *J. Dairy Sci.*, **66** (1983) 679.
10. Sawyer, W. H., *J. Dairy Sci.*, **52** (1969) 1347.
11. Baer, A., Oroz, M. & Blanc, B., *J. Dairy Res.*, **43** (1976) 419.
12. Snoren, T. H. M. & van der Spek, C. A., *Neth. Milk Dairy J.*, **31** (1977) 352.
13. Jolles, P. & Fiat, A. M., *J. Dairy Res.*, **46** (1979) 187.
14. Aschaffenburg, R., *J. Dairy Res.*, **35** (1968) 447.
15. Schaar, J., Variations in milk protein composition. Studies on κ-casein and β-lactoglobulin genetic polymorphism and on milk plasmin. PhD thesis, Swedish University of Agricultural Sciences, Uppsala, 1986.
16. Humbert, G. & Alais, C., *J. Dairy Res.*, **46** (1979) 559.
17. Reimerdes, E. H. In *Developments in Dairy Chemistry—1—Proteins*, ed. P. F. Fox. Applied Science Publishers, London, 1982, pp. 271–88.
18. Grufferty, M. B., Alkaline milk proteinase: some of its characteristics and its influence, and that of milk salts, on some processing properties of milk. PhD thesis, National University of Ireland, 1986.
19. Swaisgood, H. E. In *Developments in Dairy Chemistry—1—Proteins*, ed. P. F. Fox. Applied Science Publishers, London, 1982, pp. 1–59.
20. Creamer, L. K., Richardson, T. & Parry, D. A. D., *Arch. Biochem. Biophys.*, **211** (1981) 689.
21. Andrews, A. L., Atkinson, D., Evans, M. T. A., Finer, E. G., Green, J. P., Phillips, M. C. & Robertson, R. N., *Biopolymers*, **18** (1979) 1105.
22. Loucheux-Lefebvre, M. H., Aubert, J. P. & Jolles, P., *Biophys. J.*, **23** (1978) 323.

23. Donnelly, W. J. & Barry, J. G., *J. Dairy Res.*, **50** (1983) 433.
24. Schmidt, D. G. & Payens, T. A. J., *Surface Colloid Sci.*, **9** (1976) 165.
25. Schmidt, D. G. In *Developments in Dairy Chemistry—1—Proteins*, ed. P. F. Fox. Applied Science Publishers, London, 1982, pp. 61–86.
26. Dickson, I. R. & Perkins, D. J., *Biochem. J.*, **124** (1971) 235.
27. Dalgleish, D. G. & Parker, T. G., *J. Dairy Res.*, **47** (1980) 113.
28. Horne, D. S. & Dalgleish, D. G., *Int. J. Biol. Macromol.*, **2** (1980) 154.
29. Dalgleish, D. G. In *Food Proteins*, ed. P. F. Fox & J. J. Condon. Applied Science Publishers, London, 1982, pp. 155–78.
30. Waugh, D. F. In *Milk Proteins, Chemistry and Molecular Biology*, Vol. 2, ed. H. A. McKenzie. Academic Press, New York, 1971, pp. 3–85.
31. Schmidt, D. G., *Neth. Milk Dairy J.*, **34** (1980) 42.
32. Rose, D., *Dairy Sci. Abstr.*, **31** (1969) 171.
33. Garnier, J., *Neth. Milk Dairy J.*, **27** (1973) 21.
34. Farrell, H. M. Jr & Thompson, M. P. In *Fundamentals of Dairy Chemistry*, ed. B. H. Webb, A. H. Johnson & J. A. Alford. Avi Publishing Co. Inc., Westport, Connecticut, 1974, pp. 442–73.
35. Slattery, C. W., *J. Dairy Sci.*, **59** (1976) 1547.
36. Payens, T. A. J., *J. Dairy Sci.*, **65** (1982) 1863.
37. Pyne, G. T. & McGann, T. C. A., *J. Dairy Res.*, **27**(1960) 9.
38. Holt, C., *J. Dairy Res.*, **49** (1982) 29.
39. Holt, C. In *Developments in Dairy Chemistry—3*, ed. P. F. Fox. Elsevier Applied Science Publishers, London, 1985, pp. 143–81.
40. Holt, C., Hasnain, S. S. & Hukins, D. W. L., *Biochim. Biophys. Acta*, **719** (1982) 299.
41. Irlam, J. C., Holt, C., Hasnain, S. S. & Hukins, D. W. L., *J. Dairy Res.*, **52** (1985) 267.
42. McGann, T. C. A. & Pyne, G. T., *J. Dairy Res.*, **27** (1960) 403.
43. Walstra, P. & Jenness, R., *Dairy Chemistry and Physics*. Wiley, New York, 1984, pp. 229–53.
44. Heth, A. A. & Swaisgood, H. E., *J. Dairy Sci.*, **65** (1982) 2047.
45. *Food and Agricultural Organization Production Yearbook*, **39** (1985).
46. International Dairy Federation, The world market for cheese. *Bulletin No. 203*, 1986.
47. International Dairy Federation, Consumption statistics for milk and milk products, 1984. *Bulletin No. 201*, 1986.
48. Foltmann, B. In *Essays in Enzymology*, Vol. 17, ed. P. N. Campbell & R. D. Marshall. Academic Press, New York, 1981, p. 52.
49. Foltmann, B. In *Cheese: Chemistry, Physics and Microbiology*, Vol. 1, ed. P. F. Fox. Elsevier Applied Science Publishers, London, 1987, pp. 33–61.
50. Sardinas, J. L., *Adv. Appl. Microbiol.*, **15** (1972) 39.
51. Sternberg, M., *Adv. Appl. Microbiol.*, **20** (1976) 135.
52. Green, M. L., *J. Dairy Res.*, **44** (1977) 159.
53. Phelan, J. A., Milk coagulants—an evaluation of alternatives to standard calf rennet. PhD thesis, National University of Ireland, 1985.
54. Fox, P. F., *Neth. Milk Dairy J.*, **35** (1981) 233.
55. Dalgleish, D. G. In *Cheese: Chemistry, Physics and Microbiology*, Vol. 1, ed. P. F. Fox. Elsevier Applied Science Publishers, London, 1987, pp. 63–96.

56. Waugh, D. F. & von Hippel, P. H., *J. Am. Chem. Soc.*, **78** (1956) 4576.
57. Wake, R. G., *Aust. J. Biol. Sci.*, **12** (1959) 479.
58. Delfour, A., Jolles, J., Alais, C. & Jolles, P., *Biochem. Biophys. Res. Commun.*, **19** (1965) 452.
59. Fox, P. F. In *Developments in Food Proteins—3*, ed. B. J. F. Hudson. Elsevier Applied Science Publishers, London, 1984, pp. 69–112.
60. Fox, P. F., *Proc. 22nd Int. Dairy Congr. (The Hague)*. D. Reidel Publishing Company, Dordrecht, 1987, pp. 61–73.
61. Pyne, G. T., *Dairy Sci. Abstr.*, **17** (1955) 531.
62. Kowalchyk, A. W. & Olson, N. F., *J. Dairy Sci.*, **60** (1977) 1256.
63. Lawrence, R. C. & Creamer, L. K., *J. Dairy Res.*, **36** (1969) 11.
64. Donnelly, W. J., Barry, J. G. & Buchheim, W., *Ir. J. Food Sci. Technol.*, **8** (1984) 121.
65. Pearce, M. J., Linklater, P. M., Hall, R. J. & Mackinlay, A. G., *J. Dairy Res.*, **53** (1986) 477.
66. Green, M. L. & Grandison, A. S. In *Cheese: Chemistry, Physics and Microbiology*, Vol. 1, ed. P. F. Fox. Elsevier Applied Science Publishers, London, 1987, pp. 97–133.
67. Guthy, K. & Novak, G., *J. Dairy Res.*, **44** (1977) 363.
68. Green, M. L., Hobbs, D. G., Morant, S. V. & Hill, V. A., *J. Dairy Res.*, **45** (1978) 413.
69. Walstra, P., Bloomfield, V. A., Wei, G. J. & Jenness, R., *Biochim. Biophys. Acta*, **669** (1981) 258.
70. Payens, T. A. J., Wiersma, A. K. & Brinkhuis, J., *Biophys. Chem.*, **6** (1977) 253.
71. Payens, T. A. J., *Biophys. Chem.*, **6** (1977) 263.
72. Payens, T. A. J. & Both, P. In *Bioelectrochemistry: Ions, Surfaces and Membranes*, ed. M. Blank. Adv. Chem. Series No. 188, American Chemical Society, Washington, DC, 1980, pp. 129–41.
73. Dalgleish, D. G., *J. Dairy Res.*, **46** (1979) 653.
74. Green, M. L. & Morant, S. V., *J. Dairy Res.*, **48** (1981) 57.
75. Brinkhuis, J. & Payens, T. A., *Biochim. Biophys. Acta*, **832** (1985) 331.
76. Dalgleish, D. G., *J. Dairy Res.*, **47** (1980) 231.
77. Darling, D. F. & van Hooydonk, A. C. M., *J. Dairy Res.*, **48** (1981) 65.
78. Dalgleish, D. G., *Biophys. Chem.*, **11** (1980) 147.
79. Dalgleish, D. G., *J. Dairy Res.*, **48** (1981) 65.
80. Storry, J. E. & Ford, G. D., *J. Dairy Res.*, **49** (1982) 343.
81. Storry, J. E. & Ford, G. D., *J. Dairy Res.*, **49** (1982) 469.
82. Tusynski, W. B., *J. Dairy Res.*, **38** (1971) 115.
83. Surkov, B. A., Klimovskii, I. I. & Krayushkin, V. A., *Milchwissenschaft*, **37** (1982) 393.
84. Johnston, D. E., *J. Dairy Res.*, **51** (1984) 91.
85. Creamer, L. K., Zoerb, H. F., Olson, N. F. & Richardson, T., *J. Dairy Sci.*, **65** (1982) 902.
86. Creamer, L. K. & Olson, N. F., *J. Food Sci.*, **47** (1982) 631.
87. Hill, R. D. & Laing, R. R., *J. Dairy Sci.*, **32** (1965) 193.
88. Hill, R. D. & Craker, B. A., *J. Dairy Res.*, **35** (1968) 13.
89. Hill, R. D., *J. Dairy Res.*, **37** (1970) 187.
90. Culioli, J. & Sherman, P., *J. Texture Studies*, **9** (1978) 257.

91. Pearce, M. J., Linklater, P. M., Hall, R. J. & MacKinlay, A. G., *J. Dairy Res.*, **53** (1986) 381.
92. Foltmann, B., *Proc. 15th Int. Dairy Congr.*, **2** (1959) 655.
93. McPhie, P., *Analyt. Biochem.*, **73** (1976) 258.
94. McMahon, D. J. & Brown, R. J., *J. Dairy Sci.*, **65** (1982) 1639.
95. McMahon, D. J. & Brown, R. J., *J. Dairy Sci.*, **66** (1983) 341.
96. Bynum, D. G. & Olson, N. F., *J. Dairy Sci.*, **65** (1982) 2281.
97. Thomasow, J. & Voss, E., *International Dairy Federation Annual Bulletin*, 1977, Doc-99.
98. Okigbo, L. M., Richardson, G. H., Brown, R. J. & Ernstrom, C. A., *J. Dairy Sci.*, **68** (1985) 1893.
99. Vanderheiden, G., *CSIRO Food Res. Quart.*, **36** (1976) 45.
100. Kowalchyk, A. W. & Olson, N. F., *J. Dairy Sci.*, **61** (1978) 1375.
101. Bynum, D. G. & Olson, N. F., *J. Dairy Sci.*, **65** (1982) 1321.
102. Marshall, R. J., Hatfield, D. S. & Green, M. L., *J. Dairy Res.*, **49** (1982) 127.
103. Steinsholt, K., *Milchwissenschaft*, **28** (1973) 94.
104. Burgess, K. J., *Ir. J. Food Sci. Technol.*, **2** (1978) 129.
105. Richardson, C. H., Gandhi, N. R., Divatia, M. A. & Ernstrom, C. A., *J. Dairy Sci.*, **54** (1971) 182.
106. Scott Blair, G. W. & Burnett, J., *J. Dairy Res.*, **25** (1958) 297.
107. Burnett, J. & Scott Blair, G. W., *Dairy Industries*, **28** (1963) 220.
108. Richardson, G. H., Okigbo, L. M. & Thorpe, J. D., *J. Dairy Sci.*, **68** (1985) 32.
109. Scott Blair, G. W. & Burnett, J., *J. Dairy Res.*, **25** (1958) 457.
110. Scott Blair, G. W. & Burnett, J., *J. Dairy Res.*, **26** (1959) 58.
111. Jen, J. J. & Ashworth, U. S., *J. Dairy Sci.*, **53** (1970) 1201.
112. Keogh, M. K., MSc thesis, University College, Cork, Ireland, 1966.
113. Dill, C. W. & Roberts, W. M., *J. Dairy Sci.*, **42** (1959) 1792.
114. Erwin, R. E., Hampton, O. & Randolph, H. E., *J. Dairy Sci.*, **55** (1972) 298.
115. Tellamy, P. T., Randolph, H. E. & Dill, C. W., *J. Dairy Sci.*, **52** (1969) 980.
116. Mauk, B. R. & Demott, B. J., *J. Dairy Sci.*, **42** (1959) 39.
117. Ashworth, U. S. & Nebe, J., *J. Dairy Sci.*, **53** (1970) 415.
118. Kowalchyk, A. W. & Olson, N. F., *J. Dairy Sci.*, **62** (1979) 1233.
119. van Dijk, H. J. M. & Walstra, P., *Neth. Milk Dairy J.*, **40** (1986) 3.
120. Walstra, P., van Dijk, H. J. M. & Geurts, T. J. In *Cheese: Chemistry, Physics and Microbiology*, Vol. 1, ed. P. F. Fox. Elsevier Applied Science Publishers, London, 1987, pp. 135–77.
121. Marshall, R. J., *J. Dairy Res.*, **49** (1982) 329.
122. Guinee, T. P. & Fox, P. F. In *Cheese: Chemistry, Physics and Microbiology*, Vol. 1, ed. P. F. Fox. Elsevier Applied Science Publishers, London, 1987, pp. 251–97.
123. Prentice, J. H., *J. Texture Studies*, **3** (1972) 415.
124. King, N., *J. Dairy Res.*, **25** (1958) 312.
125. Hansson, E., Olson, H. & Sjostrom, G., *Milchwissenschaft*, **21** (1966) 331.
126. Dean, M. R., Berridge, N. J. & Mabbit, L. A., *J. Dairy Res.*, **26** (1959) 77.
127. Kimber, A. M., Brooker, B. E., Hobbs, D. G. & Prentice, J. H., *J. Dairy Res.*, **41** (1974) 389.
128. Hall, D. M. & Creamer, L. K., *N.Z. J. Dairy Sci. Technol.*, **7** (1972) 95.

129. Eino, M. F., Biggs, D. A., Irvine, D. M. & Stanley, D. W., *J. Dairy Res.*, **43** (1976) 109.
130. Eino, M. F., Biggs, D. A., Irvine, D. M. & Stanley, D. W., *J. Dairy Res.*, **43** (1976) 113.
131. Kalab, M., *Milchwissenschaft*, **32** (1977) 449.
132. Stanley, D. W. & Emmons, D. B., *J. Inst. Can. Technol. Aliment.*, **10** (1977) 78.
133. Eino, M. F., Biggs, D. A., Irvine, D. M. & Stanley, D. W., *Can. Inst. Food Sci. Technol.*, **12** (1979) 149.
134. Green, M. L., Turvey, A. & Hobbs, D. G., *J. Dairy Res.*, **48** (1981) 343.
135. Lawrence, R. C. & Gilles, J. In *Cheese: Chemistry, Physics and Microbiology*, Vol. 2, ed. P. F. Fox. Elsevier Applied Science Publishers, London, 1987, pp. 1–44.
136. Kalab, M. & Emmons, D. B., *Milchwissenschaft*, **33** (1978) 670.
137. Kalab, M., *J. Dairy Sci.*, **62** (1979) 1352.
138. Kalab, M., *Scanning Electron Microscopy III*. SEM Inc., AMF, O'Hare, Illinois, 1979, p. 261.
139. Covacevich, H. R. & Kosikowski, F. V., *J. Dairy Sci.*, **61** (1978) 701.
140. Sutherland, B. J. & Jameson, G. W., *Aust. J. Dairy Technol.*, **36** (1981) 136.
141. Bush, C. S., Coroutte, C. A., Amundson, C. R. & Olson, N. F., *J. Dairy Sci.*, **66** (1983) 415.
142. Kosikowski, F. V., *Food Technol.*, **40**(6) (1986) 71.
143. Anon., *Caseus*, December 1986, p. 11.
144. Dalgleish, D. G., *J. Dairy Res.*, **47** (1980) 231.
145. Garnot, P. & Corre, C., *J. Dairy Res.*, **47** (1980) 103.
146. Scott Blair, G. W. & Burnett, J., *Biorheology*, **1** (1963) 183.
147. Garnot, P., Rank, T. C. & Olson, N. F., *J. Dairy Sci.*, **65** (1982) 2267.
148. Green, M. L., Glover, F. A., Scurlock, E. M. W., Marhsall, R. J. & Hatfield, D. S., *J. Dairy Res.*, **48** (1981) 333.
149. Pyne, G. T., *J. Dairy Res.*, **29** (1962) 101.
150. Roefs, S. P. F. M., Structure of acid casein gels. A study of gels formed after acidification in the cold. PhD thesis, Agricultural University, Wageningen, The Netherlands, 1986.
151. Kalab, M. & Harwalkar, V. R., *J. Dairy Sci.*, **56** (1973) 835.
152. Kalab, M., Emmons, D. B. & Sargant, A. G., *Milchwissenschaft*, **31** (1976) 402.
153. Davies, F. L., Shankar, P. A., Brooker, B. E. & Hobbs, D. G., *J. Dairy Res.*, **45** (1978) 53.
154. Glaser, J., Carroad, P. A. & Dunkley, W. L., *J. Dairy Sci.*, **63** (1980) 37.
155. Heertje, I., Visser, J. & Smits, P., *Food Microstructure*, **4** (1985) 267.
156. Roefs, S. P. F. M., Walstra, P., Dalgleish, D. G. & Horne, D. S., *Neth. Milk Dairy J.*, **39** (1985) 119.
157. Walstra, P. & Jenness, R., *Dairy Chemistry and Physics*. Wiley, New York, 1984, pp. 211–28.
158. Knoop, A. M. & Peters, K. H., *Kieler Milchwirtschaftliche Forschungsberichte*, **27** (1975) 227.
159. Harwalkar, V. R. & Kalab, M., *J. Texture Studies*, **11** (1980) 35.
160. Harwalkar, V. R. & Kalab, M., *Scanning Electron Microscopy III*. SEM Inc., AMF, O'Hare, Illinois, 1981, p. 503.
161. Singh, H. & Fox, P. F., *J. Dairy Sci.*, **53** (1986) 237.

162. Chua, T. E. H. & Dunkley, W. L., *J. Dairy Sci.*, **62** (1979) 1216.
163. Emmons, D. B. & Beckett, D. C., *J. Dairy Sci.*, **67** (1984) 2200.
164. Tamime, A. Y., Kalab, M. & Davies, G., *Food Microstructure*, **3** (1984) 83.
165. Tamime, A. Y. & Robinson, R. K., *Yoghurt Science and Technology*. Pergamon Press, Oxford, 1985, pp. 7–90.
166. Harwalkar, V. R., Kalab, M. & Emmons, D. B., *Milchwissenschaft*, **32** (1977) 400.
167. Modler, H. W., Larmond, M. E., Lin, C. S., Froehlich, D. & Emmons, D. B., *J. Dairy Sci.*, **66** (1983) 422.
168. Modler, H. W. & Kalab, M., *J. Dairy Sci.*, **66** (1983) 430.
169. Harwalkar, V. R. & Kalab, M., *Milchwissenschaft*, **38** (1983) 517.
170. Walstra, P., van Dijk, H. J. M. & Geurts, T. J., *Neth. Milk Dairy J.*, **39** (1985) 209.
171. Labropoulos, A. E., Collins, W. F. & Stone, W. K., *J. Dairy Sci.*, **67** (1984) 405.
172. Grigorov, H., *Proc. 17th Int. Dairy Congr.*, **F:5** (1966) 649.
173. Kalab, M., Emmons, D. B. & Sargant, A. G., *J. Dairy Res.*, **42** (1975) 453.
174. van Vliet, T. & Dentener-Kikkert, A., *Neth. Milk Dairy J.*, **36** (1982) 261.
175. Vaikus, V., Lubinskas, V. & Mitskevichus, E., *Proc. 18th Int. Dairy Congr.*, **IE** (1970) 320.
176. Horne, D. S. & Parker, T. G., *J. Dairy Res.*, **48** (1981) 273.
177. Horne, D. S. & Parker, T. G., *J. Dairy Res.*, **48** (1981) 285.
178. Hewedi, M. M., Mulvihill, D. M. & Fox, P. F., *Ir. J. Food Sci. Technol.*, **9** (1985) 11.
179. Fox, P. F. & Morrissey, P. A., *J. Dairy Res.*, **44** (1977) 627.
180. Fox, P. F., Heat-induced coagulation of milk. In *Developments in Dairy Chemistry—1—Proteins*, ed. P. F. Fox. Applied Science Publishers, London, 1982, pp. 189–228.
181. Harwalkar, V. R., Age gelation of sterilized milks. In *Developments in Dairy Chemistry—1—Proteins*, ed. P. F. Fox. Applied Science Publishers, London, 1982, pp. 229–69.
182. Singh, H. & Fox, P. F., *J. Dairy Res.*, **52** (1985) 65.
183. Lin, C. F., Interaction of sulfated polysaccharides with proteins. In *Food Colloids*, ed. H. D. Graham. Avi Publishing Co. Inc., Westport, Connecticut, pp. 320–46.

Chapter 5

EGG PROTEIN GELS

SCOTT A. WOODWARD

Poultry Science Department, Institute of Food and Agricultural Sciences, University of Florida, Gainesville, Florida 32611, USA

1 INTRODUCTION

Eggs are capable of performing various useful functions in foods, including foaming, coagulating or gelling, and emulsifying. All liquid components of eggs have the capacity to form gels upon heating. This important property is a major factor in the daily consumption of eggs, as they are scrambled, fried, poached, hard-cooked and used in the preparation of omelettes, quiches, soufflés, custards and cakes. The purpose of this chapter is to review the gelation properties of egg components, providing a reference that will facilitate the use of eggs and egg components for their gelling function *per se*, rather than simply in preparation of traditional egg dishes.

2 STRUCTURE

The liquid portion of shell eggs consists of about 67–70% albumen and 30–33% yolk.[1-3] The typical protein compositions of egg white and egg yolk protein fractions are summarized in Tables 1 and 2. For detailed information on the chemistry and composition of eggs and egg proteins, the reader is referred to reviews on these topics.[2-7]

Egg albumen is made up of thick and thin white, with the thick white comprising about 60% of the albumen in a fresh egg.[3] The thin white has no apparent structural organization. The thick white has been described as a weak gel, with its rigidity attributed to ovomucin, a high-molecular-weight glycoprotein that is organized into fibres.[2,3] The thick white gel is broken down to thin white during normal ageing of eggs, termed egg white thinning, in which the pH of albumen rises from about 7·6 to 9·3 owing to a

TABLE 1
PROTEINS IN EGG ALBUMEN

Protein	Amount in albumen (%)	pI^a	Molecular weight	T_d^b (°C)	Characteristics
Ovalbumin	54	4·5	45 000	84·0	Phosphoglycoprotein
Ovotransferrin	12	6·1	76 000	61·0	Binds metallic ions
Ovomucoid	11	4·1	28 000	$79·0^c$	Inhibits trypsin
Ovomucin	3·5	4·5–5·0	$5·5–8·3 \times 10^6$	—	Sialoprotein, viscous
Lysozyme	3·4	10·7	14 300	75·0	Lyses some bacteria
G_2 globulin	4·0?	5·5	$3·0–4·5 \times 10^4$	92·5	—
G_3 globulin	4·0?	4·8	—	$64–66^d$	—
Ovoinhibitor	1·5	5·1	49 000	—	Inhibits serine proteases
Ficin inhibitor	0·05	5·1	12 700	—	Inhibits thioproteases
Ovoglycoprotein	1·0	3·9	24 400	—	Sialoprotein
Ovoflavoprotein	0·8	4·0	32 000	84^d	Binds riboflavin
Ovomacroglobulin	0·5	4·5	$7·6–9·0 \times 10^5$	—	Strongly antigenic
Avidin	0·05	10	68 300	85^e	Binds biotin

Adapted from Powrie and Nakai.[2]
[a] Isoelectric point.
[b] T_d at pH 7 (may vary with pH).
[c] Denaturation of ovomucoid is apparently reversible. Binding to trypsin shifts the T_d to 81°C and makes denaturation irreversible (Donovan and Beardslee[32]).
[d] Apparent T_d based on heat-induced aggregation (Matsuda et al.[63]).
[e] T_d increases to 132°C upon binding of biotin (Donovan and Ross[67]).

loss of CO_2 through the pores of the shell. Thick white is also easily broken down by mechanical means, such as blending or homogenization. Egg albumen can be regarded as a solution of globular proteins containing ovomucin fibres.[2]

Egg yolk, on the other hand, is a complex structural system in which several types of particles are suspended in a protein solution or plasma.[2,8] These particles include spheres (4–150 μm diameter), granules (0·3–2·0 μm), profiles or low-density lipoproteins (25 nm) and myelin figures (56–130 nm in length). Early studies on the size and distribution of yolk spheres were summarized by Romanoff and Romanoff.[9] Using phase-contrast and electron microscopy, Bellairs[10] verified that spheres contain subdroplets and profiles in an aqueous protein fluid. Grodzinski[11] regarded spheres as osmometers, and reported their swelling in hypotonic solutions and their shrinking in hypertonic salt solutions. He found spheres to be isotonic with a 0·16M NaCl solution. Bellairs[10] observed three types of surfaces on spheres but concluded that these surfaces differed from true cell membranes. In a later study Bellairs et al.[12] were unable to observe membranes on spheres.

TABLE 2
COMPOSITION OF EGG YOLK PROTEIN FRACTIONS

Fraction	Proportion of yolk solids (%)	Proportion of yolk protein (%)[a]	Molecular weight Lipoprotein	Molecular weight Protein moiety[b]	Protein (%)[c]	Lipid (%)[c]	Characteristics
Egg yolk	100	100	—	—	31·1	65·8	—
Plasma	77–78				17–18	77–81	
Low-density fraction	66				11	89	Density = 0·98
LDL$_1$	16	5	9·0–10·3 × 10^6	1·40 × 10^6	10·2	87–89	
LDL$_2$	50	18	3·3–4·8 × 10^6	0·55 × 10^6	12·0	83–86	
Water-soluble fraction	10·6				90	0	Serum proteins
α-Livetin	2·2	6	—	80 000	90	0	Serum albumin
β-Livetin	5·3	15	—	42 000	90	0	α$_2$-Glycoprotein
γ-Livetin	3·1	9	—	150 000	90	0	γ-Globulin
Granules	19–23				60	34	High-density fraction
α-Lipovitellin	6·0	15	4·0 × 10^5	3·1 × 10^5	78	18–22	Density = 1·27; P = 0·50%[e]
β-Lipovitellin	10·3	26	4·0 × 10^5	3·1 × 10^5	78	18–22	Density = 1·27; P = 0·27%[e]
Phosvitin	3·7	7	—	36 000–40 000[d]	90	0	Contains 10% P as PO$_4$
Low-density fraction	2·8	1	—	—	12	84	; binds iron

Summarized from Cook[22] and Powrie and Nakai.[2]

[a] Estimated by the following calculation: [[proportion of yolk solids)(protein %)]/31·1 (31·1 = the amount of total yolk protein on dry weight basis). Values should be considered relative estimates only.

[b] For lipoproteins, molecular weight of protein moiety is calculated as molecular weight × protein %.

[c] Dry weight basis.

[d] Phosvitin consists of two fractions (α and β) made up of subunits that form complexes. Reported molecular weights range from 12 400 to 190 000. See Powrie and Nakai[2] for further reference.

[e] Phosphorus.

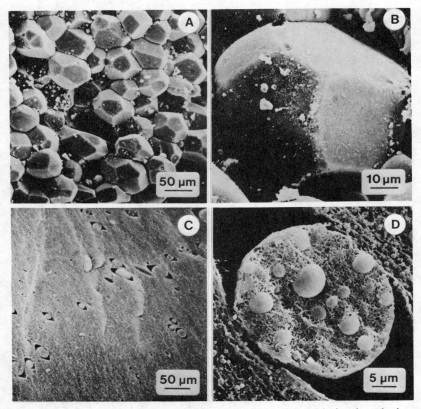

FIG. 1. Scanning electron micrographs comparing structures in hard-cooked egg yolk and heat-formed gels of stirred egg yolk. (A) Hard-cooked egg yolk containing polyhedral grains; (B) single polyhedral grain; (C) gel of stirred yolk containing a dispersion of spheres; (D) freeze-fractured sphere from stirred yolk gel, showing interior structures. (Reprinted from *Journal of Food Science*, Woodward and Cotterill.[19,77] © Institute of Food Technologists.)

Granules are more numerous than spheres and comprise about 23% of the total solids in yolk.[10,13] Chang et al.[14] disrupted granules by adding 0·34M NaCl to yolk, and reported the presence of globules, myelin figures and other microparticles. Garland and Powrie[15,16] and Kocal et al.[17] have isolated and characterized the myelin figures and low-density lipoproteins from granules.

Recent work by Fujii et al.[18] on raw yolk and Woodward and Cotterill[19] on cooked yolk has indicated that native undisturbed egg yolk consists of adjoining polyhedral 'grains' consistent in size and microstructure with yolk spheres (Fig. 1). Fujii et al.[18] observed these polyhedrons by scanning electron microscopy after chemically fixing undisturbed intact yolks. They suggested that yolk spheres may be packed tightly together in native yolk, accounting for the unusual shape (Fig. 1(A,B)). Woodward and Cotterill[19] found that gentle stirring disrupted 90–95% of these grains, presumably releasing plasma as the continuous-phase fluid and granules as part of the discontinuous phase of yolk, and leaving the remaining spheres as isolated structures suspended in plasma (Fig. 1(C)). Using scanning electron microscopy, they compared the surfaces and interiors of grains and spheres from heated samples of undisturbed and stirred egg yolk, and concluded that grains and spheres are equivalent structures. Bellairs[10,12] had observed earlier that the subdroplets, or granules, inside the spheres were identical with those in plasma, and that the fluid phase inside spheres had the appearance of plasma (see Fig. 1(D)).

Earlier researchers postulated the existence of a structural arrangement in the yolk. Maurice,[20] in observations on egg yolk conductivity, stated that a 'continuous solid phase' in yolk was easily broken down by stirring to yield a continuous aqueous phase, which was 40 times more conductive than native yolk. Shenstone,[3] reviewing the paradoxical lack of water diffusion from albumen into yolk in the shell egg, suggested that a structural arrangement limited the diffusion of water into the yolk. The ease of disruption of the grains of native yolk has made it difficult for researchers to elucidate and interpret the structure of native yolk, for in sample preparation this structure is easily destroyed.

3 SOURCE

Eggs are produced world-wide, and have been used as a dietary staple in many civilizations for centuries. Eggs are produced, processed and marketed primarily as shell eggs (84% of total production in the United States). The processing of eggs for use as a food ingredient has been developed over the past century and, at present, egg products are processed in many developed countries of the world.

4 COMMERCIAL AVAILABILITY

After being washed and graded, eggs are broken and the whites and yolks are separated by means of an automated egg-breaking/separating machine.

TABLE 3
PROXIMATE COMPOSITION OF LIQUID/FROZEN AND DEHYDRATED EGG PRODUCTS

Component	Liquid/frozen (per 100 g)				Dehydrated (per 100 g)		
	Whole	White	Pure yolk	Commer- yolk[a]	Plain whole[b]	Stab. white[c]	Plain yolk[d]
Solids (g)	24·5	12·1	51·8	44·0	96·8	93·6	97·2
Calories	152	50	377	313	600	388	692
Protein (g)	12·0	10·2	16·1	14·9	47·4	79·1	32·9
Lipids (g)	10·9	—	34·1	27·5	43·1	—	60·8
Ash (g)	1·0	0·7	1·7	1·5	4·0	5·3	3·3

Source: Cook and Briggs.[5]
[a] Commercial yolk is egg yolk diluted with egg white only.
[b] Produced from whole egg containing 24·5% solids.
[c] Produced from bacteriologically fermented egg white.
[d] Produced from yolk containing 44% solids as in footnote *a*.

Then the liquids are homogenized, pasteurized with or without added ingredients, and packaged and frozen or shipped as refrigerated liquids. Alternatively, liquids are stabilized against browning by removal of glucose, followed by spray drying.[21] Cotterill[22] listed 35 examples of egg products (whites, yolks and whole-egg blends) in four categories (refrigerated, frozen, dried and specialty products) that are available for commercial use. The compositions of typical egg products[5] are summarized in Table 3. In the United States, continuous inspection of egg products plants is mandatory, and all egg products are required to be free of viable *Salmonella* organisms.[23] Pasteurization requirements for liquid whole egg are 60°C for 3·5 min in the USA and 64°C for 2·5 min in the UK.[23]

Proteins may be fractionated from albumen by various precipitation and crystallization methods, and are available from various biochemical suppliers. Examples of available proteins or fractions include the albumin fraction, containing ovalbumin and ovotransferrin (conalbumin), a globulin fraction, purified ovalbumin, ovotransferrin, lysozyme, avidin, ovomucoid and ovoinhibitors. Phosvitin is available as a purified yolk protein. The cost of most of these preparations is prohibitive for food uses, limiting the applications of specific protein fractions to research.

An ion-exchange procedure has been patented for the removal of lysozyme from egg white,[24] and this type of process is being used commercially in some European countries. The use of cation exchangers to remove lysozyme has been reported for batch[25] and column chromatography[26] methods. The latter method appears to be superior in terms of yield and in its potential for use in an automated continuous separation process. The lysozyme so isolated has antimicrobial and antiviral properties, and the remaining 'lysozyme-free egg white' is usable as a food ingredient and is similar to egg white in its whipping and gelling properties.[26,27]

5 SOLUTION PROPERTIES

5.a Viscosity

Egg white, egg yolk and whole egg are generally regarded as non-Newtonian, pseudoplastic fluids whose behaviour can be described by the power-law and Casson models.[28-35] However, Scalzo et al.,[36] using a capillary-tube viscometer, reported that commercial liquid egg products were Newtonian fluids, within the ranges of shear rates tested. Gossett et al.[29] reported that egg white was pseudoplastic at pH values 3·0, 4·0 and from 7·0 to 10·0, but that it behaved as a Newtonian fluid at pH values 5·0, 6·0 and 11·0. Pitsilis et al.[30,31] developed rheological data that are important in the pumping of liquid egg white, egg yolk and whole egg under commercial processing conditions.

Egg white exhibits decreasing apparent viscosity with increasing shearing time and shearing rate.[34,35,37] This behaviour has been related to the phenomenon of egg white thinning. Viscosity declines as temperature increases over the range of 5°C to 45°C.[31,32,36] Egg white viscosity is also affected by pH, being maximal at pH 9 and declining as pH is increased to 11 or decreased to 4.[29,32]

Egg yolk declines in viscosity as the solids content is reduced by the addition of egg white.[28,30,38] Egg yolk becomes more viscous and whole egg less viscous when salt is added.[36,39] Both whole egg and yolk decline in viscosity with increases in temperature.[30,36] However, pasteurization of whole egg (60°C for 3·5 min in the USA and 64°C for 2·5 min in the UK) increases viscosity,[33] probably owing to protein aggregation.

Freezing and frozen storage of raw egg yolk and whole egg result in extreme increases in viscosity and an eventual loss of fluidity, referred to as freeze-induced gelation.[40] The low-density lipoprotein of yolk undergoes dehydration and an irreversible precipitation or aggregation.[40] The effects of freezing on viscosity can be prevented by the addition of NaCl, sucrose or other low-molecular-weight solutes at levels of 2–10%.

5.b Solubility

Spray-dried egg white is highly soluble, with a protein solubility of 96–98%.[41,42] Kakalis and Regenstein[41] reported that solubility is slightly decreased at pH 5, and is enhanced by the addition of NaCl and $Na_4P_2O_7$. Spray-dried egg products tend to be hard to disperse, and require some time to become fully hydrated. In studying the solubility of spray-dried egg white, Kakalis and Regenstein[41] rehydrated samples by first stirring a small amount of liquid into the solids to form a smooth paste, followed by dilution to volume, adjustment of pH and stirring for an hour. The actual time required for rehydration is likely to vary with the specific egg product, its processing history and the procedure used to rehydrate it. The successful use of egg powders as food ingredients depends to a great degree on their complete rehydration.

6 GEL PROPERTIES

6.a Preparation of Gels

Egg white has been preferred over egg yolk and whole egg for its gelling abilities in food systems for several reasons. Although all three products are available as spray-dried powders, egg white is more stable because it contains no lipids. It is colourless, with a mild flavour, whereas egg yolk and whole egg are limited by their yellow colour and more distinct flavour. The foaming ability of egg white broadens its range of gelling applications. In addition, egg yolk and whole egg contain cholesterol, which is perceived as a negative dietary factor by consumers. As a consequence of negative publicity surrounding cholesterol in foods, food manufacturers are reluctant to market cholesterol-containing products. Cost is a factor that has favoured the use of spray-dried egg white over yolk and whole egg, but increased demand for albumen since 1985 has pushed its value up to more than twice those of spray-dried yolk and whole egg.

The preparation of egg gels or food gels containing eggs consists of three steps. The egg product is brought into a liquid state by rehydrating powders or thawing frozen products as necessary. Next, any desired ingredients are incorporated, such as starches, salts, sugars, buffers, acids, etc. Finally, the liquid is poured into the desired type of container, heated to the desired temperature and cooled. Common methods of heating include immersion in a water-bath, steam injection, baking and frying. Most research on egg gels has been conducted on gels heated by immersion in a water-bath.

6.b Measurement of Gel Strength

The measurement of gel firmness may be accomplished by any of the universal texture testing instruments; Gossett et al.[43,44] reviewed a number of methods that have been applied to the measurement of egg white gels, including shear, compression, penetration, capillary extrusion and back-extrusion methods. Hamann[45] discussed the use of compression and torsion tests for determining structural failure in protein gels. Montejano et al.[46,47] compared torsion and compression testing for egg white and modified egg white gels, and additionally monitored the development of gel rigidity during heating using a non-destructive, temperature-controlled 'thermal scanning rigidity monitor' attached to an Instron Universal Testing Machine.

6.c Factors Affecting Gelation

6.c.1 Egg White Gels

It is important to understand the factors controlling the gelation of albumen in order to maximize its function in a given food product. The quality of egg white gels is greatly affected by pH and ionic conditions, similarly to the gels of most globular proteins. Gel strength and cohesiveness are minimal at pH values where net charge of proteins is minimal, in the range of pH 6–7 in egg white (Fig. 2). Increasing pH causes gel strength to increase to a maximum at pH 9, while decreasing the pH

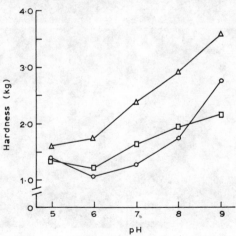

FIG. 2. Hardness of heat-formed gels of (○) egg white, (□) egg yolk and (△) whole egg as a function of pH. All gels were prepared from 11% protein samples and heated from 30 min at 85°C. (Source: Adapted from Woodward.[92])

FIG. 3. Scanning electron micrographs of egg white gels prepared at various pH values. (A) pH 9 gel; (B) pH 6 gel; (C) higher magnification of pH 6 gel, showing aggregates; (D) pH 5 gel; (E) aggregates in pH 5 gel; (F) larger aggregates in pH 5 gel. (Reprinted from *Journal of Food Science*, Woodward and Cotterill.[51] © Institute of Food Technologists.)

below 6 results in firmer but more brittle gels with low elasticity and poor water-binding properties.[48-51] Hickson et al.[52] found that the natural increase of pH in albumen due to the storage of eggs caused significant increases in gel elasticity, penetration force and viscosity index. Holt et al.[50] commented that gels below pH 7 had reduced elasticity, and that elasticity was greatest for gels at pH 9. When egg white at a pH less than 7 is heated, the gel expands, owing primarily to the liberation of CO_2,[48] which is present in albumen in equilibrium with bicarbonate.[2] This increases the brittleness of the gel and limits its water-binding ability.

The influence of pH on the gels is primarily related to the net charge of the proteins in solution. At the isoelectric point, the net charge is zero and aggregation is favoured, resulting in gels that are curdy, opaque and low in cohesiveness, lacking a continuous protein matrix.[44,51,53] As the net charge is increased by raising the pH, protein denaturation is favoured over aggregation, and resulting gels are translucent and quite elastic or rubbery, with an extensively cross-linked three-dimensional matrix (Fig. 3). The coarse, open microstructure of gels at low pH accounts for the poor water binding and texture. Enhanced firmness and water-binding ability of gels at pH 9 are due to the high degree of cross-linking, uniformity of gel networks and minimal pore size.[51]

The behaviour of albumen gels with respect to pH is of critical importance in food gels. When egg white is gelled alone, the normal pH is 9, and gels of excellent quality are obtained. However, when egg white is added to other food systems, the pH is lowered and egg white functions much less effectively in gelling. This fact must not be overlooked if egg white is to be used successfully in combination with other foods.

Salts affect the ionic nature of the gel, and NaCl tends to enhance aggregation of egg white proteins, resulting in weaker gels.[51] Beveridge et al.[48] found that low levels of NaCl did not affect gel strength, but at 0·90M NaCl gels were weakened. Holt et al.[50] found that addition of up to 0·10M NaCl caused minor increases in gel strength and elasticity. Woodward and Cotterill[51] reported that the effect of NaCl was dependent on pH, causing reductions in gel strength at pH 5, 8 and 9, and slight increases at pH 6 and 7. In a study of ovalbumin gels, Egelandsdal[54] explained that increasing the ionic strength shifted the maximum gel strength to pH values away from the isoelectric point. Hegg et al.[55] and Nakamura et al.[56] found that the addition of NaCl promoted aggregation of ovalbumin. Sodium chloride appears to expand the pH range where net charge on proteins is low and strong aggregation will occur.[51,55]

Beveridge et al.[48] evaluated the effect of various salts on gel strength, and found that $FeCl_3$ markedly toughened gels, while $CuSO_4$ and $AlCl_3$ softened gels. The binding of iron, copper and aluminium ions by conalbumin increases its heat stability;[57,58] the reason for the varying effects of these ions on egg white gel strength is unknown. It should be noted that the addition of ferric and cupric ions causes egg albumen to become pink-red and yellow-green, respectively, in the liquid state, owing to binding of these ions by conalbumin.[2] These colours change to grey-black and tan-brown, respectively, in cooked gels. The reaction of iron with H_2S, which is released in heating of egg white, forms ferrous sulphide that blackens the gel;[59] a similar reaction for copper is probably responsible for the brown discoloration of albumen gels. These discolorations would prohibit the use of iron and copper in albumen gels; however, the aluminium–conalbumin complex is colourless in albumen, and aluminium is presently used to stabilize albumen prior to pasteurization.[23]

Since heat is the driving force behind protein denaturation and gelation, the temperature of heating is of major importance in forming gels. Egg albumen begins to lose fluidity at about 60°C with the denaturation of conalbumin.[47,57] Montejano et al.[47] reported that as albumen was heated it underwent a transition from a liquid to a solid over the temperature range of 61–70°C. At 70–74°C, a large increase in elasticity signalled the development of a gel with a recoverable shape. Only minor increases in elasticity occurred as albumen was heated from 74°C to 89°C, and elasticity actually decreased from 89°C to 91°C. Gel rigidity development followed that of elasticity, being initiated at 71°C and increasing rapidly up to 83°C (Fig. 4). Montejano et al.[47] hypothesized that the development of elasticity coincided with protein denaturation and the initiation of aggregation, resulting in the establishment of a three-dimensional gel network, while subsequent heating promoted completion of protein aggregation with increased protein cross-linking and gel strength enhancement. Other workers have reported that 70°C is a minimum temperature for egg white gel formation.[50,60] As temperature is increased, the rate of gelation is also increased; Goldsmith and Toledo[60] reported that gels of maximum firmness were obtained after heating 50 min at 70°C, 18 min at 80°C and 5 min at 90°C. Over-heating, whether by high temperatures or excessive heating times, causes over-coagulation and results in gel shrinkage, syneresis and a decrease in gel strength.[49–51,61,62] The optimum temperature for gelation has been reported as 80–85°C, combined with heating times of 30–60 min.[49,50,60,62] However, some workers have found gel strength to increase as the heating temperature is increased to about 90°C.[48,51]

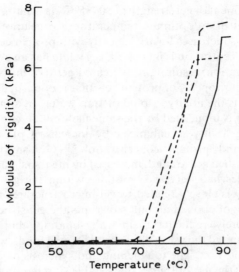

FIG. 4. Shear rigidity thermograms for native and modified egg white: (····) native; (---) oleated; (——) succinylated. Heating rate = 0·5°C/min. (Reprinted from *Journal of Food Science*, Montejano et al.[47] © Institute of Food Technologists.)

The effect of temperature on gel formation and quality is due to the thermal properties of individual proteins. In a differential scanning calorimetry study of egg white, the denaturation temperatures of conalbumin, lysozyme and ovalbumin were reported as 65°C, 72°C and 84°C, respectively.[57] Shifts in thermal stability occur with changes in pH, buffer salts, sucrose and binding of metal ions. Matsuda et al.[63] found that the temperature of aggregation of egg white proteins were pH and time dependent. At pH 9, proteins aggregated over the following temperature ranges: G_3A globulin, 54–66°C; conalbumin, 65–66°C; flavoprotein, 68–76°C; ovalbumin, 75–84°C; and globulins A_1 and A_2, 77–86°C. Watanabe et al.[64] reported that increasing the level of acidity of egg white from pH 5·5 to 2·5 caused proteins to aggregate at lower temperatures.

Thermal properties of isolated egg white proteins have been studied extensively.[54–56,58,65–75] Conalbumin, being the least heat stable, is most important in initiating gelation. Lysozyme and other globulins, having intermediate heat stability, probably play a central role in the formation of the elastic three-dimensional gel matrix. Ovalbumin, being the predominant protein in albumen as well as one of the more heat stable, contributes to the enhancement of gel strength at 80–85°C; its denaturation is related to the assignment of an optimum temperature for gelation of egg white.

Gel hardness increases logarithmically with protein concentration.[48,51] As protein concentration is increased, egg white has an enhanced tendency toward aggregation, resulting in increased gel fracturability.[51,75]

One inherent property of protein gels that is of particular importance in formed foods is the ability to bind or trap water. In albumen gels, water-binding ability is influenced by the same factors that control gel hardness. Water-binding ability is enhanced by increasing pH (Fig. 5), heating temperature and protein concentration.[51] Goldsmith and Toledo[60] reported that when gel strength increased owing to increasing temperature, water molecules had decreased motility within the gel network. Water-binding ability is closely related to gel microstructure.[51]

The freezing of cooked albumen disrupts the gel structure, toughening gels and severely reducing their water-binding ability.[40,51,76,77] This problem has limited the use of albumen in formed foods, where freeze–thaw stability is important. The toughening and syneresis of gels has generally been solved by the use of freeze–thaw stable starches or gums, or by rapid freezing.[40,78–79]

6.c.2 Modified Egg White Gels

Egg white proteins have been chemically modified in order to improve their functional properties.[81] Acylation of proteins by reagents such as acetic or succinic anhydride causes a change in the net charge of proteins by binding with amino groups, resulting in a more negatively charged protein.[72,82–84] These reactions lower the nutritional value of the albumen by reducing the bioavailability of lysine;[83] they also inhibit lysozyme activity and iron binding by conalbumin.[82] Modifications by the addition of sodium dodecyl sulphate, an anionic detergent, and oleic acid, a long-chain fatty acid, alter the ionic properties of albumen proteins, but their specific reaction with the proteins has not been fully elucidated.[55,68,69,85]

Chemical modification effectively changes the gelling characteristics of albumen. Proteins modified by sodium dodecyl sulphate are more stable to aggregation,[55,68,69] and the temperature for initiation of gelation is increased by succinylation[42] (see Fig. 4). Acetylation and oleation reduce heat stability slightly.[47,82] Oleated and succinylated albumen produce very strong and highly deformable gels, while acetylated albumen forms gels very similar in strength and deformability to native albumen gels.[46,47] The freeze–thaw stability and water-binding ability of cooked albumen gels are enhanced markedly by succinylation, oleation, addition of sodium dodecyl

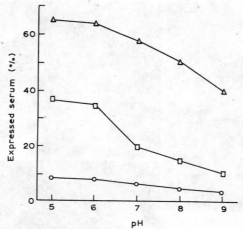

FIG. 5. Serum expressed from gels of (△) egg white, (◯) egg yolk and (□) whole egg as a function of pH. Gels were prepared from 11% protein samples and heated for 30 min at 85°C. (Source: Adapted from Woodward.[92])

sulphate and, to a lesser degree, by acetylation.[29,81,85,86] These improvements are apparently related to structural changes in the gel matrix that result in strong, elastic, translucent gels[47,61] (see Fig. 6).

No commercial applications of chemically modified egg white are presently known, although the technology for its production is available. Although chemically modified egg proteins *per se* have not been approved for food use, the chemical modifiers would be classified as food additives and would be subject to the same regulations as other additives. Oleic acid is classified as a GRAS (Generally Recognized As Safe) food additive in the United States. Sodium dodecyl sulphate (SDS) is commonly used at a level of 0·1% as a whipping aid in spray-dried egg white, while researchers have used 0·5–1·0% SDS to modify egg white. Anhydrides are used as acidulants in various food applications.

6.c.3 Egg Yolk Gels

Egg yolk contains about 16% protein, including livetins, phosvitin and various lipoproteins. In native egg yolk, the proteins are arranged structurally in yolk spheres (up to 150 μm in diameter) and granules (0·3–1·7 μm in diameter). When yolk is heated in the shell egg, the spherical structures are heat-set independently of one another, producing the familiar coarse, mealy texture of a hard-cooked egg yolk[19] (see Fig. 1(A)). However, stirring of raw yolk easily disrupts spheres, resulting in a protein

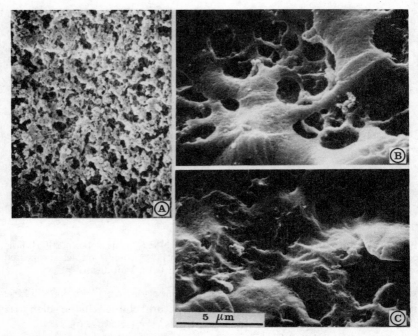

FIG. 6. Scanning electron micrographs of heat-induced gels from (A) native, (B) oleated and (C) succinylated egg white. (Reprinted from *Journal of Food Science*, Montejano et al.[47] © Institute of Food Technologists.)

solution that is capable of forming firm gels with a texture similar to that of albumen.[19,79]

Egg yolk gels are minimal in hardness at pH 6, and increase in hardness as pH is increased.[87] The effects of pH are not so pronounced in yolk as in albumen gels, however (Fig. 2). The decline in gel hardness with the dilution of yolk is logarithmic. The addition of salt causes pronounced increases in gel strength, based on the rupture of granules at 0·3M NaCl. Yolk gelation appears to be initiated at about 70°C, and gel strength is increased with increasing temperature. Egg yolk gels have a much greater capacity for binding water than do egg white gels.[29,87] Water-binding capacity is increased with increasing pH (Fig. 5), protein concentration and heating temperature.[87]

Low-density lipoprotein (LDL) is capable of producing stable gels, and apparently plays a major role in the gelation of egg yolk.[73,88] LDL solutions begin to gel at 65°C at pH 4 and 5, and at 70°C at pH 6 to 8. LDL

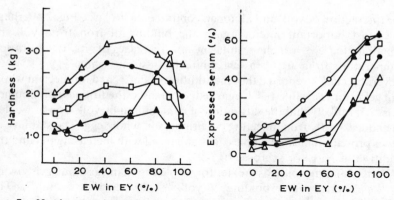

FIG. 7. Hardness and expressed serum of gels from blends of egg yolk (EY) and egg white (EW) at various pH values. (○) pH 5, (▲) pH 6, (□) pH 7, (●) pH 8 and (△) pH 9. (Source: Woodward.[92])

gels have maximum rigidity at pH 4·0 and 9·0, with a minimum at pH 6·0 to 7·0,[53] which corresponds to the reported pI range of 6·5–7·3.[88] The addition of 0·03M NaCl enhances the rigidity of LDL gels, and constant rigidity is reached at 0·05M NaCl. Other proteins in yolk are sensitive to heat; some livetins are affected by temperatures as low as 60°C, while other livetins are stable to about 80°C.[87,89,90]

6.c.4 Whole Egg Gels

Whole egg has many properties in common with egg white since it is comprised of approximately 67% albumen and 33% yolk. Beveridge et al.[91] found that whole egg gels were very similar to egg white gels. Hardness increased linearly from a minimum at pH 5 to a maximum at pH 9 (see Fig. 2). Hardness increased as temperature was increased from 77°C to 90°C and as time of heating was increased from 5 to 25 min. Dilution of whole egg caused a logarithmic decrease in hardness. Woodward and Cotterill[92,93] found that when yolk and albumen were blended together the hardness of gels was greatest for albumen:yolk ratios ranging from 40:60 to 60:40 over the pH range of 7–9, but at pH 5 the maximum hardness occurred at an albumen:yolk ratio of 90:10 (Fig. 7). They also reported increasing gel strength with increased heating temperature, protein level and pH. It should be noted that a practical upper limit for pH in whole egg gels is about 7·0, owing to the formation of ferrous sulphide during heating, termed 'greening' owing to the colour imparted in whole egg.[59,94–96]

Whole egg has greater gelling capacities than albumen or yolk based on the interaction of yolk and albumen components[92,93] (see Fig. 2). Perhaps the most important interaction is the binding of iron from yolk by conalbumin,[58,97] which results in an increase of its denaturation temperature from 61°C in egg white to 78°C in whole egg.[57,58,98,99] Beveridge et al.[91] reported that the addition of $FeCl_3$ had no effect on whole egg gel firmness, while $FeCl_3$ markedly increased the firmness of albumen gels.[48] It is likely that the saturation of conalbumin by yolk iron prevented the added $FeCl_3$ from affecting gel firmness in whole egg. The stability of other proteins in heated whole egg is enhanced with increasing pH and the addition of NaCl or sucrose.[98,99]

Water-binding ability is greater for whole egg than egg white gels, owing to the superior water binding of yolk[86,87,92] (see Figs 5 and 7). The principles that govern water binding in egg white also apply to whole egg, and pH and temperature of heating are two of the more important factors to be controlled.[92] Water-binding ability in whole egg gels or scrambled egg products is improved with increasing levels of yolk, pH, NaCl and protein.[86,92,95,100] The optimum heating temperature for maximum water binding is about 80°C;[92] excessively high temperatures or long heating times result in over-coagulation and syneresis.[61] Non-fat milk solids, starches and food gums have been used to improve water binding in whole egg gels.[79,80,101]

7 APPLICATIONS

One application of albumen in a formed food is its recent and growing use as an extender in surimi products. According to Burgarella et al.,[102] egg white modifies the thermal gelation profile of surimi and lowers its final rigidity. The two protein types do not cooperate in gelation, but interfere with each other in the gelation process. Burgarella et al.[103] suggest that egg white could be used most effectively as a gelling extender or adjunct in surimi rather than as a high-percentage ingredient. The growing use of egg white in surimi is the principle cause of the increased demand for albumen, which has made spray-dried egg white 2·5 times as expensive as egg yolk, although historically yolk products have been more expensive. The growth of the market for surimi is expected to continue to increase the demand for albumen.

Other applications of egg white using its gelling function include egg-type products, such as custards, 'long egg' products and low-cholesterol egg substitutes.[40] Wiker and Cunningham[104] developed a breaded 'egg white ring' that could be prepared, frozen and marketed as a foodservice item. Baker and Gossett[105] developed a retortable, low-cholesterol egg salad using egg white in place of hard-cooked eggs. Another area of use is in sausage or emulsion formulations as a meat extender, water-binding agent and fat stabilizer.[49,62,106–108] Egg white is also used in candy to incorporate air, to control the growth of sugar crystals, to provide structural support through heat coagulation, and to prevent syneresis.[61,109]

Although yolk gels or sets in any food in which it is cooked, it has little application outside the traditional egg cookery, such as in custards, cakes and related foods in which the flavour, texture and colour depend on the yolk. One use of egg yolk for gels has been in the formation of cooked yolk centres for 'long eggs'.[79] In those products the pre-cooked yolk centres must be milled or ground or otherwise treated to obtain the desired crumbly texture, simulating that of the yolk of a hard-cooked egg.[19,79]

The applications for whole egg are related to traditional methods of egg cookery where the heat-setting property is desirable. Examples in foodservice include scrambled egg mixes, omelettes, quiches and soufflés. The same type of products have a potential market in the pre-cooked, frozen form at the retail level. However, care must be taken to prevent the problems of gel toughening and syneresis caused by freezing.[40,95,101] Several food product concepts have been developed that take advantage of the gelling potential of whole egg; among these are an egg sausage analogue[108] and egg jerky.[110] Further development and marketing of egg-based food gels will be an important aspect in the growth of the egg and egg products industry.

REFERENCES

1. Cotterill, O. J. & Geiger, G. S., Egg product yield trends from shell eggs. *Poultry Sci.*, **56** (1977) 1027.
2. Powrie, W. D. & Nakai, S., The chemistry of eggs and egg products. In *Egg Science and Technology*, 3rd edn, ed. W. J. Stadelman & O. J. Cotterill. Avi Publishing Co., Westport, Connecticut, 1986, p. 97.
3. Shenstone, F. S., The gross composition, chemistry and physicochemical basis of organization of the yolk and the white. In *Egg Quality: A Study of the Hen's Egg*, ed. T. C. Carter. Oliver and Boyd, Edinburgh, 1968, p. 26.
4. Baker, C. M. A., The proteins of egg white. In *Egg Quality: A Study of the Hen's Egg*, ed. T. C. Carter. Oliver and Boyd, Edinburgh, 1968, p. 67.
5. Cook, F. & Briggs, G. M., The nutritive value of eggs. In *Egg Science and Technology*, 3rd edn, ed. W. J. Stadelman & O. J. Cotterill. Avi Publishing Co., Westport, Connecticut, 1986, p. 141.
6. Cook, W. H., Macromolecular components of egg yolk. In *Egg Quality: A*

Study of the Hen's Egg, ed. T. C. Carter. Oliver and Boyd, Edinburgh, 1968, p. 109.
7. Osuga, D. T. & Feeney, R. E., Egg proteins. In *Food Proteins*, ed. J. R. Whitaker & S. R. Tannenbaum. Avi Publishing Co., Westport, Connecticut, 1977, p. 193.
8. Robinson, D. S., The domestic hen's egg. In *Food Microscopy*, ed. J. G. Vaughn. Academic Press, New York, 1979, p. 313.
9. Romanoff, A. L. & Romanoff, A. J., *The Avian Egg*. Wiley, New York, 1949.
10. Bellairs, R., The structure of the yolk of the hen's egg as studied by electron microscopy. I. The yolk of the unincubated egg. *J. Biophys. Biochem. Cytol.*, **11** (1961) 207.
11. Grodzinski, Z., The yolk's spheres of the hen's egg as osmometers. *Biol. Rev.*, **26** (1951) 253.
12. Bellairs, R., Backhouse, M. & Evans, R. J., A correlated chemical and morphological study of egg yolk and its constituents. *Micron*, **3** (1972) 328.
13. Burley, R. W. & Cook, W. H., Isolation and composition of avian egg yolk granules and their constituent α- and β-lipovitellins. *Can. J. Biochem. Physiol.*, **39** (1961) 1295.
14. Chang, C. M., Powrie, W. D. & Fennema, O., Microstructure of egg yolk. *J. Food Sci.*, **42** (1977) 1193.
15. Garland, T. D. & Powrie, W. D., Isolation of myelin figures and low-density lipoproteins from egg yolk granules. *J. Food Sci.*, **43** (1978) 592.
16. Garland, T. D. & Powrie, W. D., Chemical characterization of egg yolk myelin figures and low-density lipoproteins isolated from egg yolk granules. *J. Food Sci.*, **43** (1978) 1210.
17. Kocal, J. T., Nakai, S. & Powrie, W. D., Preparation of apolipoprotein of very low density lipoprotein from egg yolk granules. *J. Food Sci.*, **45** (1980) 1761.
18. Fujii, S., Tamura, T. & Okamoto, T., Studies on yolk formation in hen's eggs. I. Light and scanning electron microscopy of the structure of yolk spheres. *J. Fac. Fish. Anim. Husb., Hiroshima Univ.*, **12** (1973) 1.
19. Woodward, S. A. & Cotterill, O. J., Texture and microstructure of cooked whole egg yolks and heat-formed gels of stirred egg yolk. *J. Food Sci.*, **52** (1987) 63.
20. Maurice, D. M., Electrical resistance and structure of the hen's egg. *Nature (London)*, **1970** (1952) 495.
21. Cotterill, O. J., Egg breaking. In *Egg Science and Technology*, 3rd edn, ed. W. J. Stadelman & O. J. Cotterill. Avi Publishing Co., Westport, Connecticut, 1986, p. 193.
22. Cotterill, O. J., Egg-products industry. In *Egg Science and Technology*, 3rd edn, ed. W. J. Stadelman & O. J. Cotterill. Avi Publishing Co., Westport, Connecticut, 1986, p. 185.
23. Cunningham, F. E., Egg-product pasteurization. In *Egg Science and Technology*, 3rd edn, ed. W. J. Stadelman & O. J. Cotterill. Avi Publishing Co., Westport, Connecticut, 1986, p. 243.
24. Ghielmetti, G. & Trinchera, C., Process for the production of lysozyme. US Patent 3 515 643 (2 June 1970).
25. Ahvenainen, R., Heikonen, M., Kreula, M., Linko, M. & Linko, P., Separation of lysozyme from egg white. *Food Process Engr*, **2** (1980) 301.

26. Li-Chan, E., Nakai, S., Sim, J., Bragg, D. B. & Lo, K. V., Lysozyme separation from egg white by cation exchange column chromatography. *J. Food Sci.*, **51** (1986) 1032.
27. Sim, J. S. & Nakai, S., Nutritional and functional property of egg albumen after lysozyme removal. *Poultry Sci.*, **64**(Suppl. 1) (1985) 182 (Abstract).
28. Chang, P. K., Powrie, W. D. & Fennema, O., Effect of heat treatment on viscosity of yolk. *J. Food Sci.*, **35** (1970) 864.
29. Gossett, P. W., Rizvi, S. S. H. & Baker, R. C., Selected rheological properties of pH-adjusted or succinylated egg albumen. *J. Food Sci.*, **48** (1983) 1395.
30. Pitsilis, J. G., Brooker, D. B., Cotterill, O. J. & Walton, H. V., Rheological properties of plain egg yolk, salted egg yolk and salted whole egg. *Trans. ASAE*, **27** (1984) 294.
31. Pitsilis, J. G., Brooker, D. B., Walton, H. V. & Cotterill, O. J., Rheological properties of liquid egg white. *Trans. ASAE*, **27** (1984) 300.
32. Pitsilis, J. G., Walton, H. V. & Cotterill, O. J., The apparent viscosity of egg white at various temperatures and pH levels. *Trans. ASAE*, **18** (1975) 347.
33. Torten, J. & Eisenberg, H., Studies on colloidal properties of whole egg magma. *J. Food Sci.*, **47** (1982) 1423.
34. Tung, M. A., Watson, E. L. & Richards, J. F., Rheology of fresh, aged and gamma-irradiated egg white. *J. Food Sci.*, **35** (1970) 872.
35. Tung, M. A., Watson, E. L. & Richards, J. F., Rheology of egg albumen. *Trans. ASAE*, **14** (1971) 17.
36. Scalzo, A. M., Dickerson, R. W. Jr, Peeler, J. T. & Read, R. B. Jr, The viscosity of egg products. *Food Technol.*, **24** (1970) 1301.
37. Beveridge, T. & Nakai, S., Effects of sulphydryl blocking on the thinning of egg white. *J. Food Sci.*, **40** (1975) 864.
38. Cunningham, F. E., Viscosity and functional ability of diluted egg yolk. *J. Milk Food Technol.*, **35** (1972) 615.
39. Jordan, R. & Whitlock, E. S., A note on the effect of salt (NaCl) upon the apparent viscosity of egg yolk, egg white and whole egg magma. *Poultry Sci.*, **34** (1955) 566.
40. Cotterill, O. J., Freezing egg products. In *Egg Science and Technology*, 3rd edn, ed. W. J. Stadelman & O. J. Cotterill. Avi Publishing Co., Westport, Connecticut, 1986, p. 217.
41. Kakalis, L. T. & Regenstein, J. M., Effect of pH and salts on the solubility of egg white protein. *J. Food Sci.*, **51** (1986) 1445.
42. Morr, C. V., German, B., Kinsella, J. E., Regenstein, J. M., Van Buren, J. P., Kilara, A., Lewis, B. A. & Mangino, M. E., A collaborative study to develop a standardized food protein solubility procedure. *J. Food Sci.*, **50** (1985) 1715.
43. Gossett, P. W., Rizvi, S. S. H. & Baker, R. C., A new method to quantitate the coagulation process. *J. Food Sci.*, **48** (1983) 1400.
44. Gossett, P. W., Rizvi, S. S. H. & Baker, R. C., Quantitative analysis of gelation in egg protein systems. *Food Technol.*, **38**(5) (1984) 67.
45. Hamann, D. D., Structural failure in solid foods. In *Physical Properties of Foods*, ed. E. B. Bagley & M. Peleg. Avi Publishing Co., Westport, Connecticut, 1983, p. 351.
46. Montejano, J. G., Hamann, D. D. & Ball, H. R. Jr, Mechanical failure

characteristics of native and modified egg white gels. *Poultry Sci.*, **63** (1984) 1969.
47. Montejano, J. G., Hamann, D. D., Ball, H. R. Jr & Lanier, T. C., Thermally induced gelation of native and modified egg white—rheological changes during processing; final strengths and microstructures. *J. Food Sci.*, **49** (1984) 1249.
48. Beveridge, T., Arntfield, S., Ko, S. & Chung, J. K. L., Firmness of heat induced albumen coagulum. *Poultry Sci.*, **59** (1980) 1229.
49. Hickson, D. W., Dill, C. W., Morgan, R. G., Suter, D. A. & Carpenter, Z. L., A comparison of heat-induced gel strengths of bovine plasma and egg albumen proteins. *J. Anim. Sci.*, **51** (1980) 69.
50. Holt, D. L., Watson, M. A., Dill, C. W., Alford, E. S., Edwards, R. L., Diehl, K. C. & Gardner, F. A., Correlation of the rheological behavior of egg albumen to temperature, pH, and NaCl concentration. *J. Food Sci.*, **49** (1984) 137.
51. Woodward, S. A. & Cotterill, O. J., Texture and microstructure of heat-formed egg white gels. *J. Food Sci.*, **51** (1986) 333.
52. Hickson, D. W., Alford, E. S., Gardner, F. A., Diehl, K., Sanders, J. O. & Dill, C. W., Changes in heat-induced rheological properties during cold storage of egg albumen. *J. Food Sci.*, **47** (1982) 1908.
53. Meyer, R. & Hood, L. F., The effect of pH and heat on the ultrastructure of thick and thin hen's egg albumen. *Poultry Sci.*, **52** (1973) 1814.
54. Egelandsdal, B., Heat-induced gelling in solutions of ovalbumin. *J. Food Sci.*, **45** (1980) 570.
55. Hegg, P. O., Martens, H. & Lofqvist, B., Effects of pH and neutral salts on the formation and quality of thermal aggregates of ovalbumin. A study on thermal aggregation and denaturation. *J. Sci. Food Agric.*, **30** (1979) 981.
56. Nakamura, R., Sugiyama, H. & Sato, Y., Factors contributing to the heat-induced aggregation of ovalbumin. *Agric. Biol. Chem.*, **42** (1978) 819.
57. Donovan, J. W., Mapes, C. J., Davis, J. G. & Garibaldi, J. A., A differential scanning calorimetric study of the stability of egg white to heat denaturation. *J. Sci. Food Agric.*, **26** (1975) 73.
58. Donovan, J. W. & Ross, K. D., Iron binding to conalbumin. Calorimetric evidence for two distinct species with one bound iron atom. *J. Biol. Chem.*, **250** (1975) 6026.
59. Tinkler, C. K. & Soar, M. C., The formation of ferrous sulphide in eggs during cooking. *Biochem. J.*, **14** (1920) 414.
60. Goldsmith, S. M. & Toledo, R. T., Studies on egg albumin using nuclear magnetic resonance. *J. Food Sci.*, **50** (1985) 59.
61. Baldwin, R. E., Functional properties of eggs in food. In *Egg Science and Technology*, 3rd edn, ed. W. J. Stadelman & O. J. Cotterill. Avi Publishing Co., Westport, Connecticut, 1986, p. 345.
62. Hickson, D. W., Dill, C. W., Morgan, R. G., Sweat, V. E., Suter, D. A. & Carpenter, Z. L., Rheological properties of two heat-induced protein gels. *J. Food Sci.*, **47** (1982) 783.
63. Matsuda, T., Watanabe, K. & Sato, Y., Heat-induced aggregation of egg white proteins as studied by vertical flat-sheet polyacrylamide gel electrophoresis. *J. Food Sci.*, **46** (1981) 1829.

64. Watanabe, K., Matsuda, T. & Nakamura, R., Heat-induced aggregation and denaturation of egg white proteins in acid media. *J. Food Sci.*, **50** (1985) 507.
65. Donovan, J. W. & Beardslee, R. A., Heat stabilization produced by protein–protein complexes. A differential scanning calorimetric study of the heat denaturation of the trypsin–soybean trypsin inhibitor and trypsin–ovomucoid complexes. *J. Biol. Chem.*, **250** (1975) 1966.
66. Donovan, J. W. & Mapes, C. J., A differential scanning calorimetric study of conversion of ovalbumin to S-ovalbumin in eggs. *J. Sci. Food Agric.*, **27** (1976) 197.
67. Donovan, J. W. & Ross, K. D., Increase in the stability of avidin produced by binding of biotin. A differential scanning calorimetric study of denaturation by heat. *Biochemistry*, **12** (1973) 512.
68. Hegg, P. O. & Lofqvist, B., The protective effect of small amounts of anionic detergents on the thermal aggregation of crude ovalbumin. *J. Food Sci.*, **39** (1974) 1231.
69. Hegg, P. O., Martens, H. & Lofqvist, B., The protective effect of sodium dodecylsulphate on the thermal precipitation of conalbumin. A study on thermal aggregation and denaturation. *J. Sci. Food Agric.*, **29** (1978) 245.
70. Holme, J., The thermal denaturation and aggregation of ovalbumin. *J. Phys. Chem.*, **67** (1963) 782.
71. Johnson, T. M. & Zabik, M. E., Gelation properties of albumen proteins, singly and in combination. *Poultry Sci.*, **60** (1981) 2071.
72. Ma, C. Y. & Holme, J., Effect of chemical modifications on some physicochemical properties and heat coagulation of egg albumen. *J. Food Sci.*, **47** (1982) 1454.
73. Nakamura, R., Fukano, T. & Taniguchi, M., Heat-induced gelation of hen's egg yolk low density lipoprotein (LDL) dispersion. *J. Food Sci.*, **47** (1982) 1449.
74. Nakamura, R., Umemura, O. & Takemoto, H., Effect of heating on the functional properties of ovotransferrin. *Agric. Biol. Chem.*, **43** (1979) 325.
75. Shimada, K. & Matsushita, S., Thermal coagulation of egg albumin. *J. Agric. Food Chem.*, **28** (1980) 409.
76. Davis, J. G., Hanson, H. L. & Lineweaver, H., Characterization of the effect of freezing on cooked egg white. *Food Res.*, **17** (1952) 393.
77. Woodward, S. A. & Cotterill, O. J., Preparation of cooked egg white, egg yolk, and whole egg gels for scanning electron microscopy. *J. Food Sci.*, **50** (1985) 1624.
78. Bengtsson, N., Ultrafast freezing of cooked egg white. *Food Technol.*, **21** (1967) 1259.
79. Hawley, R. L., Egg product. US Patent 3 510 315 (1970).
80. Ziegler, H. F. Jr, Seeley, R. D. & Holland, R. L., Frozen egg mixture. US Patent 3 565 638 (1971).
81. Ball, H. R. Jr, Functional properties of chemically modified egg white proteins. *J. Am. Oil Chem. Soc.*, **64** (1987) 1718.
82. Ball, H. R. Jr & Winn, S. E., Acylation of egg white proteins with acetic anhydride and succinic anhydride. *Poultry Sci.*, **61** (1982) 1041.
83. King, A. J., Ball, H. R. & Garlich, J. D., A chemical and biological study of acylated egg white. *J. Food Sci.*, **46** (1981) 1107.
84. Palladino, D. K., Ball, H. R. Jr & Swaisgood, H. E., Separation of native and

acylated egg white proteins with gel chromatography and DEAE-cellulose ion exchange. *J. Food Sci.*, **46** (1981) 778.
85. King, A. J., Ball, H. R. Jr, Catignani, G. L. & Swaisgood, H. E., Modification of egg white proteins with oleic acid. *J. Food Sci.*, **49** (1984) 1240.
86. Gossett, P. W. & Baker, R. C., Effect of pH and succinylation on the water retention properties of coagulated, frozen, and thawed egg albumen. *J. Food Sci.*, **48** (1983) 1391.
87. Woodward, S. A. & Cotterill, O. J., Texture profile analysis, expressed serum, and microstructure of heat-formed egg yolk gels. *J. Food Sci.*, **52** (1987) 68.
88. Kojima, E. & Nakamura, R., Heat gelling properties of hen's egg yolk low density lipoprotein (LDL) in the presence of other protein. *J. Food Sci.*, **50** (1985) 63.
89. Chang, P., Powrie, W. D. & Fennema, O., Disc gel electrophoresis of proteins in native and heat-treated albumen, yolk, and centrifuged whole egg. *J. Food Sci.*, **35** (1970) 774.
90. Dixon, D. K. & Cotterill, O. J., Electrophoretic and chromatographic changes in egg yolk proteins due to heat. *J. Food Sci.*, **46** (1981) 981.
91. Beveridge, T. & Ko, S., Firmness of heat-induced whole egg coagulum. *Poultry Sci.*, **63** (1984) 1372.
92. Woodward, S. A., Texture and microstructure of heat-formed egg white, egg yolk, and whole egg gels. PhD dissertation, University of Missouri, Columbia, Missouri, 1984.
93. Woodward, S. A. & Cotterill, O. J., Texture and microstructure of heat-formed whole egg gels. In *IFT 86 Program and Abstracts*. Institute of Food Technologists, Chicago, 1986, p. 159.
94. Baker, R. C., Darfler, J. & Lifshitz, A., Factors affecting the discoloration of hard-cooked egg yolks. *Poultry Sci.*, **46** (1967) 664.
95. Feiser, G. E. & Cotterill, O. J., Composition of serum from cooked–frozen–thawed–reheated scrambled eggs at various pH levels. *J. Food Sci.*, **47** (1982) 1333.
96. Gossett, P. W. & Baker, R. C., Prevention of the green-gray discoloration in cooked liquid whole eggs. *J. Food Sci.*, **46** (1981) 328.
97. Cunningham, F. E. & Lineweaver, H., Stabilization of egg white proteins to pasteurization temperatures above 60°C. *Food Technol.*, **19** (1965) 1442.
98. Watanabe, K., Hayakawa, S., Matsuda, T. & Nakamura, R., Combined effect of pH and sodium chloride on the heat-induced aggregation of whole egg proteins. *J. Food Sci.*, **51** (1986) 1112.
99. Woodward, S. A. & Cotterill, O. J., Electrophoresis and chromatography of heat-treated plain, sugared and salted whole egg. *J. Food Sci.*, **48** (1983) 501.
100. Feiser, G. E. & Cotterill, O. J., Composition of serum and sensory evaluation of cooked–frozen–thawed scrambled eggs at various salt levels. *J. Food Sci.*, **48** (1983) 794.
101. O'Brien, S. W., Baker, R. C., Hood, L. F. & Liboff, M., Water-holding capacity and textural acceptability of precooked, frozen, whole-egg omelets. *J. Food Sci.*, **47** (1982) 412.
102. Burgarella, J. C., Lanier, T. C. & Hamann, D. D., Effects of added egg white or whey protein concentrate on thermal transitions in rigidity of Croaker surimi. *J. Food Sci.*, **50** (1985) 1588.

103. Burgarella, J. C., Lanier, T. C., Hamann, D. D. & Wu, M. C., Gel strength development during heating of surimi in combination with egg white or whey protein concentrate. *J. Food Sci.*, **50** (1985) 1595.
104. Wiker, J. M. & Cunningham, F. E., Method of preparing high protein snack food from egg protein. US Patent 4 421 770 (20 December 1983).
105. Baker, R. C. & Gossett, P. W., Development and evaluation of a thermally-processed, low-cholesterol egg salad product. *Poultry Sci.*, **63** (1984) 2201.
106. King, A. J., Patel, S. B. & Earl, L. A., Effect of water, egg white, sodium chloride, and polyphosphate on several quality attributes of unfrozen and frozen, cooked, dark turkey muscle. *Poultry Sci.*, **65** (1986) 1103.
107. Negbenebor, C. A. & Chen, T. C., Effect of albumen on TBA values of comminuted poultry meat. *J. Food Sci.*, **50** (1985) 270.
108. Pike, O. A. & Huber, C. S., Effect of formulation on water activity and sensory attributes of egg sausage. *Poultry Sci.*, **62** (1983) 2004.
109. Cotterill, O. J., Amick, G. M., Kluge, B. A. & Rinard, V. C., Some factors affecting the performance of egg white in divinity candy. *Poultry Sci.*, **42** (1963) 218.
110. Proctor, V. & Cunningham, F., Egg jerky. *Poultry Trib.*, **90**(7) (1984) 33.

Chapter 6

GELLAN GUM

G. R. Sanderson

*Kelco Division of Merck & Co. Inc.,
8225 Aero Drive, San Diego, California 92123, USA*

1 INTRODUCTION

Food polysaccharides can be classified, simply, as thickeners or gelling agents. In most cases, they are used in foods to impart stability during transport and storage, and desirable sensory qualities; in some instances, they are used primarily as processing aids.

Although many polysaccharide thickeners are commercially available, selection of the best candidate for a particular application is usually relatively easy and only requires a knowledge of the rheological properties of the polysaccharide in solution. Such knowledge is readily available in textbooks (e.g. Refs 1 and 2) and numerous publications in the scientific literature or from the polysaccharide suppliers. It is true to say that in some applications different thickeners may be used interchangeably from a functional standpoint and, in these situations, the ultimate choice is usually determined by relative costs. In many cases, however, cost considerations are secondary to the subtle rheological benefits one material may provide relative to another.

The situation is somewhat different with hydrocolloid gelling agents. The term hydrocolloid rather than polysaccharide is used here because gelatine, one of the most important gelling agents, is a protein of animal origin. The important gelling polysaccharides are agar, high- and low-methoxy pectin, carrageenan, alginate, starch and the mixed gelling system, carrageenan/locust-bean gum. Furcellaran, xanthan gum/locust-bean gum, some cellulose ethers and konjak mannan are also used commercially to produce gels but are of lesser importance, at least on a world-wide basis. The total

'Biozan', 'Gelrite' and 'Kelcogel' are trademarks of Merck & Co. Inc. (Rahwey, New Jersey), Kelco Division, USA.

US market for gelling hydrocolloids has recently been estimated at 20·5 million kilograms excluding starch.[3] Inclusion of starch gives a figure of around 84 million kilograms but this is somewhat misleading because it includes applications in which the starch, used at relatively high concentrations, functions as a thickener rather than a true gelling agent. As in the case of polysaccharide thickeners, rheological considerations, especially in the case of starches, are important when using these gelling agents. However, other properties also assume major importance. These properties include melting and setting temperatures, setting rates, clarity, purity and suitability for in-line processing.

In molecular terms, gels are metastable, three-dimensional polymer networks formed by chemically- or thermally-induced interchain association. This intrinsic instability can lead to syneresis on storage. Syneresis may also be pronounced when the network is subjected to extreme variations in temperature such as freezing and thawing. Syneresis tends to be more of a problem in gelled than in non-gelled systems because of the reduced mobility of the polysaccharide chains. Gelling agents are generally not interchangeable, since each provides its own characteristic gel texture. For example, starch jelly candies, as the name implies, can only derive their unique texture from retrogradation of specific starches, while the standard for dessert gels is the gelatine jelly.

Each of the above gelling systems has been discussed in detail in other chapters of this book. However, a few additional comments are appropriate. In order to use these systems effectively, specialist knowledge is required. This is particularly true for the ion-sensitive materials, carrageenan, alginate and pectin. Despite the fact that they have been researched extensively and the general principles governing their use are readily available, their effective utilization still often relies heavily upon technical input from the suppliers. Gelatine and agar are simpler to use but, again, inadequate knowledge of their chemistry and properties and the differences between the various grades available commercially can lead to problems. Starch is by far the most widely used food hydrocolloid but, surprisingly, our state of knowledge of starches in their various natural and modified forms in no way reflects their importance. Thus, selection of a starch for a specific application is often made without the proper technical considerations. Because of the abundance of starch, supply is not a problem. However, the chemical modifications required to give starches the desired in-use characteristics in many applications is giving rise to increasing cause for concern. The supply situation for the other gelling hydrocolloids is much less certain. Seaweed, the source of alginate, carrageenan, furcellaran and agar, is subject to the vagaries of nature and is not always available in sufficient quantities to satisfy market demand, which can lead to hydrocolloid shortages. The scarcity and high price of agar have severely curtailed its usage in foods. Today, manufacturers of gelling agents from seaweed obtain their raw materials from areas all over the world. Political instability in some of these areas gives rise to concern for the continued availability of these materials.

Over the past 20 years, bacteria have become an increasingly important source of polysaccharides. Dextran, a blood plasma extender and a key constituent of certain types of chromatographic materials, and xanthan gum, the most extensively tested and one of the most intensively researched food polysaccharides, are good examples of bacterial polysaccharides. Kelco Division of Merck & Co. Inc. has pioneered the development of bacterial polysaccharides and has now isolated in excess of 900 gum-forming bacteria. Although all of the gums produced by these bacteria are of scientific interest, most are not of commercial value because they do not offer significant advantages over existing alternatives. However, there are a few notable exceptions, namely rhamsan gum (S-194), a more efficient suspending agent than xanthan gum; Biozan welan gum (S-130), a thickener for oil-field applications with superior thermal stability to xanthan gum; and gellan gum (S-60), a novel gelling polysaccharide.

Gellan gum offers a potential solution to many of the problems that exist with current gelling agents. Being a fermentation product, it can be produced on demand and with consistent quality. Availability and variability are, thus, not concerns. It is functional at very low use levels and is, therefore, very efficient in many applications. Gellan gum can provide a range of gel textures as opposed to a single characteristic texture. Consequently, it can be used to mimic the texture of existing gelling agents or to create new textures. In today's food industry, a tool to create new textures and hence permit the creation of new, differentiated food products would be highly desirable. Gellan gum is not a difficult product to use and, although it is anticipated that eventually various forms of gellan gum will be available to industry, selection of the best form for a particular application will be determined by textural rather than by complex functional considerations. It should be stressed that gellan gum is in its early development stage, and much still has to be learned about its basic chemistry, properties, applications and, indeed, its ultimate utility in the market place. In this respect, it is encouraging that, for the first time in the history of hydrocolloids, commercial development is proceeding in parallel with the generation of fundamental scientific data on the product. The 'state of the art' on gellan gum is discussed in detail in the following sections.

2 STRUCTURE

The monosaccharide building units of gellan gum are glucose, glucuronic acid and rhamnose in the molar ratios of 2:1:1. The primary structure, a tetrasaccharide repeating unit, is shown in Fig. 1.[4,5] In the native form of the polysaccharide, there are approximately one and a half *O*-acyl groups per repeating unit. Originally the *O*-acyl substituent was thought to be *O*-acetyl, resulting in the various forms of gellan gum being referred to as high- and low-acetyl, and so on. Recent studies by Kuo *et al.*[6] suggest that gellan gum contains both *O*-acetyl and *O*-L-glyceryl substituents on the 3-linked glucose unit, the former tentatively assigned to the 6-position and the latter to the 2-position. Analysis of samples in our laboratory following this work has indicated that glycerate substitution predominates over acetate. Undoubtedly, these bulky glycerate groups hinder chain association and account for the significant changes in gel texture that accompany deacylation. Inclusion of bulkier substituents on the gellan gum backbone has an even more dramatic impact on properties. For example, welan gum and rhamsan gum, which have the same backbone as gellan gum but are substituted by mono- and disaccharide side-chains, respectively, have no similarity to gellan gum in solution behaviour.

The shape or conformation adopted by the gellan gum molecule as a result of this primary structure has been under investigation for a number of years using X-ray crystallography. The quality of early diffraction patterns[7,8] was not adequate to permit a detailed structural analysis. Subsequent work at Bristol,[9] although resulting in high-quality diffraction patterns from well-oriented polycrystalline samples, also failed to produce

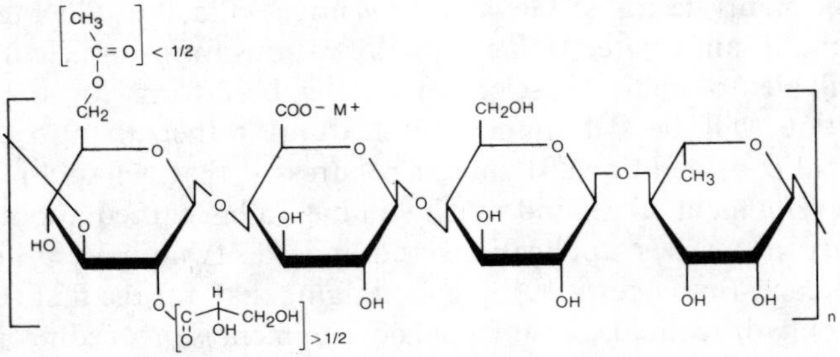

FIG. 1. Gellan gum tetrasaccharide repeat unit.

a structure consistent with the X-ray data. A recent re-examination of the Bristol data[10] has indicated that gellan gum forms an extended, intertwined, three-fold, left-handed parallel double helix. Molecular shape in the solid state is usually an indicator of molecular association in solution. The mechanism whereby gellan gum molecules associate in solution is believed to involve ion-mediated aggregation of double helices.[11,12]

3 SOURCE AND PRODUCTION

Gellan gum is produced by the bacterium *Pseudomonas elodea*. The gum is formed by inoculating a carefully formulated fermentation medium with this organism. The medium consists of a carbon source such as glucose, a nitrogen source and a number of inorganic salts. The fermentation is allowed to proceed under sterile conditions with strict control of aeration, agitation, temperature and pH.[13] When fermentation is complete, the viscous broth is pasteurized to kill the viable cells and then subsequently processed to recover the polysaccharide in either the fully acylated native form or the deacylated form, as shown in Fig. 2. Gelrite—gellan gum for microbiological media and related applications—and Kelcogel—food-grade gellan gum—are low acyl products. Gels from the native material can loosely be described as cohesive and elastic, while those from the deacylated materials are strong and brittle. Materials of intermediate acyl content,

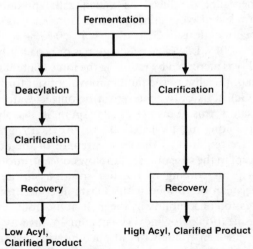

FIG. 2. Recovery procedures for gellan gum.

which can be obtained by careful control of the deacylation step, provide gel textures intermediate between those of the native and fully deacylated products.

4 COMMERCIAL AVAILABILITY

4.a Supply

Gellan gum is proprietary to Kelco.[14-18] Gelrite and various grades of low-acyl gellan gum are commercially available and are supplied as 60-mesh free-flowing powders. Native gellan gum is currently not available, whereas Kelcogel food-grade gellan gum is available for testing by the food industry.

4.b Regulatory Status

Gellan gum, at the time of writing, is not approved for use in foods. However, extensive safety studies have been conducted to secure this approval. Results from these studies (summarized in Table 1) have not given any cause for concern.

5 SOLUTION PROPERTIES

5.a Solubility

To date, most of the studies on gellan gum have focused on the low-acyl materials. These are produced as mixed salts, predominantly in the potassium form but also containing divalent ions such as calcium. Typical levels of the major cations in Gelrite are: Ca^{2+}, 0.75%; Mg^{2+}, 0.25%; Na^+, 0.70%; and K^+, 2.0%. Low-acyl gellan gum is only partially soluble in cold water. Solubility is increased by reducing the ionic content of the water and by conversion of the gum to the pure monovalent salt forms, but complete solubility of Gelrite is only achieved in deionized water using the pure monovalent salt forms. Low-acyl gellan gum is dissolved by heating aqueous dispersions to at least above 70°C. Progressively higher temperatures are required as the ionic strength of the aqueous phase is increased. Except in the case of Gelrite at low concentrations in the absence of ions, subsequent cooling of the hot solutions always results in gel formation. Gels can be formed with Gelrite in concentrations as low as 0.05%. Suppression of solubility by the inclusion of ions is a useful tool for the practical utilization of low-acyl gellan gum. In this way, the gum can be easily pre-dispersed in water without encountering hydration problems, and can be 'activated' simply by heating. Use of gellan gum in this manner is

TABLE 1
GELLAN GUM SAFETY STUDIES

Animal	Test	Level of safety
Rat	Oral LD_{50}	>5000 mg/kg
	4-hour dust inhalation, LC_{50}	>6 mg/litre nominal
	3-month dietary	No signs of toxicity at up to 6% of diet
	2-generation reproduction	No adverse effects noted at up to 6% of diet
	Teratology	No dose-related effects noted at up to 5% of diet
	In utero/chronic toxicity/ carcinogenicity	
	F_0 generation, 63 days	Possible effect on rate of weight gain of males: no significant other findings. Dietary up to 5% of diet
	F_1 generation, 104 weeks	Treatment produced no overt signs of toxicity and no effect on spontaneous tumor profile. Achieved intake ~3 g/kg/day in high dose group
Mouse	Chronic dietary, 96–98 weeks	No adverse reaction to treatment noted. Produced no adverse signs of toxicity and no effect on the spontaneous tumor profile. Fed up to 3% of diet, with achieved intake being ~5 g/kg/day in high dose group
Dog	Chronic dietary, 52 weeks	No toxicological effect noted after 52 weeks of feeding at concentrations of 3%, 4·5% or 5% (representing daily intakes of ~1·0, 1·5 or 2·0 g/kg/day) for both males and females
Monkey	28-day oral	No clinical signs or changes in blood chemistry noted at doses up to 3 g/kg
Rabbit	Skin/eye irritation	No skin irritation noted. No eye irritation noted
In vitro	Ames	No negative effect
	UDS	No negative effect
	V-79 cell mutagenesis	No negative effect
	Clinical organisms	No significant differences in growth noted when 50 organisms were grown on plates gelled with gellan gum and plates gelled with agar
Humans	23-day dietary	No adverse effects on plasma biochemistry, haematology or urinalysis. Dosed at 200 mg/kg/day

analogous to the use of native starches, which, being cold-water-insoluble granules, can be conveniently slurried in water prior to cooking. Solutions of gellan gum will react in the cold with mono- and divalent ions to form gels and, depending on the types and levels of ions, the resulting gels may not melt on heating. To circumvent this usually undesirable situation, it is recommended that, in applications where partial or complete pre-solution of gellan gum is unavoidable, the gellan gum be incorporated above 70°C. Bearing in mind the above considerations, there are a number of alternative ways of incorporating low-acyl gellan gum into a given system. It may be added alone or in combination with other dry or liquid ingredients to a cold mix that is then heated and cooled to induce gelation. Alternatively, it may be added to a mix that has been pre-heated above 70°C. The preferred method of addition is best determined by consideration of the ingredients in the formulation and processing conditions. The ions present in the system have a major impact on the quality of the final gel and, for best results, ions additional to those inherently present in the system may be required. These can also be added in the cold or after heating.

5.b Rheology of Solutions

Native gellan gum on heating and cooling in the presence of cations forms cohesive, elastic gels similar to those obtained by heating and cooling mixtures of xanthan gum and locust-bean gum. Since this texture does not appeal to most consumers, native gellan gum alone is not expected to see widespread utility as a gelling agent. However, when dispersed in cold water, it provides extremely high viscosities. A possible limitation to its use as a thickener is high sensitivity to salt. This effect is shown in Fig. 3, which compares the viscosities of 0·3% solutions of xanthan gum and native gellan gum at different concentrations of salt. The viscosities recorded are K values derived from the 'power-law' equation, $\eta = K\dot{\gamma}^{n-1}$, and are approximations of the viscosities at one reciprocal second. The well-known

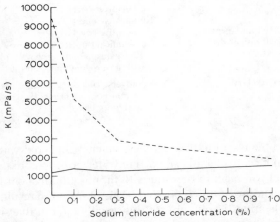

FIG. 3. K values for 0·3% (---) native gellan gum and (——) xanthan gum.

stability of xanthan gum viscosity to changes in salt concentration is apparent. In contrast, the viscosity of the native gellan gum displays a strong dependence on salt concentration. The native gellan gum solutions are highly thixotropic and the apparent high viscosities appear to be the result of the formation of a gel-like network. Similar thixotropic behaviour is observed when low concentrations of xanthan gum/locust-bean gum are dispersed in cold water.

6 GEL PROPERTIES

6.a Measurement of Gel Texture

Before considering the influence of ions in greater detail, a discussion of gel texture and gel texture measurement is necessary.[19,20] A wide array of terms, including elastic, brittle, hard, firm, cohesive and rubbery, are used to describe the sensory attributes of gels. In contrast, objective measurement of texture is largely restricted to measurement of gel strength. This parameter is widely used as an indicator of the quality of a particular gelling agent and is obtained using a variety of instruments. Although providing a gel strength value, different instruments do not always measure the same textural parameter. A Bloom gelometer, for example, provides an indication of the perceived firmness or modulus of a gel, while a Marine Colloids Gel Tester determines the force required to rupture the gel. Clearly, these parameters are not the same in textural terms. Gel strength measurements are thus only of limited value from a textural standpoint and, even when the conditions and instrument for measurement are specified, a full description of gel texture is not realized. For this reason, comparison of different gelling agents on the basis of gel strength can lead to highly misleading conclusions. Only in cases where a comparison of the qualities of different grades of a specific gelling agent is required is gel strength of real value. Gel strength is normally used, for example, as an indicator of the quality, or Bloom strength, of gelatine.

In the early 1960s the problem of correlating objective measurements with the sensory perception of texture in foods was addressed by General Foods.[21-23] The technique of texture profile analysis (TPA) that evolved from these studies has been further developed in our laboratory to permit routine measurement of gel texture. These developments consisted of interfacing an Instron 4201, used for compressing the gel, with a Hewlett–Packard 86B computer specially programmed for rapid acquisition and analysis of the data generated.

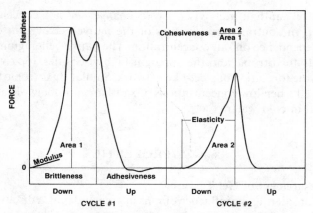

FIG. 4. Idealized texture profile for a gel.

As practised in our laboratory, the technique involves preparing gel discs by allowing hot solutions to cool in plastic ring moulds. The gel disc, after removal from the mould, is placed on a plastic plate with a roughened surface and compressed by the Instron crosshead, normally to 30% of its original height (70% compression), using a compression rate of 2 in/min. The crosshead is then withdrawn and the compression cycle is repeated. The textural parameters measured are shown on the idealized force–deformation curve in Fig. 4.

The modulus is the initial slope of the force–deformation curve. This is a measure of how the sample behaves when compressed a small amount. The modulus usually correlates very closely with a sensory perception of the sample's firmness. The measurement is analogous to squeezing a piece of fruit to determine its ripeness. Modulus is expressed in units of force per unit area (usually in pounds force per square inch or newtons per square centimetre).

Hardness is defined as the maximum force that occurs at any time during the first cycle compression. It may occur when the gel initially breaks (as is shown in the 'idealized' plot) or it may occur later in the test as the sample is flattened and deformed. In most cases, the hardness is correlated to the rupture strength of the material. It is similar to gel strength measured on a Marine Colloids Gel Tester and is expressed in units of force (pounds force or newtons).

The first significant drop in the force–deformation curve during the first compression cycle is defined as the brittleness. It is the point of first fracture or cracking of the sample. A gel that fractures very early in the compression cycle is considered to be more brittle or fragile than one that breaks later. Brittleness is measured as the percentage strain required to break the gel. As the number becomes smaller, it indicates a more brittle gel because the gel breaks at a lower strain level. Since the gels are normally compressed to a 70% strain level, the maximum value for brittleness is 70%. Such a high value would represent a gel, such as xanthan gum/locust-bean gum, that is not brittle at all.

Following the first compression cycle, the force is removed from the sample as the Instron crosshead moves back to its original position. If the material is at all sticky or adhesive, the force becomes negative. The area of this negative peak is taken as a measure of the adhesiveness of the sample. Obviously, the more sticky the product, the higher the adhesiveness value. There are no real units for this parameter. It is expressed in the internal integrator units of the computer.

As the second compression cycle is begun, the sample's elasticity is determined. By noting where the force begins to increase during this second compression cycle, a measure of the sample height may be obtained. If the sample returned to its original height, the elasticity would be 100%. The elasticity is a measure of how much the original structure of the sample was broken down by the initial compression. In sensory terms, it can be thought of as how 'rubbery' the sample will feel in the mouth. The units are dimensionless (a length divided by another length) and are usually expressed as a percentage.

Cohesiveness is measured by taking the total work done on the sample during the second cycle and dividing it by the work done during the first cycle. Work is measured as the area under the respective curves. This ratio is usually expressed as a percentage. Samples that are very cohesive will have high values and will be perceived as tough and difficult to break up in the mouth. As with elasticity, there are no units, since the quantity is dimensionless.

6.b Factors Affecting Gel Texture

Figure 5 shows the hardness and modulus of Gelrite (0·25%) as a function of calcium ion concentration at neutral pH. Gels were prepared by dispersing Gelrite in distilled water, heating to 70–75°C, adding the appropriate concentration of ions, and cooling. Although, in the absence of ions, only a very weak gel is formed, the hardness rapidly increases to a maximum at low Ca^{2+} levels and then gradually decreases as the ionic concentration is increased. The modulus shows a somewhat more symmetrical increase and decrease, and between 0·016% and 0·05%

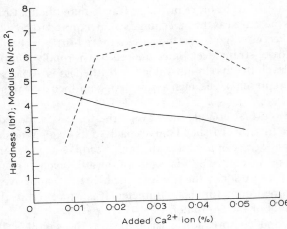

FIG. 5. Gelrite (0·25%); (——)) hardness and (---) modulus versus Ca^{2+} concentration ('as is' pH).

calcium (400 and 1250 ppm $CaCO_3$) modulus values between 5·5 and 6·5 N/cm² are obtained. Within this same range, the hardness varies between around 3·0 and 4·0 lb. Thus, provided products are formulated so that the concentration of free ions is in this range, modulus and hardness are fairly insensitive to changes in ionic composition brought about by variations in raw materials. It is noteworthy that since the bottom end of this range (400 ppm $CaCO_3$) is higher than the hardness encountered in most waters, some products may require the addition of ions for best results.

Figure 6 shows that Gelrite gels are brittle, with brittleness values between 30% and 40%. Some reduction in brittleness (higher values correspond to lower brittleness) is seen at lower ion concentrations. In contrast, elasticity (Fig. 6) is highest under these conditions and drops rapidly as ion level is increased to an almost constant value of around 10%. Other divalent ions, notably Mg^{2+}, have similar effects on gel texture. These trends are also observed with monovalent ions such as K^+ and Na^+. However, with these latter ions much higher concentrations are required. For example, maximum hardness requires an approximate 25-fold increase in the molar concentration of Na^+ and K^+ relative to Ca^{2+} or Mg^{2+}.[24]

Comparative studies in our laboratory on the influence of ionic concentration on κ-carrageenan (1·5%), agar (1·5%) and gelatine (5·2%) revealed, not unexpectedly, that the gel textural parameters for κ-carrageenan were strongly dependent on K^+ concentration. Potassium had

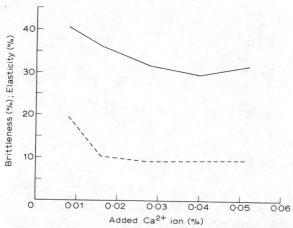

FIG. 6. Gelrite (0·25%); (——) brittleness and (– – –) elasticity versus Ca^{2+} concentration ('as is' pH).

less of an effect on gelatine texture and essentially no influence on agar texture. The level of potassium required to give the maximum modulus value for each of these hydrocolloids was then used in determining the relationship between modulus/hardness and gum concentration. The results are shown in Figs 9 and 10. Figure 7 compares the same parameters for Gelrite. As is indicated, Gelrite, at a given concentration, provides much higher hardness than the alternative hydrocolloids at the same use level.

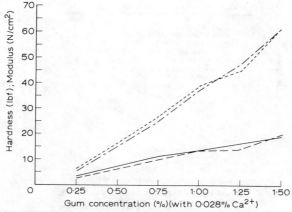

FIG. 7. Gelrite—Hardness (——) as is, (– – –) pH 4·0 and modulus (···) as is, (—·—) pH 4·0 versus gum concentration.

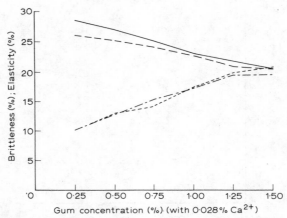

FIG. 8. Gelrite—Brittleness (——) as is, (– – –) pH 4·0 and elasticity (· · ·) as is, (—·—) pH 4·0 versus gum concentration.

The difference is substantially more striking when the moduli are compared. Gel strength measurements based on hardness or modulus would thus clearly reflect the greater efficacy of gellan gum. Figures 7, 9 and 10 also show the influence of pH on modulus and hardness. On reducing the pH to 4·0 by the addition of citric acid, no change in modulus is observed for all of the gelling agents. This is also true for Gelrite hardness and, at lower concentrations, for gelatine hardness. At higher concentrations of

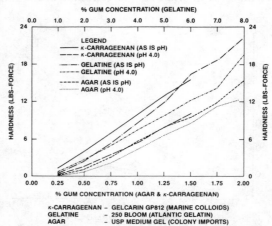

FIG. 9. Hardness versus gum concentration for κ-carrageenan in 0·63% K^+, agar in 0·70% K^+ and gelatine in 0·35% K^+.

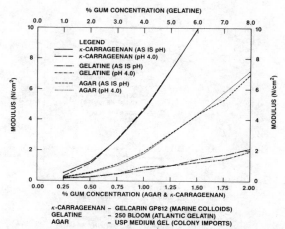

FIG. 10. Modulus versus gum concentration for κ-carrageenan in 0·63% K⁺, agar in 0·70% K⁺ and gelatine in 0·35% K⁺.

gelatine, lowering pH causes a reduction in hardness. Reduced hardness at pH 4·0 is observed for all concentrations of κ-carrageenan and agar, particularly in the case of the former.

The dependence of brittleness/elasticity on Gelrite concentration is shown in Fig. 8. The gels become slightly more brittle and elastic as concentration is increased. Brittleness of κ-carrageenan, agar and gelatine

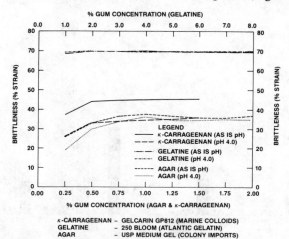

FIG. 11. Brittleness versus gum concentration for κ-carrageenan in 0·63% K⁺, agar in 0·70% K⁺ and gelatine in 0·35% K⁺.

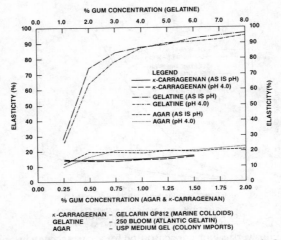

FIG. 12. Elasticity versus gum concentration for κ-carrageenan in 0·63% K⁺, agar in 0·70% K⁺ and gelatine in 0·35% K⁺.

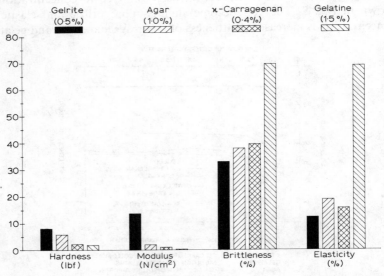

FIG. 13. Textural characteristics of Gelrite compared to agar, κ-carrageenan and gelatine.

remains almost constant above a threshold gum concentration, as shown in Fig. 11. Except in the case of carrageenan, where a reduction in pH increases brittleness by around 10%, change in pH has little effect on brittleness. Figure 12 shows that the elasticity of κ-carrageenan is largely independent of concentration and pH. The same figure indicates that pH has a minor influence on the elasticity values of agar and gelatine gels, which show a fairly rapid increase at low gum concentrations but tend to level off at higher concentrations.

The textural characteristics of gels made from typical use levels of agar (1·0%), κ-carrageenan (0·4%) and gelatine (1·5%) are compared to those of a 0·5% Gelrite gel in Fig. 13. The data indicate that Gelrite gels can be qualitatively described as hard, firm, brittle and non-elastic, while a totally opposite description would be appropriate for gelatine. The fact that gelatine has a brittleness of 70% indicates that the gel did not break under the 70% maximum compression employed in the texture profile test procedure. Agar and κ-carrageenan are slightly less brittle and more elastic than Gelrite but lower in hardness and, in particular, modulus.

6.c Melting and Setting Points

Key properties of gels are melting and setting points. For gellan gum, these strongly depend on ion concentration and type, and, to a lesser extent, on gum concentration, as shown in Figs 14 and 15. Most Ca^{2+} gels set between

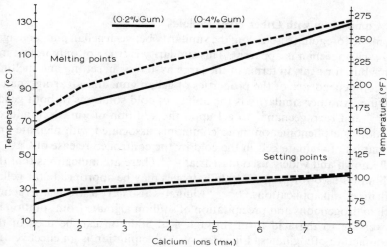

FIG. 14. Melting and setting points of Kelcogel gellan gum gels containing Ca^{2+}.

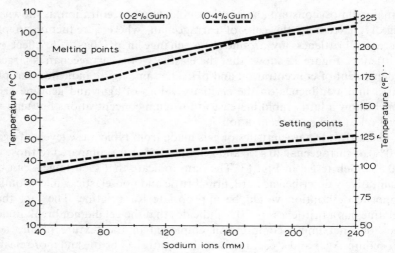

FIG. 15. Melting and setting points of Kelcogel gellan gum gels containing Na^+.

25 and 40°C while those with Na^+ set in the range 40–50°C. Ionic concentration has a marked influence on melting temperature and at lower ion levels the gels re-melt on heating, while at higher levels the gels do not melt below 100°C. This latter property is useful in applications in which heat-stable gels are required.

6.d Comparison with Other Hydrocolloids

From the foregoing discussion, the similarity between gellan gum and agar and κ-carrageenan is apparent. This similarity is true not only in textural terms but, for agar, in terms of the large hysteresis in setting and melting. The ion dependence of the properties of gellan gum and κ-carrageenan is striking. Another similarity is the ability of cold solutions of both gellan gum[24] and carrageenan[25] to gel upon the addition of ions. Ion-induced gelation is a phenomenon more commonly associated with alginate and formation of alginate gels in the cold by the controlled release of Ca^{2+} is well known and widely used industrially.[26] There are indications that the techniques used for preparing alginate gels may be appropriate for gellan gum in certain applications.[24,27,28] Alginates also have a strong interaction with hydrogen ions and precipitation of sodium alginate from solution by conversion to the acid form through acid addition can be used in the manufacture of alginates. Likewise, acid precipitation is an effective and alternative means of isolating gellan gum. The gels produced from gellan gum by the addition of hydrogen ions are extremely strong. The fact that gelation of gellan gum is induced by ions both upon cooling like carrageenan and in the cold like alginate has led to its being described as the 'missing link' in gelling hydrocolloids.[29] The application of this statement is that understanding the molecular basis of gellan gum gelation may help resolve the controversy over the mechanism of carrageenan gelation, which is generally, although not universally, believed to proceed through double-helix formation upon cooling,[30] in contrast to alginate, where the accepted mechanism is ion-induced dimeric association of polymer chains in the cold.[31,32] (See Chapter 2 and Chapter 3, Section 6.)

6.e Blends with Other Hydrocolloids

Combinations of more than one hydrocolloid are widely used in foods. In fact, use of blends rather than a single hydrocolloid could be considered standard practice. Non-gelling hydrocolloids are normally used together to obtain optimal rheology. In some cases, xanthan gum/guar gum being a good example, the combination is used to obtain a synergistic increase in viscosity. The changes in viscosity that result from blending non-gelling hydrocolloids can be predicted using the so-called log-mean blending law. Viscosity values that differ from those predicted indicate synergistic (greater than expected) or anti-synergistic behaviour. Combinations of gelling and non-gelling hydrocolloids or two or more gelling hydrocolloids are much more complex, and various possible network structures have been proposed.[33] Since not a great deal is known about mixed polysaccharide gelling systems, the concept of synergism applied to these systems is generally inappropriate. Despite this lack of fundamental knowledge, mixed polysaccharide gels are well established commercially. Combinations of κ-carrageenan or agar and locust-bean gum are perhaps the best examples. The inclusion of locust-bean gum provides textural modification and allows reduction of the total polymer concentration required for gel formation. Use of the gelling system xanthan gum/locust-bean gum to modify the textures of gels such as those from agar and carrageenan has also been suggested.[34] (See Chapter 8 for more details.)

The effect of both gelling and non-gelling hydrocolloids on the texture of low-acyl gellan gum gels has been extensively studied. Commonly used thickeners such as guar gum, locust-bean gum, xanthan gum, carboxymethylcellulose and tamarind gum, when added to gellan gum in progressively increasing amounts while maintaining a constant total gum concentration, cause a progressive reduction in hardness and modulus. Brittleness remains essentially constant, with an accompanying slight

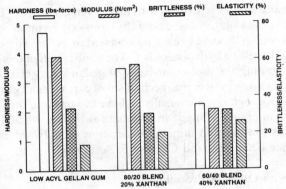

FIG. 16. Textural changes on blending low-acyl gellan gum with xanthan gum.

increase in elasticity. These effects are shown for low-acyl gellan gum/xanthan gum combinations in Fig. 16. For the key textural parameters, hardness, modulus and brittleness, these thickeners function essentially as inert diluents and the texture of the resulting blends is similar to the texture of low-acyl gellan gum alone at a concentration equivalent to that in the blend. It is common practice to include thickeners in gelled systems to reduce syneresis, improve freeze–thaw stability and, in some cases, eliminate unfavourable interactions between ingredients. Thus, some products formulated with gellan gum also require the presence of thickener.

The textural similarity between gels from low-acyl gellan gum, κ-carrageenan and agar has already been mentioned. Blends of low-acyl gellan gum and agar (0·50% and 0·25% total gum concentration) provide gels in 4 mM Ca^{2+} that show a decrease in hardness and modulus as the blend becomes richer in the agar component, the decrease being more pronounced at the higher gum concentration. Brittleness and elasticity values remain virtually constant around 34% and 14%, respectively. Similar blends of κ-carrageenan and low-acyl gellan gum in 0·16 M K^+ show a rapid drop in hardness (4·5 to 2) in going from 0·5% low-acyl gellan gum to 0·5% of an 80:20 blend containing 80% gellan gum. The hardness remains approximately constant at around 2 up to 20:80 gellan gum:κ-carrageenan and then rises to around 4 for carrageenan alone at 0·5%. Modulus falls sharply from 4·6 in going from 0·5% low-acyl gellan gum alone to the 80:20 blend and remains between 1·5 and 2 thereafter. At 0·25% total gum concentration the same trends are apparent but less pronounced. As in the case of the low-acyl/agar blends, the low-acyl/κ-carrageenan blends have a fairly constant brittleness and elasticity

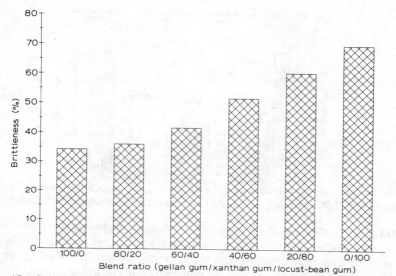

FIG. 17. Brittleness values for low-acyl gellan gum/xanthan gum/locust-bean gum gels in 0·004 M Ca^{2+}. Total gum concentration = 0·5% for all gels; xanthan gum: locust-bean gum ratio = 1:1.

irrespective of blend composition. The values for these latter textural parameters are almost identical to those for the low-acyl/agar blends. These data indicate that the characteristic brittle texture of low-acyl gellan gum gels cannot be substantially changed by progressive substitution with other brittle gelling agents.

This is not the case when low-acyl gellan gum is used in combination with the xanthan gum/locust-bean gum gelling system. As can be seen in Fig. 17, the gellan gum gels become less brittle as the blend becomes richer in xanthan gum/locust-bean gum.[35,36] Figure 18 indicates the other textural changes that take place. Hardness and modulus are reduced, while elasticity is increased. Similar textural changes are induced when locust-bean gum is replaced by other hydrocolloids, such as *Cassia* gum and konjak mannan, both of which are capable of interacting with xanthan gum to form gels with texture similar to those obtained from xanthan gum and locust-bean gum. The striking similarity between the textures of xanthan/locust-bean gum gels and native or high-acyl gellan gum is evident by comparing Figs 18 and 19. Consequently, it is perhaps not surprising, as indicated in Fig. 19, that blends of high- and low-acyl gellan gum provide textural variations similar to those obtained by blending different ratios of low-acyl gellan gum

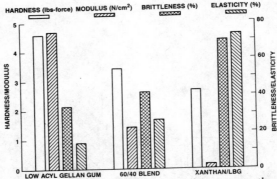

Fig. 18. Comparison of the textures of low-acyl gellan gum, xanthan gum/locust-bean gum (LBG) and low-acyl gellan gum/xanthan gum/locust-bean gum gels.

and xanthan gum/locust-bean gum. For labelling purposes, achievement of textural modification by blending gellan gum alone would clearly be preferred. Figure 20 demonstrates the range of different textures that can be obtained simply by using different concentrations of low-acyl gellan gum by itself or blended with different proportions of the high-acyl form.

If starch is excluded, gelatine is the most widely used gelling agent. In contrast to the strong, brittle, non-elastic gels produced by low-acyl gellan gum, gelatine gels have a low modulus or perceived firmness and are highly elastic and not brittle. Combinations of low-acyl gellan gum and gelatine offer another avenue to textural diversity. For example, addition of progressively increasing amounts of 250 Bloom type A gelatine to 0·25% low-acyl gellan gum, a typical in-use concentration, causes a gradual

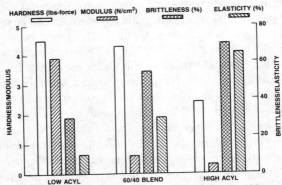

Fig. 19. Comparison of the textures of low-acyl, high-acyl and low-acyl/high-acyl gellan gum gels.

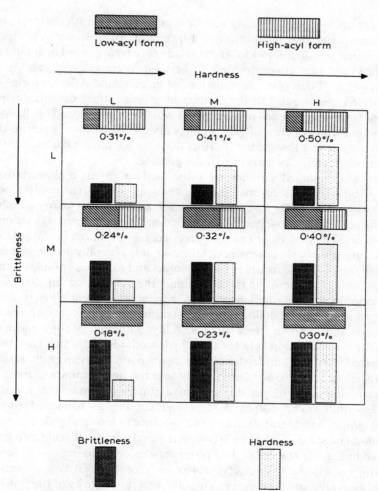

FIG. 20. Textural variations using the high-acyl/low-acyl gellan gum gelling system.

increase in hardness, modulus and elasticity and a gradual reduction in brittleness. Conversely, addition of low levels of low-acyl gellan gum, up to 0·05%, to replace up to around 1% gelatine in a 5% gelatine gel has no marked influence on the characteristic gelatine texture. It is conceivable, however, that the melting/setting temperatures of the gelatine may be advantageously increased by inclusion of low levels of the higher melting and setting low-acyl gellan gum. Low-acyl gellan gum/gelatine combinations may also help prevent toughening of gelatine gels upon refrigerated storage, allow low-grade gelatines to be upgraded in quality or permit lower gum concentrations to be used in certain applications. A recent patent[37] describes combinations of gelatine and different forms of deacylated gellan gum. In the context of gelatine, the excellent flavour release from low-acyl gellan gum gels is worthy of mention. This flavour release is a consequence of gellan gum's ability to structure water in the gelled state at very low use levels rather than the gels having the 'melt in the mouth' characteristics associated with gelatine.

Raw and modified starches are often used to impart a characteristic heavy-bodied consistency and, in some cases, a gel-like structure to foods. Cold-water-dispersible instant starches are available but many starches require cooking to cause gelatinization and generate the desired functional properties. The molecular changes that occur when starch is cooked and cooled are still poorly understood. However, it is generally accepted that a cooked starch paste consists of swollen intact and ruptured granules within a continuous aqueous phase containing the solubilized amylose and amylopectin, the two component polysaccharides of starch. When starch is used, additional hydrocolloids are often required to modify texture, reduce syneresis and improve freeze–thaw stability. Although possible mechanisms for the interaction between these hydrocolloids and starch have been suggested,[38] current understanding is again poor. Consequently, starch/hydrocolloid combinations are usually selected on an empirical basis. A standard indicator of performance of a starch system is the viscosity changes that occur during heating and cooling as measured on an amylograph. Amylograph data on the influence of low-acyl gellan gum on the modified starch, Col-Flo 67 (National Starch and Chemical Corp.), are shown in Fig. 21. The gellan gum produces a more rapid increase in initial build-up of viscosity. Viscosity subsequently remains fairly constant on further cooking and then rises less rapidly than the viscosity of the Col-Flo 67 alone upon cooling. These limited results are difficult to interpret on a fundamental basis but suggest that starch/gellan gum combinations are worthy of more detailed study. In practical terms, it has already been shown that in certain pudding and pie fillings the levels of modified starch can be reduced by a half by inclusion of around 0·1% Kelcogel gellan gum. In these products, the structure imparted by the gellan gum results in a firmer, shorter texture.

The compatibility of polysaccharides with proteins depends on a number of factors such as relative concentrations, pH, ionic strength, temperature,

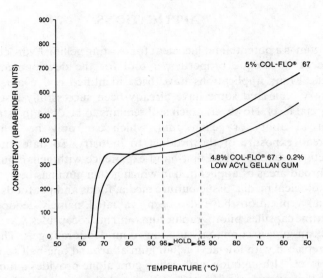

FIG. 21. Brabender amylographs for Col-Flo 67 with and without low-acyl gellan gum.

time and, in the case of foods, the nature of the other ingredients present. From a compatibility standpoint, model studies are thus of limited value and, as for starches, incompatibility problems that occur in products are most quickly solved by empiricism. Our experience with the interactions between gellan gum and proteins has also been similar. Limited model studies indicated that while low-acyl gellan gum was compatible at neutral pH with milk proteins, soy, egg albumen, whey and sodium caseinate, precipitation with all of these proteins occurred around pH 4. In contrast, it has been shown possible to produce a number of direct acidified and cultured dairy products using low-acyl gellan gum in combination with protective colloids such as guar gum and carboxymethylcellulose. Low-acyl gellan gum also shows good compatibility with proteins in non-acidified milk systems. The need to study protein/polysaccharide interactions under specific in-use conditions is emphasized by the fact that, although low-acyl gellan gum and gelatine combinations are potentially useful as already discussed, precipitation can occur under certain conditions. The observation that sodium caseinate and soy protein can prevent gelation of low-acyl gellan gum without causing precipitation also requires further investigation to define more fully the conditions under which this occurs.

7 APPLICATIONS

Gellan gum is a potential replacement for existing gelling hydrocolloids or, because of its unique properties, a tool for the development of new products. Many applications have been identified and evaluated on a laboratory scale and some have already been successfully tested at the commercial level. However, much still remains to be done to establish the commercial utility of gellan gum, which can only be achieved by widespread exposure of the product to industry. To date, only a few prospective users have had first-hand experience with gellan gum.

The broad areas of application in which gellan gum has been tested are microbiological media, tissue-culture media, foods and pet foods. Selected industrial applications have also been evaluated, namely deodorant gels, soft gelatine capsules, photographic films and microcapsules. Carrageenan/locust-bean gum is commonly used in room-deodorant gels. This system can be replaced with low-acyl gellan gum at around one-half to one-third the use level. Although low-acyl gellan gum alone provides a more brittle gel than the carrageenan/locust-bean gum gel, the texture produced by the latter can be simulated by inclusion of xanthan gum/locust-bean gum with the low-acyl gellan gum. When freeze–thaw stability is required, propylene glycol may be included in the gellan gum gel.[39] In photographic film and soft gelatine capsules, blends of low-acyl gellan gum and gelatine have been shown to function as replacements for gelatine alone. Although further work is required to substantiate more fully the benefits of these blends, the fact that the soft gelatine capsules do not dissolve in the gut when gellan gum is used is seen as an advantage in certain situations. This property has led to the suggestion that gellan gum may be a suitable vehicle for controlled release of certain drugs.[40] Microcapsules are usually formed by coacervation of gelatine and gum arabic. The suitability of low-acyl gellan gum and gelatine for coacervation and microencapsulation has been demonstrated in our laboratory[41] and, more recently, by other studies.[42]

The similarity between agar and Gelrite has led to the commercialization of Gelrite as an alternative to agar for microbiological media. Studies using 50 different bacterial species have shown that Gelrite can not only be substituted for agar in many routine media applications but can also give a higher degree of cell growth in certain situations.[43] Gelrite is particularly useful for the culture of thermophilic microorganisms, as Gelrite gels are thermostable and can withstand prolonged incubations at high temperatures.[44] In addition, acceptable gel strengths can be obtained using Gelrite at a lower level than agar and spreader colonies do not become too large. It has been demonstrated that Gelrite is superior to agar for cultivation of mesophilic *Methanobacterium* and *Methanobrevibacter* organisms, the advantages being greater gel strength, reduced preparation time of plates, drier media and, in the case of mesophilic *Methanobacterium* species, reduction of the extended incubation times required.[45] In all of these microbiological media applications, the high purity of Gelrite and the water-like clarity of the resulting gels are distinct additional advantages. To maintain perspective, it should also be pointed out that a few problems have also been encountered in certain media when Gelrite is used. Some media are difficult to re-melt, the plates are sometimes more difficult to streak than are agar plates, and the high setting temperatures can cause haemolysis of blood for blood plates.[43] Although failure to solve these problems may prevent the use of Gelrite in some situations, the benefits obtained from Gelrite should lead to its establishment as a better alternative to agar for many microbiological media.

Agar is currently the gelling agent of choice for plant tissue culture. However, the presence of impurities or inhibitory factors, including sulphur, in agar can adversely affect growth.[46] Gelrite offers a promising alternative to agar in this application because of its high purity. Recent studies on a limited number of plant species have shown that growth is indeed improved on media containing Gelrite.[47] In addition, Gelrite could be used at one-fifth of the agar use level, showed good resistance to contamination by moulds, was easily washed from the plant tissue for transplantation, and allowed clear observation of root and tissue development. Despite these encouraging results, the full potential of Gelrite for plant tissue culture requires a more extensive study involving a larger number of plants.

Pet foods are a major application for alginate and carrageenan. The types of pet foods using these materials consist of comminuted meat bound in a gel matrix and are widely available in Europe and Australia, usually in canned form. In the United Kingdom and Australia, these products are also sold in non-sterile containers such as sausage casing and are sometimes referred to as 'dog-brawn'. These brawns, which contain preservatives, have a short shelf life and are produced by a number of small manufacturers supplying local markets. A desirable feature of the canned pet foods is the presence of meat-like chunks. These are sometimes produced in the cold by the reaction of alginate with calcium ions to produce a gel structure that provides structural integrity to the chunks during filling and in the early stages of retorting. Alternatively, the alginate may be used to bind the comminuted meat into a continuous gelled block. Again the gel is formed in

TABLE 2
MAJOR APPLICATION AREAS FOR GELLAN GUM

Major food area	Typical products	Gelling agents/thickeners currently used
Confectionery	Starch jellies, pectin jellies, fillings, marshmallow	Pectin, starch, gelatine, agar, xanthan gum/locust-bean gum
Jams and jellies	Reduced-calorie jams, imitation jams, bakery fillings, jellies	Pectin, alginate, carrageenan
Fabricated foods	Fabricated fruits, vegetables, meats	Alginate, carrageenan/locust-bean gum
Water-based gels	Dessert gels, aspics	Gelatine, alginate, carrageenan
Pie fillings and puddings	Instant desserts, canned puddings, pre-cooked puddings, pie fillings	Starch, carrageenan, alginate
Pet foods	Canned meat chunks, gelled pet foods	Alginate, carrageenan/locust-bean gum
Icings and frostings	Bakery icings, canned frostings	Agar, starch, pectin, xanthan gum/guar gum
Dairy products	Ice-cream, gelled milk, yoghurt, milkshakes	Carrageenan, gelatine, alginate, guar gum, locust-bean gum

the cold by standard techniques.[26] Despite the advantage of cold make-up, use of alginate in pet foods is now largely restricted to the production of chunks and other shaped pieces. Low grade κ-carrageenan is now the product of choice for binding the meat in a gelled matrix. More specifically, blends of κ-carrageenan, locust-bean gum and potassium chloride are used, the carrageenan/locust-bean gum system providing a less brittle gel than carrageenan alone. Many varieties of pet food exist, each with its own particular blend of gelling agents carefully selected to provide the desired gel texture. When strong, robust gels are required, the level of carrageenan/locust-bean gum/potassium chloride may be as high as 1·3%; levels of around 0·4% produce a weak, easily broken gel, which is preferred for some varieties. Low-acyl gellan gum has been successfully evaluated at plant scale in a number of canned pet foods and in dog-brawn. As for the carrageenan system, the gellan gum use levels vary depending on the final texture required. Although gellan gum can be used alone, it is more commonly used with other hydrocolloids. For example, guar gum and xanthan gum are useful for providing syneresis control, while xanthan gum and locust-bean gum combinations or high-acyl gellan gum reduce gel brittleness. As a general rule of thumb, low-acyl gellan gum blends can

TABLE 3
KELCOGEL VERSUS AGAR IN JAPANESE FOODS

Product	Kelcogel use level (%)	Agar use level (%)	Texture[a]								
			Kelcogel					Agar			
			h	m	b	e	c^b	h	m	b	e	c^b
Mitsumame jelly cubes	0.4	1.4	4.0	6.2	27.3	16.5	4.7	3.2	2.2	30.7	15.1	6.8
Hard red bean jelly	0.5	1.5	6.0	13.0	26.8	25.1	9.6	7.6	6.0	37.0	41.6	10.4
Soft red bean jelly	0.1	0.35	1.0	2.1	24.4	42.5	12.2	1.0	1.3	34.0	46.9	13.5
Tokoroten noodles	0.4	0.8	3.1	2.4	34.6	22.9	8.3	3.1	1.4	33.9	24.6	9.7

[a] Texture determined by the texture profile analysis procedure described in this chapter.
[b] h = hardness (lb), m = modulus (N/cm^2), b = brittleness (%), e = elasticity (%), c = cohesiveness (%).

replace κ-carrageenan blends in gelled pet foods at around half the use level. Thus, in a laboratory dog-brawn formulation, a blend of Kelcogel gellan gum, xanthan gum and locust-bean gum at 0.45% in combination with potato starch gave a similar product to a carrageenan/locust-bean gum blend at 1.0%, while in a prototype canned pet food 0.75% of a Kelcogel gellan gum/guar gum blend was able to replace carrageenan/locust-bean gum/potassium chloride at 1.3%.

The potential food uses of gellan gum have already been described[24,48] and the major areas of application are listed in Table 2. Previously published food formulations[48] indicate that low-acyl gellan gum is functional at very low use levels. These, however, were based on low-acyl, unclarified gellan gum. More recent studies have shown that Kelcogel food-grade gellan gum can be used in these formulations at around two-thirds of the low-acyl, unclarified gellan gum use level. In products using κ-carrageenan or agar as the primary gelling agent, Kelcogel can be used at around one-half to less than one-third of the use level to provide similar texture. The relative performance of Kelcogel and agar in selected Japanese foods is shown in Table 3. The lower use levels of Kelcogel and the textural similarities of the end products are apparent. The data show that the Kelcogel products are, in fact, firmer than their agar counterparts. This increased firmness is organoleptically desirable. The tokoroten noodles with Kelcogel also contain 0.4% of a 50:50 blend of xanthan gum and locust-bean gum to improve resilience. These examples illustrate that, while low-acyl gellan gum can be used alone in many products, further textural variations may be achieved by inclusion of xanthan gum/locust-bean gum combinations. One of the major advantages of gelatine is its so-called 'melt-in-the-mouth' characteristics. Low-acyl gellan gum does not melt in the mouth but, as already mentioned, the low use levels and the particular nature of the gel network result in outstanding flavour release. Thus, water dessert gels made with low-acyl gellan gum, although unlike the traditional gelatine gels in texture, are excellent appealing products that, upon consumption, resemble a beverage rather than a solid. The ability to produce such novel products suggests that gellan gum will be a valuable tool for the creation of new product concepts as well as a replacement for current gelling agents. The excellent clarity of gellan gum is also a major advantage in certain applications.

REFERENCES

1. Glicksman, M., *Gum Technology in the Food Industry*. Academic Press, New York, 1969.
2. Whistler, R. L. & BeMiller, J. N., *Industrial Gums*, 2nd edn. Academic Press, New York, 1973.
3. Ulrich, P. T., unpublished data.
4. O'Neill, M. A., Selvendran, R. R. & Morris, V. J., Structure of the acidic extracellular gelling polysaccharide produced by *Pseudomonas elodea*. *Carbohydrate Res.*, **124** (1983) 123.
5. Jansson, P. E., Lindberg, B. & Sandford, P. A., Structural studies of gellan gum, an extracellular polysaccharide elaborated by *Pseudomonas elodea*. *Carbohydrate Res.*, **124** (1983) 135.
6. Kuo, M. S., Dell, A. & Mort, A. J., Identification and location of L-glycerate, an unusual acyl substituent in gellan gum. *Carbohydrate Res.*, **156** (1986) 173.
7. Carroll, V., Miles, M. J. & Morris, V. J., Fibre-diffraction studies of the extracellular polysaccharide from *Pseudomonas elodea*. *Int. J. Biological Macromolecules*, **4** (1982) 432.
8. Carroll, V., Chilvers, G. R., Franklin, D., Miles, M. J., Morris, V. J. & Ring, S. G., Rheology and microstructure of solutions of the microbial polysaccharide from *Pseudomonas elodea*. *Carbohydrate Res.*, **114** (1983) 181.
9. Upstill, C., Atkins, E. D. T. & Atwool, P. T., Helical conformations of gellan gum. *Int. J. Biological Macromolecules*, **8** (1986) 275.
10. Chandrasekaran, R., Millane, R. P., Arnott, S. & Atkins, E. D. T., The crystal structure of gellan gum. *Carbohydrate Res.*, **175** (1988) 1.
11. Rinaudo, M., Gelation of ionic polysaccharides. In *Gums and Stabilisers for the Food Industry—4*, ed. G. O. Phillips, D. J. Wedlock & P. A. Williams. IRL Press, Oxford, 1988, p. 119.
12. Chandrasekaran, R., Puigjaner, L. C., Joyce, K. L. & Arnott, S., Cation interactions in gellan: an X-ray study of the potassium salt. *Carbohydrate Res.*, **181** (1988) 23.

13. Kang, K. S., Veeder, G. T., Mirrasoul, P. J., Kaneko, T. & Cottrell, I. W., Agar-like polysaccharide produced by a *Pseudomonas* species: production and basic properties. *Appl. Environ. Microbiol.*, **43** (1982) 1086.
14. Kang, K. S., Colegrove, G. T. & Veeder, G. T., Heteropolysaccharides produced by bacteria and derived products. European Patent 0 012 552 (1980).
15. Kang, K. S., Colegrove, G. T. & Veeder, G. T., Deacetylated polysaccharide S-60. US Patent 4 326 052 (1982).
16. Kang, K. S. & Veeder, G. T., Polysaccharide S-60 and bacterial fermentation process for its preparation. US Patent 4 326 053 (1982).
17. Kang, K. S. & Veeder, G. T., Fermentation process for preparation of polysaccharide S-60. US Patent 4 377 636 (1983).
18. Kang, K. S., Veeder, G. T. & Colegrove, G. T., Deacetylated polysaccharide S-60. US Patent 4 385 125 (1983).
19. Mitchell, J. R., Rheology of gels. *J. Texture Studies*, **7** (1976) 313.
20. Mitchell, J. R., The rheology of gels. *J. Texture Studies*, **11** (1980) 315.
21. Szczesniak, A. S., Classification of textural characteristics. *J. Food Sci.*, **28** (1963) 385.
22. Friedman, H. H., Whitney, J. E. & Szczesniak, A. S., The texturometer—a new instrument for objective texture measurement. *J. Food Sci.*, **28** (1963) 390.
23. Szczesniak, A. S., Brandt, M. A. & Friedman, H. H., Development of standard rating scales for mechanical parameters of texture and correlation between the objective and the sensory methods of texture evaluation. *J. Food Sci.*, **28** (1963) 397.
24. Sanderson, G. R. & Clark, R. C., Gellan gum, a new gelling polysaccharide. In *Gums and Stabilisers for the Food Industry—2*, ed. G. O. Phillips, D. J. Wedlock & P. A. Williams. Pergamon Press, Oxford, 1984, p. 201.
25. Whitt, H. J., Carrageenan, Nature's most versatile hydrocolloid. In *Biotechnology of Marine Polysaccharides: Third Annual Massachusetts Institute of Technology Sea Grant College Program Lecture and Seminar Series*, ed. R. R. Colwell, E. R. Pariser & A. J. Sinskey. Hemisphere Publishing Corp., New York, 1984.
26. Bryden, W. & Sanderson, G. R., Structured foods with the algin/calcium reaction. *Kelco Bulletin F-83*. Kelco Division of Merck & Co. Inc., San Diego, California, 1982.
27. Baird, J. K. & Shim, J. L., Non-heated gellan gum gels. US Patent 4 503 084 (1985).
28. Baird, J. K. & Shim, J. L., Non-heated gellan gum gels. US Patent 4 563 366 (1986).
29. Smidsrod, O., personal communication, 1984.
30. Rees, D. A., Williamson, F. B., Frangou, S. A. & Morris, E. R., Fragmentation and modification of ι-carrageenan and characterization of the polysaccharide order–disorder transition in solution. *Eur. J. Biochem.*, **122** (1982) 71.
31. Grant, G. T., Morris, E. R., Rees, D. A., Smith, P. J. C. & Thom, D., Biological interactions between polysaccharides and divalent cations: the egg-box model. *FEBS Letters*, **32** (1973) 195.
32. Morris, E. R., Rees, D. A., Thom, D. & Welsh, E. J., Conformation and intermolecular interactions of carbohydrate chains. *J. Supramolecular Structure*, **6** (1977) 259.

33. Brownsey, G. J. & Morris, V. J., Mixed and filled gels—models for foods. In *Food Structure—Its Creation and Evaluation*, ed. J. M. V. Blanchard & J. R. Mitchell. Butterworth, London, 1988.
34. Kovacs, P., Useful incompatibility of xanthan gum with galactomannans. *Food Technol.*, **27** (1973) 26.
35. Sanderson, G. R. & Clark, R. C., S-60 in food gel systems. Japanese Patent 193535/82.
36. Clare, K., Clark, R. C., Pettitt, D. J. & Sanderson, G. R., Low acetyl gellan gum blends. US Patent 4 647 470 (1987).
37. Shim, J. L., Gellan gum/gelatin blends. US Patent 4 517 216 (1985).
38. Ring, S. G. & Morris, V. J., Starches—useful interactions. *Food*, **8** (1986) 19.
39. Colegrove, G. T., K9A50 for use in aqueous gel systems. *Commercial Development Bulletin CD-30*. Kelco Division of Merck & Co. Inc., San Diego, California, 1983.
40. Zatz, J. L., personal communication, 1985.
41. Lindroth, T. A., Gellan gum in microencapsulation. *Commercial Development Bulletin CD-31*. Kelco Division of Merck & Co. Inc., San Diego, California, 1983.
42. Chilvers, G. R. & Morris, V. J., Coacervation of gelatin–gellan gum mixtures and their use in microencapsulation. *Carbohydrate Polymers*, **7** (1987) 111.
43. Shungu, D., Valiant, M., Tutlane, V., Weinberg, E., Weissberger, B., Koupal, L., Gadebusch, H. & Stapley, E., Gelrite as an agar substitute in bacteriological media. *Appl. Environ. Microbiol.*, **46** (1983) 840.
44. Lin, C. C. & Cassida, L. E. Jr, Gelrite as a gelling agent in media for the growth of thermophilic microorganisms. *Appl. Environ. Microbiol.*, **47** (1984) 427.
45. Harris, J. E., Gelrite as an agar substitute for the cultivation of mesophilic *Methanobacterium* and *Methanobrevibacter* species. *Appl. Environ. Microbiol.*, **50** (1985) 1107.
46. Colegrove, G. T., Agricultural applications of microbial polysaccharides. *Ind. Eng. Chem. Prod. Res. Dev.*, **22** (1983) 456.
47. Shimomura, K. & Kamada, H., The role of medium gelling agents in plant tissue culture. *Plant Tissue Culture*, **3** (1986) 38.
48. Sanderson, G. R. & Clark, R. C., Gellan gum. *Food Technol.*, **37** (1983) 63.
49. Anderson, D. M. W., Brydon, W. G. & Eastwood, M. A., The dietary effects of gellan gum in humans. *Food Additives and Contaminants*, **5**(3) (1988) 232–49.

Chapter 7

GELATINE

F. A. JOHNSTON-BANKS

*Gelatine Products Limited,
Sutton Weaver, Runcorn, Cheshire WA7 3EH, UK*

1 INTRODUCTION

Of all the hydrocolloids in use today surely none has proved as popular with the general public and found favour in as wide a range of food products as gelatine. A sparkling, clear dessert jelly has become the archetypal gel and the clean melt-in-the-mouth texture is the characteristic that has yet to be duplicated by any polysaccharide. Despite its apparently unfashionable status, more gelatine is sold to the food industry than any other gelling agent. It is relatively cheap to produce in quantity, and there is a ready supply of suitable raw material.

Being a protein it is nutritious, containing essential amino acids not found in quantity in many foods (e.g. lysine); this was confirmed by the European Commission, who decided gelatine was to be classed as a food in its own right and should not be subject to an 'E' number registration.

It possesses a relatively low melting point when compared to polysaccharides and thus it is not often used in dilute solution at ambient temperatures; instead, it is normally reserved for ready-to-eat food products residing in the refrigerator or cooling cabinet (e.g. mousse, trifles, etc.). It can be sold as a dry ingredient (crystal table jellies) or used in a more concentrated gel form (fruit pastilles, marshmallow).

Unfortunately the analysis and an understanding of gelatine has proved difficult owing to its heterogeneity, and this has therefore led to the study of its precursor, the ubiquitous protein collagen.

Gelatine is commercially derived from collagen by a controlled acid or alkaline hydrolysis; therefore, the source, age and type of collagen each influence the properties of the gelatines derived from them. This chapter will attempt to relate the functional properties of gelatine with its known structural features, and highlight how these can be selected by the gelatine manufacturer to optimize the gelatine's effectiveness in particular product applications.

2 STRUCTURE

2.a Collagen Structure

It is little appreciated that collagen constitutes as much as 30% of total human protein, with similar proportions being found in many other animals. It is widespread in both vertebrates and invertebrates, from primitive marine worms to mammals, differing somewhat in amino-acid composition but providing the same function—strength and support for the tissues and organs of the animal concerned.

The collagen molecule exists as a triple helix, comprising three discrete α-chains that adopt a three-dimensional structure in which the compact glycine residues occupy every third site in the chain, leaving a relative space that hydroxyproline and proline residues on the other two chains can occupy: an ideal geometry for interchain hydrogen bonding. The imino acids also impart a considerable degree of rigidity to the molecule encouraging the stability of long-range order. It is also probable that interstitial water molecules play a crucial role, acting as hydrogen-bond bridges in order to increase the stability of the helix.[1]

A cooperative effort by many laboratories established the primary structure of the $\alpha 1(I)$ chain and described its sequence of 1044 residues. (The sequence of the $\alpha 1(III)$ chain was later established using DNA sequence data.) The collagen molecule can be characterized by a number of distinguishing features:

(a) It has a high percentage (33%) of glycine (Gly).
(b) It has a high proportion of the imino acids proline (Pro) and hydroxyproline (Hyp) (22%).
(c) It contains the rare amino acid hydroxylysine (Hyl) (1%).
(d) The molecule features repeating (Gly-X-Y) triplets,[2] where a high proportion of 'X' and 'Y' are the imino acids proline and hydroxyproline, although this does not apply to the short non-helical terminal regions.

The X- and Y-positions are not occupied randomly but are subject to a definite distribution. The non-helical terminal regions (telopeptide zones) at each end of the molecule are implicated in cross-linking the single α-chains, both inter- and intramolecularly, giving rise to the observed three-dimensional structure.

Hydroxyproline is only found in position Y, as is hydroxylysine, while proline can be found in either. Several other amino acids are predominantly found in either the X- or Y-positions.[3] Examples of this are glutamic acid and leucine, which are associated with the X-position, while, for example, arginine favours the Y-position in the Gly-X-Y triplet. Both the polar and non-polar amino acids tend to be distributed preferentially in certain areas of the chain, producing distinct hydrophobic and hydrophilic regions that give rise to surface-active properties.

The triple-helix model for collagen may lead to the assumption that the three α-chains are identical, but for most types of collagen this is not the case. For example, type (I) consists of two identical α-chains termed α1(I), and one other of a slightly different composition termed α2(I). Type (I) and other collagens can be distinguished by variations in their amino-acid composition.

At least 10 collagen types have been identified in the last decade, and another four or so have tentatively been recorded. Collagen types I, II, III and V exist as continuous helices with three α-chains. Collagen types IV and VI–IX also have triple-helical regions but these are separated by extensive linear domains.

The two types occurring in sufficient quantity to be important to gelatine technology are listed in Table 1.[4] The best known are types I and III, from which all commercial gelatines are produced. Ossein gelatines are derived from bone sources and so consist solely of type I collagen.[5] Hide and pigskin gelatines are derived from collagens that will contain a small proportion of type III. However, in most skin tissues, this level is approximately 10–20%, and for most applications it is of little significance to the user.

It is known that the collagens in bone are cross-linked to form a three-dimensional network, while those from skin tend towards a two-dimensional structure. This can affect pretreatment methods for gelatine extraction, and thus the physical properties of the resultant products.

TABLE 1
α-CHAIN COMPOSITION OF COLLAGEN TYPES USED IN GELATINE MANUFACTURE

Type	Molecular chain composition and notation	Distribution
I	$\{2[\alpha1(I)], [\alpha2(I)]\}$	Skin, bone, cartilage, etc.
III	$3[\alpha1(III)]$	Skin (not present in bone)

TABLE 2
AMINO ACID DISTRIBUTION FOR THE THREE α-CHAINS FOUND IN THE COLLAGEN PRECURSORS OF GELATINE

(After Miller, E. J., in *Chemistry of the Collagens and Their Distribution*, from Extracellular Matrix Biochemistry, ed. K. Piez & M. Reddi. Elsevier Science Publishers, Amsterdam, 1988, p. 66.)

Residue	Residues per 1000		
	α1 (type I)	α2 (type I)	α1 (type III)
3-Hydroxyproline	1	2	0
4-Hydroxyproline	108	93	125
Proline	124	113	107
Lysine	26	18	30
Hydroxylysine	9	12	5
Glycine	333	338	350
Cysteine	0	0	2
Serine	34	30	39
Alanine	115	102	96
Histidine	3	12	6
Valine	21	35	14
Methionine	7	5	8
Isoleucine	6	14	13
Leucine	19	30	22
Arginine	50	50	46
Phenylalanine	12	12	8
Aspartic acid	42	44	42
Threonine	16	19	13
Glutamic acid	73	68	71
Tyrosine	1	4	3

Although the α-chains are quite similar regarding their physical properties, small but significant differences in chemical composition can be observed. Table 2 gives the differences in amino-acid content between the three single α-chains found in types I and III collagen. As can be seen, there are substantial differences between the three α-chains, especially regarding those amino acids present in lower proportions. It is not surprising, therefore, that variations in amino-acid analyses occur from one gelatine to another. This can be due to:

(1) differences in the degree of hydroxylation of lysine groups;
(2) differences in levels of hydroxyproline between α1(I) and α2(I);
(3) differences in levels of valine, methionine, leucine, isoleucine and

histidine (which can increase the hydrophobic nature of zones in the molecule);
(4) the presence of cysteine in $\alpha 1(III)$.

When collagens have been subjected to a long chemical pretreatment, they give rise to gelatines relatively low in amino acids originating from the telopeptide residues. This has important implications for photographic gelatine manufacture, where even traces of some telopeptide material (e.g. cysteine) can have dramatic effects on its characteristics. The presence of cysteine cross-linkages in type III collagen compensates for the low hydroxylysine content in the terminal regions (normally utilized in cross-link formation), ensuring the stability of the fibrils.

2.b Collagen Cross-linking

In most tissues, collagen is arranged in bundles of four or five molecules to form structures known as fibrils. These are further associated with other fibrils nearby to form further bundles or larger diameter. The collagen chains are stabilized by the formation of intermolecular cross-links formed from lysine or hydroxylysine residues in the telopeptide zones interacting with adjacent lysine or hydroxylysine residues in similar regions, or on the helix, to produce stable covalent cross-links; this ensures insolubility in water and immunity from attack by most enzymes. (Collagenase is the only common enzyme able to degrade collagen under mild conditions, while gelatine is attacked by most proteases.) Various interchain bonds have been proposed for collagen, e.g.

(1) head to tail bonds, which fix the overlap of α-chains;
(2) reducible cross-linkages, which stabilize the aggregation of these chains; also termed side-to-side bonding.

The degree of cross-linking has profound effects on the pretreatment process required for gelatine manufacture. In young foetal tissues, the concentration of cross-links is at a minimum, and much of the collagen is found to be soluble in mild acid or salt buffer. As the collagen matures, the proportion of the soluble fraction (procollagen) reduces markedly as the cross-linking progresses. In mature collagens, the pretreatment process proceeds at a considerably slower rate. Young, immature pigskin collagen is much easier to convert than that found in mature bovine hide, and this is reflected in the different approach to pretreatment taken by the manufacturer.

The nature of these mature cross-links has been under dispute for some time, although recently a new amino acid derived from lysine has been tentatively implicated as one constituent.[6]

2.c Gelatine Structure
2.c.1 Primary Structure
The primary structure of gelatine closely resembles the parent collagen. Small differences are due to raw material sources together with pretreatment and extraction procedures. These can be summarized as follows.

(a) Partial removal of amide groups of asparagine and glutamine, resulting in an increase in the aspartic and glutamic acid content. This increases the number of carboxyl groups in the gelatine molecule and thus lowers the isoelectric point. This is described in detail later, but it is sufficient to say here that the degree of conversion is related to the severity of the pretreatment process. The mildly acidic pretreatment required for immature pigskin will remove only a few amide groups, while liming treatment will result in almost complete conversion of amide.

(b) Conversion of arginine to ornithine in more prolonged treatments experienced during long liming processes. This takes place by removal of a urea group from the arginine side-chain. The slow rate of this reaction can be illustrated by examining gelatines produced by liming for 8–10 weeks, where only 3% of the arginine is converted. Gelatines pretreated for longer periods, e.g. 4–8 months, are shown to have a higher rate of conversion, about 34%.[7] This feature has been proposed as one explanation of the greater surface-active properties found in pigskin gelatines, as the guanidino groups of arginine seem to be related to these functional properties.[8]

(c) There is a tendency for trace amino acids, such as cysteine, tyrosine, isoleucine, serine, etc., to be found in lower proportions than in their parent collagens. This is due to the inevitable removal of some telopeptide during cross-link cleavage, which is then lost in the pretreatment solutions.

Comparisons of the amino-acid profiles of gelatines from various sources were described by Ward and Courts.[9]

2.c.2 Secondary Structure and Molecular Weight
For many years the molecular weight has been quoted to explain the various aspects of gelatine behaviour in solution and in gels. Relating

physical properties to the average molecular weight of a polydisperse colloid is, at best, difficult. Gelatine is not completely polydisperse, but has a definite molecular weight distribution pattern corresponding to the α-chain and its oligomers. One to eight oligomers may be detected in solution, but it is possible that higher numbers exist. Doublets, known as β-chains, are formed from both α1- and α2-chains, giving rise to β11- and β12-molecules. Oligomers of three α-chains will mainly exist as intact triple helices, but a certain proportion will exist as extended α-polymers bonded randomly by end-to-end or side-to-side bonds. The structure of oligomers of greater than four α-chain units obviously becomes increasingly more complex and difficult to interpret.

It is necessary to separate these molecular-weight fractions and analyse them quantitatively in order for a comparison with physical properties to be obtained. Polyacrylamide gel electrophoresis (PAGE) can be used to obtain highly accurate molecular-weight spectra of both commercial and laboratory gelatines, giving quantitative separation. The study of these spectra can begin to relate physical properties to molecular structure.[10-14] Other analytical methods have also recently been tried with some success.[12,15,16]

A modern picture of gelatine structure can be seen in Fig. 1, a simulated

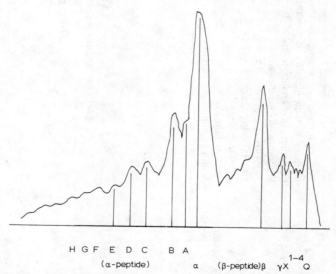

FIG. 1. Molecular weight distribution showing the major structural components of gelatine (see also Table 3). Electrophoresis method after Tomka.[13]

TABLE 3
THE MAJOR MOLECULAR FRACTIONS OF GELATINE AS SEEN IN FIG. 1

Molecular fraction	Description
Q	Very high molecular weights, of $15-20 \times 10^6$ daltons and thought to be branched in character owing to their inability to penetrate the gel successfully
1–4	Oligomers of α-chains, levels of five to eight
X	Oligomer of four α-chains
γ	285 000 daltons, i.e. $3 \times \alpha$-chain
β	190 000 daltons, i.e. $2 \times \alpha$-chain
α	95 000 daltons
A-peptide	86 000 daltons
α-, β- and γ-peptides	Seen as tailing their parent peaks

TABLE 4
MOLECULAR FRACTIONS FOUND IN HIGH-GRADE GELATINES

Source	Fractions (%)									Bloom (g)	Viscosity (mpoise)
	<A	A	α	β-pept.	β	γ-pept.	γ–X	1–4	Q		
Limed ossein	20	9	28	5	11	4	12	6	5	250	65
Limed hide	28	4	30	6	13	3	8	4	3	250	57
Acid pigskin	35	4	10	12	11	6	11	7	4	250	55

electrophoretogram of a typical high-bloom, high-viscosity gelatine (and see Table 3).

Differences can be detected between commercial gelatines from the different raw materials and these are summarized below. In general terms, the sum of the α- and β-fractions, together with their larger peptides, will be proportional to the bloom strength, and the percentage of higher-molecular-weight material to the viscosity. The setting time is increased by peptide fractions below α-chain, but a certain proportion of the very high molecular-weight 'Q' fraction can reduce it markedly. The melting point also increases in line with higher molecular weight content.

2.c.2.1 OSSEIN

The electrophoretograms of ossein gelatine are distinctive in a number of features, the most important being the amount of A-peptide component present (6–15%) compared with hide and pigskin gelatines (< 5–6%).

The general level of peptides is also lower in limed ossein gelatines.

2.c.2.2 HIDE

Hide gelatines are generally characterized by a higher level of peptides, in excess of 20–25%, but without the A-peptide content found in ossein gelatines. The peptide fraction tends to increase with gelatines of lower viscosity together with a corresponding reduction of the high-molecular-weight components.

2.c.2.3 PIGSKIN

Pigskin gelatines are in a class of their own, since the acid conditions encountered during manufacture result in the production of a much higher percentage of lower peptides. Oligomers of α-chain are also more likely to exist at the expense of free α-chain material, and in greater complexity, resulting in a greater number of individual peaks that are seen between the γ- and Q-fractions.

Typical results for high grade gelatines are shown in Table 4.

In more recent high-resolution work,[17] both $\alpha 1$- and $\alpha 2$-peaks can be seen, but whether the $\alpha 1$(III) chains can be separated is not yet clear, owing to their low concentration. Often the α-peptides are found to display a range of peaks of diminishing size, indicating that these peptides also possess some degree of distribution.

Two-dimensional electrophoresis has also been performed, and this clearly shows the range of peptides obtained and also their PI distribution.[18]

2.c.3 *Molecular Shape*

The Marc–Houwinc parameter x, derived from intrinsic viscosity measurements, is related to the overall molecular shape. For pigskin gelatines x is found to be in the range 0·6–0·8, while for a typical limed ossein gelatine the value is found to be 0·45–0·65. A lower value indicates a more spherical molecule, while higher values indicate an increasing amount of rod-like character necessary for superior surface activity. These results agree with the practical observations quite well.

3 SOURCES AND PRODUCTION

3.a Manufacturing Sources

Gelatine is produced from sources of collagen that are available both in quantity and at a reasonable price for the manufacturer. Commercially viable sources are pigskin, cattle skin and cattle bone.

Pigskin is obtained directly from abattoirs and other processing plants in the form of frozen blocks and so requires thawing prior to use. This material can contain up to 30% of fat, which can be sold as a by-product.

Cattle (bovine) hide is obtained as surplus material from leather tanneries. The hides are washed, the adhering flesh and hair being removed, and generally cleaned up before being sent to the gelatine manufacturer.

Processing cattle bone is a complex and expensive operation involving degreasing and demineralization of the raw material to produce the product known as ossein, which results in these commanding higher prices compared to skin gelatines. This is an obvious drawback from the food processor's point of view, and thus they find their greatest use in photographic and pharmaceutical applications.

3.b Pretreatment Processes

To convert insoluble collagen into soluble gelatine, two processes are in current use:

(i) acid pretreatments leading to acid process or type A gelatines;
(ii) alkaline or basic pretreatments leading to alkaline process or type B gelatines.

The pretreatment process is designed to convert the collagen into a form suitable for extraction. In order to achieve this a sufficient number of the covalent cross-links in the collagen must be broken in order to enable the release of free α-chains. The process is also designed to remove other organic substances, such as proteoglycan, blood, mucins, sugars, etc., that also occur naturally in the raw material. It is optimized by each manufacturer to give the required physical and chemical properties to the gelatines that are produced.

The effect of pretreatment on the collagen concerned is proportional to the degree of cross-linking in the raw material. Acid pretreatment processes, which are less aggressive than those using alkali, are applied to pigskin and fresh ossein, which are both relatively young and immature forms of collagen. Fresh pigskin is approximately nine months old, and thus possesses a lower concentration of non-reducible cross-linkages. For this reason, an 18–24 h soak in dilute acid is generally sufficient to bring about the conversion. Sulphuric and hydrochloric acids are used, often with the addition of phosphoric acid to retard colour development.

Alkaline pretreatment processes are normally applied to bovine hide and ossein. Lime is most commonly used for this purpose; it is relatively mild and does not cause significant damage to the raw material by excessive hydrolysis. Unfortunately, the reaction is a slow one, and up to 8 weeks or more are required for complete treatment. Concentrations of up to 3% lime are used in conjunction with small amounts of calcium chloride or caustic soda. Frequent renewal of the liquors is practised in order to remove extracted impurities and to maintain the degree of alkalinity present. If caustic soda is used, then with care a 10–14 day pretreatment is possible.[19]

3.c Manufacture: Extraction

The conversion of pretreated raw material into gelatine takes place in five basic stages: (i) washing, (ii) extraction, (iii) purification, (iv) concentration, and (v) drying.

The extraction process is designed to obtain the maximum yield in combination with the most economic of physical properties, i.e. to optimize the balance between pH, temperature and the extraction time. Figure 2 gives an indication of the relationship between these parameters.[20]

In practice, gelatine is obtained from the raw material in three or four separate extractions, each at an increasing temperature. Typical values are 55°C for the first extraction, 60°C for the second, 70°C for the third and 80–90°C for the final extraction, each giving gelatines of decreasing bloom strength and increasing colour.

The evolution of this process has been well summarized.[21]

The pH of extraction can be selected either for the maximum extraction rate (low pH) or for the maximum in physical properties (neutral pH), or, as is usual, some compromise between the two. To extract older collagens at

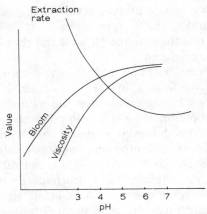

FIG. 2. Graph showing factors determining optimum gelatine extraction conditions. Abscissa represents the pH value of the extracting collagen; ordinate represents the bloom or viscosity value of the resultant gelatine, or the extraction rate as weight/h.

neutral pH, a substantial proportion of the cross-links need to be cleaved, necessitating a longer liming pretreatment. If shorter liming times are used, then a lower extraction pH is necessary in order to achieve acceptable conversion rates. However, owing to the acidity present, the resultant gelatines will have lower viscosities (lower molecular weight) than those extracted at neutral pH. More efficient pretreatment conditions also allow the manufacturer to use lower extraction temperatures, resulting in gelatines of greater gel strength (bloom). Shorter treatments generally require higher extraction temperatures if neutral pH levels are chosen, resulting in gelatines of lower gel strength.[22] Recent work shows clearly the influence of collagen age on gelatine extraction and subsequent properties.[23]

Mention should be made of certain gelatines extracted from chrome-tanned leather waste in the form of shavings or powder. In order to make use of this material, the process must break the stable chromium/carboxyl tanning cross-links, in addition to the existing interchain bonds. The process, developed in the 1930s by the then British Glues and Chemicals, involves a high-temperature extraction in the presence of magnesium oxide to precipitate the chromium as hydroxide. These gelatines, known as 'spa gelatines', are characterized by a low bloom strength (50–100) but have the colour and appearance equivalent to gelatines of much higher bloom strength. Other methods have also been developed commercially.[24]

Following extraction, the gelatines are filtered to remove suspended insolubles such as fat or unextracted collagen fibres. This is usually performed using materials such as diatomaceous earth to give solutions of high clarity. The gelatine is further purified by deionization, which both removes inorganic salts left from pretreatment and also adjusts the pH to that suitable for sale. Most commercial gelatines are sold in the pH range 5·0–5·8. An exception to this is found in certain pigskin gelatines, where no deionization takes place and then the pH of extraction (4·0–4·5) is the same as the pH of sale.

The final stage is evaporation, sterilization and drying. These are performed as quickly as possible to minimize loss of properties, after which the gelatines are subjected to laboratory testing for the physical and bacteriological characteristics required.

4 COMMERCIAL AVAILABILITY

From the laboratory analyses, the most important of which is the 'bloom' or gel strength figure, the various gelatines produced can be blended into larger quantities to meet customer specifications. Normally these will be delivered in palletized 25 or 50 kg polythene-lined paper sacks, although drums or other containers are occasionally specified. The manufacturer will thus have to carry sufficient stocks in order to blend to the customers' characteristics. If the gelatines are reasonably close in bloom strength, then they are blended arithmetically. However, if the bloom strengths are further apart, an allowance is made for a certain drop in the expected value; this becomes larger the higher the proportion of low-bloom material present.

4.a Commercial Characteristics

Gelatine owes its importance as a hydrocolloid to its unique rheological characteristics, and there can be few people who cannot appreciate the 'melt in the mouth' texture so characteristic of a gelatine gel. The rheological properties are usually summarized by manufacturers in

(1) the bloom value (which is a function of the gel strength);
(2) the viscosity (which gives a measure of the solution properties).

These arbitrary data can define the gelatine properties quite well under most circumstances, especially if comparisons are made from the same manufacturer. It is useful to realize the implications of these terms and their relationship to each other, together with the principles that govern these properties on the molecular scale.

It is known that the gel strength properties are related to the α- and β-chain components, and that the viscosity is related to the average molecular weight, in particular the degree of oligomerization of the α-chains. Setting time is also strongly influenced by the high-molecular-weight fractions, as these are associated with the initial conformations necessary for the onset of gelation. Melting point is related to several features, one being the proportion of low-molecular-weight peptides that cannot take part in forming the gel network.

The α-chain and its oligomers are the most important features in the molecular-weight spectrum (molecular weights of 95 000 and multiples). Many rheological studies in the past have been carried out using much lower average molecular weights, i.e. 40 000–80 000 daltons were commonly quoted figures. These would obviously indicate degraded gelatines of low physical properties, and conclusions from these results must therefore be treated with care before extrapolating to higher grades.

It is well known in the food industry that the bloom and viscosity can act as a good guide to the behaviour of a gel under most conditions. However, bloom is a function of the gel strength at $10°C$ (for 18 h) and the viscosity is quoted using the same concentration at $60°C$ ($6\frac{2}{3}\%$). Therefore, a user requiring a gel at, e.g., 16–20°C must exercise caution. The gel strength will also be dependent on the viscosity under these conditions, higher viscosities giving higher gel strengths, while for temperatures lower than 10°C the reverse is often true.

The following viscosity ranges are available for type A and type B gelatines:

High bloom	30–70 mpoise—limed hide/ossein
(250–300)	30–60 mpoise—acid pigskin/ossein
	70–130 mpoise—specialized hide/ossein
Medium bloom	20–80 mpoise—limed hide/ossein
(150–200)	20–40 mpoise—acid pigskin/ossein
Low bloom	15–30 mpoise—limed hide/ossein and spa processes
(50–100)	15–30 mpoise—acid pigskin/ossein

4.b Particle Size and Solubility

Gelatine, although insoluble in cold water and other liquids such as milk, sugar solution, dilute food acids, etc., will swell and absorb up to 10 times its weight in water, the rate of which is dependent on the particle size—or, more accurately, the surface area of the gelatine per unit weight. The swelling characteristics in cold water are determined by temperature and the salt or sugar content of the liquid, all of which can affect the rate of water uptake.

The particle size distribution chosen for a particular process should enable the gelatine to dissolve quickly and efficiently. To prepare solutions of high concentration, e.g. for those required in low-water-content foods such as gelatine confectionery and table jelly tablets, particle sizes from 14 to 20 mesh (BSI) are usual, although some processors prefer larger sizes of 6–8 mesh (BSI) when there is only a small volume of water available for solution.

Often both maximum and minimum particle sizes are specified in order to reduce undissolved particles and 'balling up' of fines, thus reducing possible wastage. Gelatine dissolution can often prove troublesome when employing modern continuous techniques if insufficient care is given to the initial design of the plant concerned. In practice, a compromise is often made and the gelatine powder is given a separate cold soak, and then introduced into the process, dissolving in the hot sugar or starch solutions. Exceptions to this are found in some batch processes in which the dissolution time is unimportant; for example, coarse, dry gelatines are added directly to tinned hams immediately prior to sealing and sterilizing. In this way the tendency to clump is reduced and an even dispersion is ensured throughout the can.

4.c Isoelectric Point
4.c.1 Definitions
Before we go further, the distinction needs to be made between isoelectric and isoionic points, both of which are sometimes quoted.

The isoelectric point (pI) is defined as the pH at which no net migration takes place in an electric field,[25] while the isoionic point is defined as the pH at which the number of protons lost by the gelatine, compared to its state when carrying the maximum possible charge, is just equal to that maximum charge,[26] i.e. the pH at which there is no net charge on the molecule. In a deionized solution, the isoelectric and isoionic points are for most purposes identical. Acid-processed gelatines with a greater distribution of isoelectric points can differ in their isoelectric and isoionic points in solutions of ionic strengths similar to those found in most foods.

The isoelectric point distribution of gelatine is dependent on the type of pretreatment applied during manufacture. Type A or acid-processed gelatines have isoelectric points that can vary from 6·5 to 9·0. Acid ossein gelatines, being at the lower end of the range with pI values of 6·5–7·5, while acid pigskin is more likely to have values from 7·5 to 9·0. Limed or alkaline-process gelatines have isoelectric points over a narrower pH range, typically 4·8–5·0. These differences are caused by side-reactions occurring during the pretreatment process. The amide groups of asparagine and glutamine are hydrolysed, releasing ammonia (which can be readily detected near any pretreatment area!). Acid-processed gelatines have a pI closer to collagen (pH 9·4), owing to limited chemical alteration of their side-groups; the small changes which do take place lowers the pI to 7–9, and are influenced by the effects of production. The lower pI of acid ossein is due to the acid demineralization step prior to extraction. A fall in pI also occurs over the four extraction liquors in all type A gelatines due to the increase in extraction conditions.

4.c.2 Influence of Isoelectric Point on the Properties of Gelatine

The isoelectric point can have important effects on the properties of gelatines used for particular purposes that are dependent on the overall ionic charge. Many physical properties of gelatine have either a minimum or a maximum at the isoelectric point, and this can clearly influence the choice.

Commercial gelatines are normally sold at pH 5·2–5·5, and so those made from limed ossein or hide are close to the isoelectric point. These possess a low negative charge compared to acid pigskin gelatines, which possess a much higher positive charge. For example, when preparing an aspic, the gelatine chosen must have a high degree of clarity at pH levels near 5–6. The high charge on pigskin gelatines will give an acceptable product in these situations. Type B gelatines, owing to the proximity of the pI to this pH, will tend to be more turbid in dilute solution. Lowering the pH to 4·5, together with the addition of salts, should solve the problems, while the slightly negative charge on type B gelatines enables them to be compatible (at pH levels greater than 5·5) with negatively charged polysaccharides, e.g. carrageenan.

4.c.3 Methods of Determining the Isoelectric Point
These are based on the following.
(1) Total deionization using strong acid and base resins.
(2) Changes in clarity of dilute gels over a pH range. Figure 3 shows the variations of turbidity due to pH of 2% gelatine gels of limed ossein, acid ossein and acid pigskin. The relative molecular pI distribution can be clearly seen.

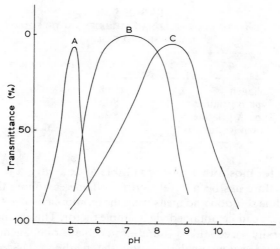

FIG. 3. Isoelectric point distribution for different gelatines, variation of turbidity with pH of 2% solution. A, limed ossein or hide, isoelectric point 4·8–5·2. B, acid ossein, isoelectric point 6·0–7·5. C, acid pigskin, isoelectric point 7·5–9·0.

(3) Isoelectric focusing.[27-29]
(4) Absorption of gelatine onto monodisperse latex particles.[30]

4.d Analysis and Characterization of Gelatines

Commercial gelatines consist of gelatine protein in a high state of purity; the non-proteinaceous matter present is mainly mineral ash and residual moisture. The only additive sometimes incorporated during manufacture is sulphur dioxide, used as a colour inhibitor during extraction and evaporation and not as a preservative. There is also a small proportion of carbohydrate (1–1·5%) in the form of glucose and galactose units bonded to the gelatine at hydroxylysine residues.

4.d.1 Analysis of Gelatine Content

The gelatine content of foods can be determined in several ways, each depending on other known constituents present.

(1) Kjeldahl nitrogen digestions, or biuret reagent determinations, can only be applied if it is certain that there are no other proteinaceous compounds present. The procedure for Kjeldahl determinations is well known and suitable conditions have been published.[31] The conversion factor for gelatine determinations is considerably lower

TABLE 5
COMPARISON OF KJELDHAL NITROGEN FACTORS FOR GELATINE AND CASEIN

	N (%)	Protein factor
Collagen	18·68	5·36
Type B gelatine	18·15	5·51
Type A gelatine	18·30	5·46
Casein (comparison)	15·73	6·36

than for most other proteins (Table 5). Recently, a rapid Kjeldahl procedure has been developed that involves boiling the sample in alkaline solution and measuring the ammonia evolved. The protein content can be obtained via a simple factor. This method, although generally successful, is found unsuitable for collagenous proteins.

(2) Hydroxyproline analysis using the method of Stegemann and Stakder.[32] This is the definitive method, applicable in nearly all situations with a high degree of accuracy.

4.d.2 Amino-acid Composition

As described earlier, this can vary with the source of collagen used to manufacture the gelatine and also the production process followed. However, a typical analysis is shown in Table 6 for comparison.

TABLE 6
TYPICAL AMINO-ACID COMPOSITION OF A GELATINE

Residue	(%)	Residue	(%)
Hydroxyproline (3- and 4-)	22·6	Methionine	0·8
Proline	13·8	Isoleucine	1·3
Lysine	3·7	Leucine	2·9
Hydroxylysine	0·7	Arginine	7·6
Glycine	22·6	Phenylalanine	2·0
Cysteine	0	Aspartic acid	5·7
Serine	3·5	Threonine	1·9
Alanine	9·1	Glutamic acid	9·5
Histidine	0·7	Tyrosine	0·2
Valine	2·3		
Total			100·0

4.e Gel Strength
The gel strength of gelatine is nearly always quoted bloom strength. The 'bloom' value of gelatine is c required to push a cylindrical plunger, of 13 mm di previously prepared gel of $6\frac{2}{3}$% w/w concentration 16–18 h. Gel strength determinations not performed conditions cannot be quoted as 'bloom strength'.

4.f Viscosity
This test can be performed in U-tube viscometers because of the Newtonian behaviour of gelatine solutions. A BSI U-tube viscometer is used (size c is most suitable) with the gelatine concentration set at $6\frac{2}{3}$% as for the bloom test. Viscosity is determined at $60 \pm 0.1°C$ and should be reproducible to within 0.2 of a second. Viscosity is measured in millipoise.

4.g pH
This is measured using a 1% solution cooled to 45°C, using a standard combination glass electrode.

4.h Moisture
The BSI method states that 1 ml of water should be added to 1 g of gelatine in a weighed dish and placed in an oven at 105°C for 18 h.

4.i Ash
For ash content, 5 g of gelatine is placed in a previously cleaned crucible and held at 550°C to constant weight. Higher temperatures may volatilize ammonium salts, whilst a lower temperature would lead to difficulty in complete combustion.

4.j Colour
Colour is usually measured subjectively, by comparison with standards, while the clarity is often found by optical techniques such as transmission factors.

4.k Sulphur Dioxide
Sulphur dioxide is measured by the method of Monier and Williams,[33] which involves a lengthy steam-distillation step during which sulphur dioxide is volatilized from a refluxing acidic gelatine solution (HCl or H_3PO_4 are normally used). A slow stream of carbon dioxide or nitrogen gas purges the solution of the sulphur dioxide, which is absorbed into either

...de or hydrogen peroxide. It is debatable whether sufficient accuracy can be maintained below 50 ppm.

A quicker method, but one that may be subject to interferences, is via iodometric titration of an acidified gelatine solution. Addition of alkali is necessary in order to release the bound SO_2 from aldehyde groups as bisulphite complexes. A measure of the 'free' SO_2 can be made without the use of alkali, and this can be compared to the total SO_2 content using the Monier–Williams method, as a measure of SO_2 activity.

4.l Melting and Setting Points

Melting point, setting time and setting temperature are tests that can be performed accurately but are strongly dependent on the method used. Ward and Courts[34] mention the more commonly used procedures. The concentration, pH and salt content all affect the results.

4.m Other Non-gelatine Components

Other constituents of collagen relevant to gelatine manufacture and utilization are the non-collagenous proteins and carbohydrates. Non-collagenous proteins, e.g. albumins, enzymes, globulins, etc., tend to be precipitated out during manufacture from good-quality raw materials, but if poorer material is used some substances will be partially solubilized by the pretreatment process and extracted together with the gelatine, causing an increase in colour, turbidity and odour. Certain carbohydrate compounds are always present together with collagens *in vivo*. These are composed of certain polysaccharides as well as the natural hexose content of collagens. Of the polysaccharides found in collagen tissues, two are important enough to be mentioned here. One is hyaluronic acid, a highly acidic long-chain polymer of molecular weight 10^6. The other is dermatan sulphate, which is a proteoglycan covalently bonded to the chain via serine residues. These two polysaccharides are instrumental in maintaining the structure of collagen in living tissue.[35]

The hexose content of collagen is found to consist of glucose and galactose units bonded to certain hydroxylysine residues.[36] These units, and a portion of the polysaccharides, are carried through extraction, and are found in commercial gelatines, although in small quantities.

5 SOLUTION PROPERTIES

The typical commercial gelatines tend not to find use as thickening agents owing to their relatively low specific viscosity (<60 mpoise at $6\frac{2}{3}\%$) as there are more efficient polysaccharides available for this purpose. Instead, gelatines in the liquid phase can be used for their surface-active properties, either as stabilizers and emulsifiers or as polyelectrolytes.

It is known that some situations require the use of both high- and low-viscosity gelatines, and so manufacturers are able to supply a wide range. Higher-viscosity gelatines will give higher melting temperatures and somewhat quicker setting times, while those of lower viscosity can be prepared in a much higher concentration without causing problems due to tailing when depositing in moulds, etc. Also, lower-viscosity gelatines dissolve at a faster rate.

5.a Dissolving Gelatine

To dissolve gelatine, two methods are in current use: (a) indirect solution and (b) direct solution.

(a) *Indirect solution.* Add the gelatine to cold water (never the reverse!), ensuring all of the particles are evenly wetted and leave to swell for a sufficient time until a soft, friable mass is formed. This can simply be performed by an agitator in a tank to which the correct amount of cold water has previously been added. The solution is then formed when the mass is heated to a temperature of 50–60°C. Constant stirring will ensure complete dissolution.

(b) *Direct solution.* This method eliminates the cold soaking stage but requires higher water temperatures (60–80°C are typical) and also high-speed agitation to ensure an absence of clumping when the gelatine is introduced to the liquid. The rate of dissolution is determined by the particle size distribution. This procedure involves heating the water to about 60–80°C and introducing a high level of agitation, preferably creating a vortex in the mixing tank. The powdered gelatine is then introduced as a slow stream, pausing now and then to allow the powder to be wetted before further additions. When all of the gelatine has been added, the stirrer is left on for a few minutes to complete the dispersion. As stated earlier, the dissolution rate is dependent on the particle size, but unfortunately finer particles have an increasing tendency to clump, and thus the optimum particle size to specify for this method is 16–28 mesh BSI.

5.b Viscosity of Gelatine Solutions

A solution of gelatine in water will possess a viscosity or thickening power proportional to its concentration, pH, ionic strength and the specific

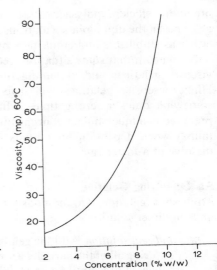

FIG. 4. Variation of viscosity with concentration for gelatine solutions of 2–12%.

viscosity of the gelatine itself. The relationship of viscosity to concentration is not directly proportional but is approximately logarithmic. Croome[37] found a straight-line dependence for log viscosity on the reciprocal of absolute temperature.

The specific viscosities of gelatine solutions are measured kinematically using BSI U-tube viscometers. The justification for their use is based on an analysis of stress–strain curves with temperature using Brookfield (or similar) viscometers. The viscosity of gelatine is normally quoted in millipoise.

At most temperatures gelatine acts as a Newtonian fluid, but just above the setting point the viscosity becomes markedly time dependent owing to a degree of aggregation taking place. This is accentuated by higher concentrations and high-molecular-weight content (i.e. high-viscosity values), as shown in Fig. 4.

As may be expected, the effect of pH is influenced by the ionic strength of the solution. This effect is not as noticeable in concentrated solutions, but increasingly in more dilute solutions the pH–viscosity curve is affected by the ionic strength. The effect of pH on viscosity is reduced by the addition of salt (Fig. 5). This is due to both inter- and intrachain repulsion effects.

The viscosity is always at a minimum at the isoelectric point, and increases as the overall charge on the molecule is increased. A maximum viscosity value (pH 3·5) can be seen, also a second maxima at higher pH

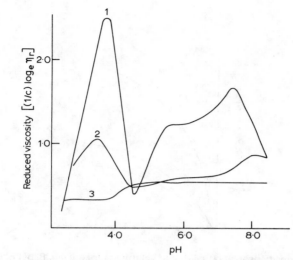

FIG. 5. Change in viscosity with pH for 0·2% gelatine solutions of varying ionic strength. 1, deionized gelatine; 2, gelatine in 0·017M NaCl; 3, gelatine in 0·50M NaCl. (From Stainsby, G., *GGRA Res. Rep. No. A2*, 1951. Reproduced by permission of Leatherhead Food Research Association.)

levels. Towards the extreme values of pH the viscosity drops to below that seen at the isoelectric point, but this is of no consequence in food processing. The differences between the two gelatines are related to variations in charge density distribution along the molecular chains owing to changes in amide content loss during processing.

5.c Viscosity Drop

Unlike loss in gel strength, which is usually a result of elevated temperatures, viscosity loss during processing is a function of both temperature and pH. The relationship for a typical limed hide gelatine of medium viscosity (40 mpoise) is shown in Fig. 6. It can be seen that minimum viscosity loss occurs at pH 6, the same point as is found when studying bloom loss. Normally the degradation is at an acceptable level when pH is held at 5·3–8·0. This is one of the reasons why food acids are among the last of the ingredients to be added to low-pH jelly products during manufacture, and ideally they should be added during cooling. The loss in viscosity can be compensated for by increasing the gelatine concentration, but this is an expensive method of rectifying poor processing technique. The rate of bloom strength loss, although lower, is also not

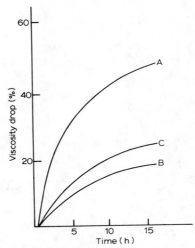

FIG. 6. Viscosity drop at 60°C for a $6\frac{2}{3}$ gelatine. A, pH of 4·0; B, pH of 6·0; C, pH of 8·0.

insignificant enough to be neglected here. Viscosity is considered important in this situation owing to its relationship with the gel strength at temperatures higher than 10°C.

The rate of viscosity loss can be influenced by the type of food acid present in the product. For example:

	Least degradation	
Citric	Most commonly used.
Malic	Used to a lesser extent to achieve a different flavour profile from that of citric.
Tartaric	Used to a lesser extent to achieve a different flavour profile from that of citric.
Adipic	Used mainly in dry mixes and jelly crystals.
Fumaric	Often used in a 1:4 ratio with adipic.
	Most degradation	

Lactic acid may be present when gelatine is added to fermented milk products, and acetic acid is sometimes encountered as a preservative.

As fumaric acid has the greatest effect on viscosity of all the common acids used, it is rarely added in solution, but instead as a dry form in products such as jelly crystals. The 'milder' food acids such as lactic or adipic have a reduced effect on the rate of loss; this should be kept in mind during product development. Additionally, the acid concentration may be raised by the presence of buffer salts.

5.d Surface Activity and Related Sol Properties

Gelatine is often utilized for its surface-active properties, altering the rate of crystal growth in supersaturated solutions, e.g. marshmallow and ice-cream, where the sugar and ice crystal growth is suppressed, giving the desired product. It also finds applications in stabilizing other emulsions such as mayonnaise, paté and meat homogenates. Gelatine is added to various types of yoghurt formulations to stabilize the product. It forms a weak gel structure that imbibes free water, thus inhibiting separation of the whey, especially after pasteurization.

Another feature of its surface activity is its polyelectrolyte behaviour. Gelatine solutions are widely used in this respect, i.e. for clearing wine and apple juice musts of yeast suspensions, tannins and other polysaccharides formed during fermentation. It can also be used in the modern 'hot fining process' designed for improved cider and apple juice production. These properties are related on the molecular level to three parameters:

(i) the overall charge, and its distribution along the chain;
(ii) the distribution of non-ionic groups;
(iii) the molecular weight (more correctly the mean chain length).

Recently, Japanese workers have modified the structure of some gelatines by the addition of the hydrophobic group L-leucine dodecyl ester, using papain as a catalyst.[38] This group was found to promote excellent emulsification properties to the protein; its characteristics are shown in Table 7. Comparing this to a food-grade emulsifier, e.g. Tween 80,[39] using oil–water mixtures, it was found that the emulsifying activity of the modified gelatine depended on its concentration in solution, but was not affected by temperature. The activity of Tween 80, on the other hand, was quite sensitive to temperature, but was resistant to changes in concentration.

TABLE 7
PROPERTIES OF LEUCINE DODECYL MODIFIED GELATINE

Molecular weight	$7\,500 \pm 2\,500$
Leucine dodecyl content	$1 \cdot 2 \pm 0 \cdot 1$ mol/7 500 g
Surface tension	$0 \cdot 7$ N/m^2

Reproduced with permission of Pergamon Press.

Recent work[40] has described this product as an inhibitor for ice nucleation in frozen water products. It is likely that these systems may well become popular in the future if approval can be obtained for their use in foods.

6 GEL PROPERTIES

6.a Formation of the Gelatine Gel

Since the pioneering work of Eagland,[41] Harrington,[42] Rao[43] and Finer[44] during the 1970s, some excellent reviews on the research performed in this area have been published that have now come close to explaining the mechanism involved in gelation: see Ward and Courts,[27] Ledward[45] and Stainsby.[46]

The major components present in gelatine solution are the α, β, γ chains and higher α-chain oligomers and their respective peptides. Included with these is a small proportion of very high-molecular-weight Q fraction.

The stability of collagen can be shown to be proportional to its total pyrolidine (Pro + Hyp) content.[47] From these early studies it was found that the same trends were found in gelatines prepared from a wide range of collagens. The ultimate strength of the gel can therefore also be related to its pyrolidine content, and it was shown quite early in gelatine research that gelatines derived from fish collagens are weak, and of low melting point, compared to mammalian gelatines, which have a higher proportion of proline and hydroxyproline.[48] (The manufacture and properties of fish gelatines were reviewed recently by Norland.[49])

The pyrolidine content is therefore implicated in the gelling mechanism, the sites most likely to be involved in the formation of junction zones being those that are rich in pyrolidine imino acids. It has also been found that the optical rotation of gelatine solutions changed during gelation and that the final value varied with the rigidity of the gel. This suggests that a continuous increase in the number or extent of the junction zones is occurring. In dilute solution, at concentrations below 0·5%, the system will gel unimolecularly, while at concentrations above this the mechanism tends increasingly towards a bi- or trimolecular one[50-53] (Fig. 7).

French workers have described three mechanisms of molecular transformations from dilute to more concentrated solutions:

(1) From $<10^{-2}$ g/100 g—individual coils < internal bonds
(2) From 0·2 to 0·5 g/100 g— aggregate chains that tend to be discrete
(3) From >1 g/100 g—interbonded aggregates forming a gel

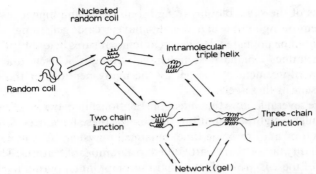

FIG. 7. Some possible conformations leading to gelation of single α-chains. (From Finer,[44] reproduced with permission of John Wiley and Sons, Inc.)

They describe gelation as a two-step process:

(a) A fast initial process
(b) A log phase, >1000 h[54]

Since the junction zones are rich in (Gly-Pro-Hyp) triplets, it would be expected that the zones once formed would also show similar helical structures to those found in native collagen, and this was found to be the case. These helical regions are stabilized by hydrogen bonding to form domains of long-range rigidity.

Hydrogen bonding takes place between the imino acids and nearby C—H groups, e.g. on glycine found on adjacent chains. (It is well known that hydrogen-bond breakers such as thiocyanate or urea can prevent gelation.) It is now considered that such a regime akin to collagen is likely in which it is proposed that some association of water molecules takes place with additional hydrogen bonding. The equilibrium of junction zone formation and breakage moves to create a dynamic gel structure as the temperature of the solution approaches the setting point. At this stage the larger branched molecules begin to aggregate via junction zone formation with other smaller α-chains or, as more likely at this stage, with other large molecules present.

The Q fraction has a very profound effect on the rate of this initial aggregation, thus influencing the setting time,[14] but has no significant effect on the gel strength after the onset of gelation.

The viscosity has a considerable effect on the properties near the gelling point. Melting temperature, setting temperature, setting time and gel strength at temperatures near the melting point depend on the content of higher-molecular-weight molecules, as would be expected.

Gelatines of the same bloom strength but possessing higher viscosities (high content of high molecular weights) have greater gel strengths above 10°C, as would be predicted. However, at temperatures below 10°C, lower-viscosity gelatines can have a higher gel strength. This is due to the increased reinforcement of the gel by the many peptides of the shorter chains present in these gelatines.

There are several factors that affect the relationship between gel strength and molecular weight distribution. In the past these have been known as 'rigidity factors' and for some time remained a mystery. It is now known that the 'rigidity factor' can be attributed to a number of features. The most influential of these is probably the distribution of intact pyrolidine groups along the chain.[44,45] There is also the effect of the proportion of $\alpha 1$- to $\alpha 2$-chains. $\alpha 2$-Chains are thought to have a lower tendency towards gel formation, but one author at least speculated that a certain proportion of $\alpha 1$ to $\alpha 2$ may possess some extra stability.[52]

A most novel approach has been taken by Heideman[55] in some extremely elaborate and detailed work. He has described the construction and reproducible functioning of a microgelometer with which to analyse the gel strength of chomatographically separated components (i.e. α- and β-chains, etc.) and so determine their contribution to gel strength. Theoretical estimates of blends of these components are able to be reproduced. The relative strength of cyanogen bromide (CB) peptides (of known sequence) were also determined, and the gel strengths obtained were in close agreement with the 'rigidity factors' discussed here (Fig. 8).

The other contributing cause is the location of junction zone-forming

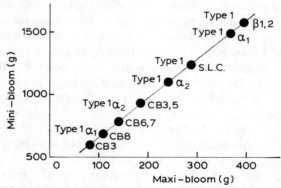

FIG. 8. Relative bloom strengths (maxi-bloom) of collagen fragments separated via chromatography determined from microgelometer studies (mini-bloom). (From *Das Leder*, **40**(5) (1989) 87–91; reproduced with permission.)

areas in the gelatine molecule, which can vary depending on the type of hydrolysis used to form the gelatine;[56] those situated at the ends of the molecule will not be as efficient and are normally shorter in length compared to those found in the central regions. Grose[57] found that acid hydrolysis left many of these zones intact, while alkali hydrolysis, giving a more random attack, cleaved the chain at these sites to a greater extent, explaining the relative resistance to bloom drop at mildly acid pH levels (see Fig. 15).

Recent work has confirmed that acid hydrolysis of collagen follows certain rules. Peptides were demonstrated to be obtained reproducibly, after hydrolysis under set conditions, and these were subsequently sequenced to confirm this.[58] The fraction of large molecules, i.e. $\sum(\alpha+\beta+\gamma)$ as percentage of peptides, has been termed the ϕ value.[59]

Rose[60] points out that α-chain peptides of certain lengths can possess an effective concentration of imino acid residues higher than other sizes of chain. A computer program was developed to calculate the extent of this change with respect to peptide length.

When a gel is in the process of formation, the more the chains can align themselves, before or during maturation, the greater the number of junctions that will form. If gelatines are snap-chilled, they are found to have significantly lower gel strengths than expected. The sites involved in hydrogen bonding cannot align themselves efficiently enough to produce the expected number of junctions, whereas if gels are 'tempered' at a temperature just above the setting point, the gel strength is above that which would be expected.[61]

Japanese workers have recently described the phase diagrams that result through both heterogeneous and homogeneous associations in solutions over a wide range of temperatures.[53]

Another approach was outlined by workers at the General Foods Corporation, in which gelatine is treated as a classical partially crystalline glassy polymer (PCGP). Many of the varied properties displayed by gelatine in solid, sol and gel forms are explained well using this line and it is probable that this will be followed further in the future.[62]

6.b Measurement of Gel Strength

To determine the gel strength, one of the three following instruments can be used:

(a) The Lead Shot gelometer, also known as the Bloom gelometer
(b) The Boucher gelometer
(c) The Stevens–LFRA Texture Analyser

The gelatine bloom test has a poor reputation for repeatability between manufacturers and their customers. (How much of this is due to over-enthusiasm on both sides for value for money is not recorded here!) Therefore, the following comments, which relate to the BSI bloom test, should be borne in mind if reproducible results are to be consistently obtained.

(1) THE WATER CHILL BATH

The temperature of the chill bath should be sufficiently well controlled to have ideally a maximum variation of $\pm 0.02°C$ when held at the correct temperature of $10.00°C$. A variation of $0.05°C$ may affect the value by 2–3 bloom points when using high-strength gelatines.

The perforated platform upon which the test jars are placed must be perfectly flat and horizontal. Any tilt or bowing under the weight of jars will lead to an angle between the gel surface and the plunger, and a lower bloom value.

To obtain the required 16–18 h at $10°C$ necessary for the test, it is important that the water bath is of sufficient capacity to reduce the water temperature to $10.00°C$ in under 1 h after the gelatine samples have been placed in the bath. With smaller-capacity baths, the water temperature may well rise to over $15°C$ and may not return to the correct temperature for over 2–3 h. It is also essential that the water level in the chill bath should be at least 1 cm above the surface of the jelly.

(2) THE GELOMETER
 (a) Type of gelometer used
  (i) *Lead Shot gelometer*. This gelometer is still in use but is often mistreated. The advisory comments in the manual (photocopies can be obtained, e.g. from the Leatherhead Food Research Association) should be followed closely, especially those regarding the mesh range of the shot flow rate, cleaning of the gold disc, etc.
  (ii) *Boucher gelometer* (discontinued production). This instrument is more reliable, but it is essential to keep it in a room in which the air temperature is controlled to $20 \pm 2°C$, otherwise the torsion characteristics of the wire will be altered; eventually it will be subject to fatigue, giving rise to erratic results. The Boucher tends to give low results on gelatines above 250 bloom, and higher results on gelatines below 100 bloom.
  (iii) *Texture Analyser*. Accurate results can be achieved quite simply on this instrument; the only comment to make apart from the initial

setting of the plunger travel using dial gauges is to ensure the instrument has thoroughly warmed up before performing the tests so as to ensure a good thermal distribution in the load cell and other important components.

(b) The plunger. The plunger itself probably has the most consistent influence over the result, more so perhaps than differences in gelometer type. There are two plungers in use—the BSI plunger, which is popular in Europe and which has a radiused edge of 0·35–0·43 mm. It is important that this is adhered to and any plungers having a different radius or possessing a rough chipped edge or face should be rejected. The American Organization of Analytical Chemists' plunger, widely used in America, is of the same dimensions but with a sharp edge; this gives it a larger surface area, and it is for this reason that American bloom values are slightly higher than those from Europe. A factor of 1·03 should be used when specifying equivalent bloom values.

(3) TEST JARS

The shape of the bloom test jar is of importance, in that in an empirical test all outside influences must be standardized. The relationship between the surface area and depth, and also the material of construction, may affect the final result. The correct jars are available from C. Stevens & Sons Ltd (Loughton, Essex, UK), the manufacturers of the Texture Analyser.

(4) STANDARDS

Most manufacturers adhere closely to all these details, but still experience day-to-day variations in their results. Gelatines of known bloom strength should also be included in the test sequence, to enable any other influences to be taken into account. Standards of 100, 200 and 300 bloom are often used to check the samples at the start of testing, and sometimes at the end of the session.

It is also beneficial to cool the plunger prior to testing. A warm, dry plunger may give erroneous results until it has reached a temperature similar to that of the sample.

6.c Factors Affecting the Properties of Gelatine Gels
6.c.1 Concentration

The exact relationship between concentration and gel strength depends on the type and origin of the gelatine itself, but using gelatines commonly found in the food industry, i.e. 100–250 bloom, the relationship becomes

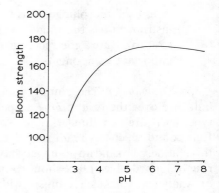

FIG. 9. Effect of the solution pH on the bloom strength of gelatine.

$(C_1)^n \times (B_2) = (C_2)^n \times (B_1)$, where C is the concentration of the gel concerned, B is the bloom of the gel concerned, and n is equal to 1·7 for high-bloom gelatines and 1·8–1·9 for lower strength gelatines of 150–100 bloom. Some authors, using degraded gelatines of low average molecular weight, quote $n = 2$: see, e.g., comments in Ref. 63.

6.c.2 Effect of pH

The gel strength is seriously affected by the solution pH only at the extremes of the range; from pH 4 to pH 9 it is not affected to any significant extent. Figure 9 shows the influence of pH on the bloom test. Dilute gels (<2%) are more affected, while stronger gels (>10%) are relatively insensitive to pH.

6.c.3 Time and Temperature of Set

The gel strength will depend on the time and temperature of set. This will vary from gelatine to gelatine and is also dependent on the proportions of molecular fractions present and thus the viscosity. Figure 10 shows the

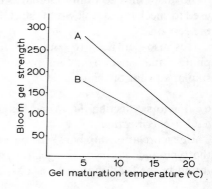

FIG. 10. Variation in gel strength (at $6\tfrac{2}{3}\%$ concentration) with temperature of maturation for 275-bloom (curve A) and 175-bloom (curve B) gelatines.

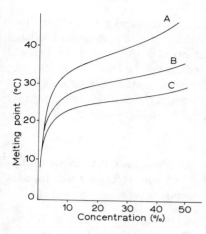

Fig. 11. The relationship between melting point and concentration for gelatines of different bloom strength. A, 250 bloom; B, 175 bloom; C, 100 bloom.

relationship between gel strength and the maturation temperature for dilute gels.

6.c.4 Effect of Melting Point

The unique organoleptic qualities exhibited by gelatine are strongly dependent on its melting point, which in turn is influenced by:

(i) the bloom value and viscosity of the gelatine;
(ii) the concentration of the gel.

These in turn can be altered by other ingredients incorporated in the food concerned, e.g. salts, sugars and other gelling or thickening agents, etc. Figure 11 shows the influence of concentration on melting point for 250, 175 and 100 bloom gelatines.

6.c.5 Effect of Low-Molecular-Weight Compounds

The effect of other dissolved ingredients in the gel can be either to weaken the gel or to strengthen it. Most simple sugars, glycerol and other non-electrolytes can contribute to an increase in gel strength (fructose and sorbitol being an exception), while the addition of most electrolytes has the opposite effect.

6.c.6 Compatibility with Other Food Gels

Gelatine is incorporated as a functional ingredient in a wide variety of foods. When doing so, it is important to have knowledge of its compatibility with other substances with which it is likely to be blended in order to form the final food product.

A good example is the reaction between gelatine and gum arabic in low-pH solutions. When using moderate or dilute concentrations, these colloids of opposing charge will coacervate when mixed. If this behaviour occurred under all conditions, the confectionery industry would lose one of its more useful combinations. However, in a high-sugar-solids mix, gelatine is found to be perfectly compatible with gum arabic, and the pastilles and gums that are formed are part of many confectioners' product ranges.

6.c.6.1 INTERACTIONS WITH HYDROCOLLOIDS

Polysaccharide hydrocolloids can be classified into two main groups depending on their structural features. Gelatine solutions will interact either constructively or destructively depending on the environment.

(1) Non-charged polysaccharides: these include locust-bean gum, guar gum, starches and sugar or glucose syrup solutions.
(2) Charged polysaccharides: examples are pectins and alginates, agar, carrageenan and gum arabic.

(1) Non-charged polysaccharides
(a) *Locust-bean gum and guar gum.* These two galactomannans are characterized by their non-gelling properties and are used as viscosity improvers. Although they are non-ionic in structure, they do possess a certain degree of polarity, and have molecular weights reaching 1 000 000 in some cases. There has been little published work concerning combinations of gelatine with these two galactomannans, but in dilute solution some degree of coacervation possibly exists. This is probably due to phase separation of the large galactomannan molecules from the gelatine. Similar reactions are shown by the non-charged glucose syrups. Careful addition of these products to certain foods, especially dairy products (e.g. 'fromage frais' or yoghurt preparations), can result in a successful combination. To reduce interactions, it is necessary to add these ingredients separately.
(b) *Starch.* Starch producers have for some time been attempting to find a replacement for gelatine in confectionery products, using either thin boiling starch or other modified products. Gelatine has not been replaced in these foods, but can be supplemented with a certain proportion of starch without significant alteration of the required properties. In dilute solution, gelatine can be added to dairy desserts incorporating starch effectively, and without interactions.

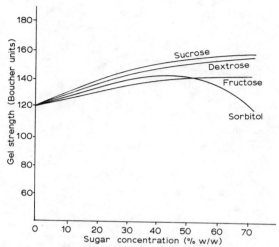

FIG. 12. Effect of some sugars on the gel strength of 3% gelatine gels. (From Marrs and Weir,[64] reproduced by permission of Leatherhead Food Research Association.)

(c) *Reactions with sucrose and simple monosaccharides.* The incorporation of these simple sugars is fundamental to many of the products in which gelatine is found. Table jellies, desserts and confectionery all include mixtures of these sugars and gelatine in varying proportions and prepared under various conditions. Therefore, it is useful to have information regarding the gelatine–water–sugar system in order to predict the physical properties of the mixture concerned. Marrs and Weir (see, e.g., Refs 64–66) have studied extensively the compatibility between gelatine and various sugars, including glucose syrups and sorbitol. The relationship between simple sugars was shown[64] to be a beneficial one. Setting time, melting point and gel strength all improved with increasing sugar content. Figure 12 shows the effect of dextrose, fructose, sorbitol and sucrose on the gel strength of a 3% limed hide gelatine.

(d) *Reactions with glucose syrups.* Glucose syrups are used for both textural and economic reasons; knowledge of the glucose syrup–gelatine–water relationship is particularly relevant during product development. Glucose syrups are manufactured by controlled hydrolysis of starches, using enzymes, acids or combinations of both. The degree of hydrolysis is given commercially as its dextrose equivalent (DE), a measure of the syrup's reducing power compared

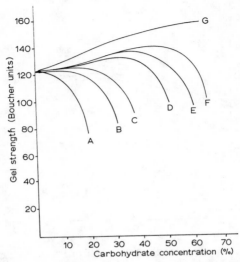

FIG. 13. Effect of glucose syrups on the gel strength of 3% gelatine solutions at 10°C. A, maltodextrin 10–13; B, maltodextrin 17–20; C, glucose syrup 26–32; D, glucose syrup 36–40; E, glucose syrup 39–43; F, glucose syrup 60; G, glucose monohydrate. (From Marrs and Weir,[65] reproduced by permission of Leatherhead Food Research Association.)

to that of glucose. These are commonly produced in ranges from 100 DE (dextrose) to 10 DE; syrups with DE values below 25 are termed maltodextrins, owing to their lack of sweetness and greater percentage of high-molecular-weight components. Reactions between gelatine and glucose syrups at low concentration are similar to those exhibited by simple sugars. However, in greater concentrations, the high-molecular-weight fraction of both the gelatine and the glucose syrup react to influence the solution and gel properties dramatically. This was confirmed[65,66] when specific molecular weight ranges of polyethylene glycols (PEG) were used as model systems. These were found to exhibit parallel effects to glucose syrups of equivalent molecular weight and related the loss in clarity and gel strength to the degree of coacervation between both the high-molecular-weight (HMW) fractions of gelatine and PEG. An acid pigskin gelatine was used, but type B gelatines may be more sensitive to this effect owing to their greater HMW content. Higher ash levels sometimes found in type B gelatines can also increase the likelihood of incompatibility. Figure 13 shows the effects of glucose syrups of differing dextrose equivalent and concentration on the strength of a 3% gelatine gel at 10°C. The gel strength is very much reduced when syrups of less than 20 DE are used at concentrations of over 30%. In confectionery formulations, it is usual to use syrups of higher DE owing to their textural properties (normal grades are 40 and 60 DE); even so, the maximum content of these in a low-ionic-strength mixture should be no greater than 55% and less if inorganic components such as ash or buffer salts, etc., are present.

(2) Charged polysaccharides. This class of hydrocolloid with its negative charge will interact with gelatine, coacervating in a joint phase from solution when the gelatine has a net positive charge and acting synergistically when the gelatine has a negative charge. The degree of synergism is dependent on the negative charge density, and its distribution on the hydrocolloid. Interactions between gelatine and anionic polysaccharides have been studied in detail over the last decade by Russian workers, notably Tolstoguzov, who has over a number of years researched the fundamental properties of soluble protein–anionic polysaccharide complexes (see, e.g., Refs 67–70).

The sulphated carrageenans are more highly charged than the carboxylated polysaccharides and exert strong synergistic effects with gelatines showing negative charge. This is found when type B gelatines are considered, possessing isoelectric points of 4·8–5·0. Above these pH levels, gelatine and carrageenan interact to form solutions and gels whose clarity and gel strength increase as the pH is raised. Satisfactory gels are found above pH 6·0, such as those found in non-acidic food products like meat jellies, hams or pork pies, and also for neutral cream-based desserts. Tolstoguzov[67–70] has tried to provide an explanation for this effect in terms of the formation of strong complexes that are soluble and stabilized by salts.

Carboxyl-containing polysaccharides show much less pronounced non-equilibrium effects than sulphated polysaccharides, and so are generally more incompatible.

In practice, gelatine and charged polysaccharides are not normally intentionally combined below the isoelectric point, except in high-sugar-solids confectionery and with low-charge-density polysaccharides under a very low-water-activity regime. In these conditions, perfectly clear acid jellies or gums can be made.

(a) *Pectins.* Pectins are sometimes used in conjunction with a small proportion of gelatine to provide special textures in confectionery jellies, providing a softer and cleaner chew.

(b) *Substituted celluloses.* Substituted celluloses are occasionally added to gelatines to improve the viscosity and melting point when a neutral product is required. They are utilized in the meat industry, where gelatine for use in, e.g., pork pies, may have a small proportion included; carboxymethylcellulose is the best derivative for this purpose.

(c) *Alginates.* Mixtures of gelatine and alginates in food products are not normally encountered, but, in common with many other soluble proteins, gelatine will react in alkaline solution with propylene glycol alginate, forming a non-reversible gel.[71] The mechanism involves covalent bonding of the carboxyl groups on the alginate to the ε-amino groups on lysine in the gelatine molecule. The reaction takes place in alkaline solution ($>$pH 9·5), forming gels that are non-elastic and stable to both freezing and heating. When washed with water or phosphate buffer in order to reduce the pH, the gels swell to an extent dependent on the ionic strength and pH of the gel. At pH levels below 3, the gels become cloudy owing to alginic acid formation. These gels were found to be as digestible as gelatine itself;[72] however, approximately 20% of the previously available lysine remains covalently bonded to the alginate and is therefore unable to be digested.

(d) *Agar.* Gelatine is not often mixed with agar in confectionery, and if at all then in small proportions in order to modify the texture created by the polysaccharide.

(e) *Carrageenan.* The three types of carrageenan may be used in dilute solutions with type B gelatines, provided the pH is kept above 6. κ-Carrageenan is by far the most important and is often found with gelatines in meat or savoury products. It cannot be used in confectionery, however, owing to its high viscosity and handling difficulties. It forms synergic mixtures with gelatine, and can usefully be blended up to about 3–4%, where the soft gelatine texture is still apparent.

(f) *Gum arabic.* Gelatine is used extensively with this gum in high-solids products such as fruit pastilles or gums to provide textures somewhere between that expected of the two hydrocolloids when they are used individually. The coacervation between type A gelatines and gum arabic was the first example of microencapsulation technology, developed and fully patented many years ago by the NCR Co. Despite other more refined systems in use today, the original method is still in existence because the ingredients used are permitted additives for consumption and are completely 'natural'. This technology is applied, for example, in the production of encapsulated flavour oils.

(g) *Gellan gum.* Gellan gum is the latest microbial polysaccharide to be introduced into the food arena by Kelco International Ltd. At the time of writing, it is being tested for FDA and EEC approval, for which no adverse toxicological effects have been found to date. Gellan gum is an efficient gelling agent, forming both neutral and acid-type gels with calcium at concentrations around 0·2%. The gels, however, possess melting points of 80°C and over, and therefore do not melt in the mouth. However, it has been demonstrated that small quantities of gellan gum can increase the melting point and setting time so that shelf-stable and 'instant' gels can be formulated. The concentration of gellan gum is below the critical concentration at which anionic polysaccharides coacervate out when the pH of the gel is dropped below the isoelectric point. The acid regime recommended consists of a sequestrant (sodium citrate) in order to quickly absorb calcium to allow the gellan to dissolve, together with a slowly dissolving acid (e.g. adipic acid) that also releases the bound calcium and which then complexes with the gellan, forming a calcium–gellan gel in a similar way to formation of calcium alginate gels. Using gellan gum concentrations as low as 0·05%, together with the usual gelatine concentrations, the setting time of a typically prepared gel is reduced to less than 15 min. Unfortunately, the gels have a 'melting point' of about 80°C. If lower concentrations of gellan are used, the rate of set is reduced but the apparent melting point is brought back to the area where a typical 'melt-in-the-mouth' sensation is optimized. It is expected that gelatine/gellan blends will become of importance following toxicological clearance and further research (see Chapter 6 for more detail).

6.c.7 *Aspects of Filled Gels*

These have been well covered by Ring and Stainsby,[73] who found two distinct groups depending on their effect in the gel. Rigid particles increased the gel strength in proportion to the amount of filler added independently of the original gel strength (Fig. 14).

Soft, deformable fillers interacted to a greater extent, their contribution

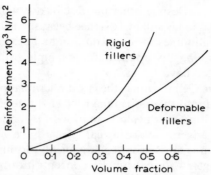

Fig. 14. Reinforcement of gelatine gels using deformable and rigid fillers (gelatinized starch and charcoal). (Adapted from Ref. 73, reproduced by permission of Pergamon Press.)

to the total texture increasing with decreasing system gel strengths. Boyar *et al.*[74] compared textures of gels with varying proportions of TiO_2, starch and ground almonds as typical examples of insoluble fillers. All gave a transition from a brittle to a more plastic gel with increasing filler concentration. The change in texture was roughly proportional when using ground almonds, while for starch higher concentrations were necessary before a degree of plastic behaviour was observed. However, with TiO_2 dispersions, only when concentrations approached 50% w/w was there any

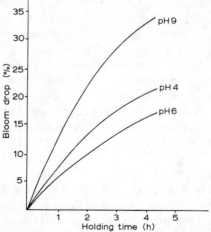

Fig. 15. Effect of pH on the thermal degradation of gelatine at 70°C. (Monitored by viscosity decrease, i.e. molecular weight.)

significant change in texture. This could be explained by the increased solid–solute interactions between the gelatine molecules and the dispersed ingredients. Ground almonds contain a significant proportion of oil and would show little interaction, while the starch granules, being liable to deformation, would not support the gel to a great extent. TiO_2, a rigid particle, supports the gel and associates with it via strong particle–gelatine interactions and a degree of hydrogen bonding.

6.c.8 Characterization of Gel Properties

Because gelatine gels owe public popularity to their unique texture and sensory characteristics, various workers have attempted to correlate results from instrumental tests with standardized sensory profile data. Henry,[75] Szczeniak,[76] Munoz[77] and Larson-Powers[78] found that gelatine gels could be distinguished from other model gel systems. They both identified and defined the parameters of firmness, cohesiveness and elasticity, etc. Christenson and Trudoe[79] also used sensory evaluation techniques for several hydrocolloids in an attempt to duplicate the gelatine gel. Recently, the Leatherhead Food Research Association has attempted to use electromyography to characterize gel systems, including gelatine gels.[80]

6.c.9 Effect of Processing Conditions

Gelatine will degrade and lose its gelling properties when subjected to conditions of heat, extremes of pH and exposure to enzyme attack.

The differing mechanisms for acid and alkali hydrolysis can be seen in Fig. 15, in which the hydrolysis is plotted against the holding time and percentage loss in gel strength.

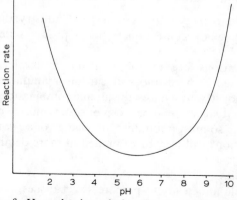

FIG. 16. Effect of pH on the degradation (loss in gel strength) of gelatine.

Acid hydrolysis is somewhat selective, tending more to break cross-links between the chains; the peptide bonds are attacked less often. Alkaline hydrolysis, however, tends to give a wider pattern of cleavage, preferring the more numerous peptide bonds rather than selecting the chain cross-links. This has the effect of reducing the α- and β-chain size very quickly and it is interesting to note that the minimum point in Fig. 16 is not at pH 7 as may perhaps be expected but at a pH of 5·5–6. This stability of the gel strength to acid conditions explains the success of acid extraction methods when using pigskin; bovine hide and ossein can also be extracted in the same way.

7 APPLICATIONS

Edible gelatine is widely used as an ingredient in the food industry owing to the following functional properties.

(1) It forms high-quality gels in dilute solution with typical clean melt-in-the-mouth textures.
(2) It forms elastic gum-type textures in the concentrated gel, slowly dissolving in the mouth.
(3) It produces emulsification and stabilization of immiscible liquid–liquid, liquid–air or liquid–solid mixtures.
(4) In very dilute solutions, it acts as a polyelectrolyte that will flocculate suspended particles or unstable colloids.
(5) In a mixture with other powdered ingredients, usually as a minor component, gelatine acts as an efficient tableting aid and binder.

Gelatine is sold to many sections of the food industry; the major areas of use are in table jellies, confectionery, meat products and chilled dairy products.

Other applications are in the pharmaceutical industry—a large proportion is used in hard and soft capsule manufacture—and in the photographic industry, which uses the unique combination of gelling agent and surface activity to suspend particles of silver chloride or light-sensitive dyes but in doing so must not induce any degree of agglomeration. These general applications will now be examined in more detail.

7.a Dessert Jellies

Of all the simple and ready-to-prepare sweet dishes available from the supermarket shelf there are few that can match the brilliance, colour, texture and price of a table jelly. Table jellies have been popular in the UK for 50 years now and, although sales have been declining recently, there is still a strong demand for this type of product. The market is split into several areas, with basic jellies available at a lower price while the higher-price products incorporate spray-dried fruit juices and often added vitamin C. Other novelty jellies aimed at children, using fluorescent colourings, have been developed.

There are two ways of presenting a jelly to the consumer, in a powder or tablet form. In the former, all of the ingredients are dry-blended and packaged; in the traditional tablet, the ingredients are processed to give a concentrated block that requires melting and diluting to the correct concentration.

The crystal jelly is less costly to produce and package. Dry ingredients are used, although some extra cost is incurred in using spray-dried or encapsulated flavours, but a cheaper gelatine of lower gel strength can be incorporated.

The jelly tablet, on the other hand, requires a gelatine of higher gel strength because of the processing involved, which requires it to be dissolved in hot syrup, mixed with acid and buffer solutions, and deposited in moulds after de-aeration. In spite of the additional costs, the tablet jelly remains the most popular version in the UK. Outside the UK, notably in warmer climates where a tablet would melt or soften to an unacceptable degree, it is virtually unknown and the jelly dessert is supplied in crystal form. This predominates in countries such as the USA, where 'Jello' is the famous market leader.

7.a.1 Crystal Jellies

A typical recipe for a crystal jelly is given below:

Sugar	250 g
Gelatine*	57 g
Fumaric acid	8·5 g
Sodium citrate	2·4 g
Flavour	sufficient
Colour	sufficient

*The gelatine to be 200 bloom, 30 mesh (BSI)

Mix well: 50 g to make one pint jelly

It is critical that all of the ingredients are of the same particle size in order to ensure even distribution of all the ingredients during blending. Further cost reductions can be achieved by substituting glucose and/or saccharin to replace a portion of the sugar. The gelatine used is usually set at 200–220 bloom and has been ground to pass through a 30-mesh sieve in order to minimize dissolution time in hot water. The food acid may be citric acid; however, fumaric or malic acids can also be used to give a different accent to fruit flavours and also to prevent clumping on storage.

Colours added are normally absorbed onto a carrier or encapsulated to avoid specking—shown as undispersed dye particles around the edge of the mixing bowl.

Flavours may be of the basic type or sophisticated encapsulated concentrates of natural fruit to satisfy various consumer tastes and demands.

In some export markets, the customs of the region have an influence over the type of gelatine to be used in the product. For example, in the Middle East, where jelly is popular in the hot climate, the religious laws stipulate that the consumption of pigskin gelatine is forbidden, and the use of alternative types is necessary.

7.a.2 Jelly Tablets

Jelly tablets are probably purchased by the majority of British families at some time during the year. They are produced in as many different flavours and colours as marketing people can dream up, but the 'reds and purples' are still the most popular with adults and children alike. The tablet form is manufactured in order to provide the consumer with the pre-dissolved ingredients of a table jelly ready to prepare by steeping in hot water followed by cooling in the refrigerator.

7.b Confectionery Jellies and Gums

Confectionery is normally taken to indicate foods consisting of high sugar solids. Here, in a reduced water activity environment, the properties of gelatine are substantially modified. The compatibility of gelatine with other anionic polysaccharides has been covered elsewhere in this chapter and reference was made to the incompatibility of these hydrocolloids in dilute solutions. In confectionery systems, however, acidic mixtures are used extensively with no coacervation taking place. The manufacturing procedures are similar to that of tablet jellies in that a gelatine solution is introduced into a sugar syrup and moulded, usually by depositing in starch. After cooling, the jellies, or gums in particular, are often 'stoved' to reduce the water content. Stoving normally involves drying for several hours, or even days at temperatures of 35–55°C, depending on the required residual

TABLE 8
GELATINE GRADES REQUIRED FOR VARIOUS CONFECTIONERY PRODUCTS

Product	Bloom	Product	Bloom
Set marshmallow	240	Jelly babies	160
'Gummie bears'	240	Fruit chews	150
Foamed products	200	Jelly shapes	150
Nougat	175	Tableted mints	125
Wine gums	175		

water content and thus the desired texture—the gelatine present retarding sugar crystallization.

The grades used for confectionery jelly articles are more variable than for dessert jellies owing to the wide range of textures required. A company buying gelatine for its confectionery production would probably be purchasing up to four grades of gelatine, and these are shown in Table 8.

The boiling temperature can exert an influence on the texture, as can the percentage of glucose syrups and their DE value. Glucose syrups of low DE can coacervate with gelatine under certain conditions. Therefore, care must be taken when developing recipes for production, as it is quite common for systems that do not coacervate in the laboratory to do so under the non-ideal conditions present during processing. Gelatine manufacturers have also developed new gelatine types and mixtures that have properties to improve the texture or aid processing inside the factory. For example, in recent years, the use of hydrolysed gelatines for confectionery has grown despite their higher price, because they can give a variety of new textures to some confectionery products.

An example is given below of a typical confectionery product in which a successful combination of gelatine/polysaccharide is utilized—in this case, for gelatine pastilles. Gum arabic is added to provide a firmer texture that will last for longer in the mouth.

PASTILLES WITH GUM ARABIC

Sugar	100 g
Glucose syrup	75 g
Gum arabic, prepared	100 g
Gelatine, 160 bloom	10 g
Water	50 g

(1) Boil the sugar and syrup to 127°C using 30 parts water.
(2) Stir in the gum arabic solution.

(3) Cool to 80°C and add the pre-dissolved gelatine.
(4) Add colour, flavouring and acid.
(5) Deposit in starch and stove for 2–3 days depending on required texture.

Further alterations in texture can be achieved in these products by altering the bloom strength of the gelatine. Higher-bloom gelatines can be used in lower concentrations and are thus associated with additional volumes of water, producing a light, elastic product. Lower-bloom gelatines require a higher concentration and so give rise to a tougher, more resilient product typical of higher-total-solids confectionery. Much of the knowledge regarding this subject is probably acquired over many years of experience as the 'confectioner's art'.

When gelatine is used in combination, notably with anionic polysaccharides, coacervation is not apparent. This is due to the low water activity of the system reducing interactions between the two phases. However, in order to achieve this compatibility it is essential to incorporate each ingredient into the sugar syrup separately. Mixtures with polysaccharides added as a blend will generally give rise to coacervation in the product.

7.c Gelatine as a Stabilizer in Marshmallow Foams

The name marshmallow was given to this example of a sugar foam confection when the stabilizer used historically was derived from the roots of a common herb, the marsh mallow. The stabilizers in use today are gelatine and egg albumin, more suitable perhaps than the original! Gelatine is the more widely used ingredient, especially as its price, availability and ease of use make it ideal for modern production methods. Egg albumin gives a lighter, shorter texture compared to gelatine. Low-bloom gelatine gives a more delicate texture, whereas high-bloom gives a more elastic chew. Aids to enhance foaming can be used in addition to gelatine; one of the most popular is Hyfoama (hydrolysed casein) or in some countries hexametaphosphates are used.

Gelatine stabilizes the sugar foam in five important ways:

(1) The surface tension of the syrup is lowered, enabling a higher whip to take place.
(2) The viscosity of the syrup is increased and the foam is less likely to collapse.
(3) The gelatine stabilizes the syrup–air emulsion formed during beating.

(4) Gel formation allows a permanent foam insensitive to stress.
(5) The gelatine prevents crystallization of the saturated sugar solutions, to give a smooth product.

To achieve the differing textures, i.e. toughness, moistness and lightness, four different styles have evolved in the confectionery and biscuit industry.

(a) *Extruded marshmallow.* This recipe is designed to be used in modern extruders, e.g. Johnson's Votator, etc. The extruded marshmallow is required to have a very quick set on leaving the extruder in order to retain its shape. Therefore, a high-bloom/viscosity gelatine is used with values up to 280 bloom/55 mpoise viscosity. The final product tends to be drier than the other types, although the use of invert sugar can minimize this.

(b) *Deposited marshmallow.* At the other end of the scale, a light and delicate marshmallow can be obtained by depositing in starch. The time left in the starch moulds is not critical but a rapidly setting gelatine may be specified.

(c) *Cutting marshmallow.* This type is formed by pouring the beaten syrup onto cooled slabs or trays and allowing it to set, after which it is cut into squares and packaged. Cutting marshmallow is moist in texture and softer in the mouth, although it tends to be more expensive because of the nature of the production process. Gelatines of 240 bloom are often used.

(d) *Biscuit topping.* As the name suggests, this variety is deposited onto a biscuit base and enrobed with chocolate or other coatings. The necessity here is for a fast set, therefore high viscosity gelatines are used in slightly higher concentration than in extruded marshmallows.

It is generally found that high-grade pigskin gelatines tend to have the best combination of net molecular charge and molecular chain length for use in marshmallow production. Type B gelatines are more likely to have inferior foam-stabilizing properties, although limed hide and ossein producers have had limited success in producing gelatines with similar properties. Incorporation of additives mentioned previously can help considerably. The increased charge on pigskin gelatine molecules at the pH levels normally used for marshmallow (pH 5·5–6·5) gives a higher degree of rigidity to the chains (owing to the increased repulsion of the positively charged groups on the molecule) compared to type B gelatines, which are close to the isoelectric point. Acid ossein gelatines, because of their relatively lower average pI when compared to acid pigskin gelatines, are not as efficient in their stabilizing capacity.

7.d Polyelectrolyte Properties

The gelatine molecule, with its mixture of acidic and basic amino-acid side-chains, has polyelectrolyte characteristics and is described as amphoteric. At pH levels lower than the isoelectric point the molecule will have a net positive charge, and above this point it will have a net negative charge. Since the isoelectric point in gelatine can vary considerably, the extent of net charge is dependent on the environmental pH. A pigskin gelatine molecule will have a higher net charge at pH levels below 7 than limed hide or ossein, and it would be expected to show enhanced reactions with negatively charged particles as found in wines, fruit juices and other products. As a result of this polyelectrolyte behaviour, gelatine is used extensively in the clarification of wines and fruit juices in order to remove the unstable and undesired levels of polyphenolic compounds, which cause cloudiness and sediments on storage and an unpalatable astringency if in excess.

Once the suspended matter has been removed, the resulting solution is visibly suitable for bottling or further storage. However, if polyphenols and other unstable colloids are present they will slowly precipitate with age; also, high levels of tannin are undesirable to the palate. Heatherbell[81] states that these colloidal systems can vary in size from 10^{-4} to 10^{-3} μm. Two mechanisms are responsible for the action of gelatine on these components.

(a) Absorption of a gelatine layer on the particle will act to neutralize the surface charge, and thus the mutual repulsion that had been responsible for the colloidal system is removed. The system can now agglomerate and so aid flocculation.

(b) The positively charged gelatine molecules and the negatively charged tannin molecules will coacervate in dilute solution and flocculate.

The reaction between gelatine and tannins in wine closely parallels that between collagen and vegetable tanning agents (e.g. mimosoa bark extract) in leather manufacture. This similarity demonstrates the bond stability of these components once formed. The following comments can be made in respect of the fining of wine or juice by the action of gelatine.

(1) Type A gelatines with a higher isoelectric point will be more highly charged at pH 3·5, the average pH found in wines. The use of this

gelatine would therefore be expected to achieve more acceptable results than those with type B gelatines.
(2) Extensive research has been performed to find the grade of gelatine that gives the best results. Unfortunately, the published information gives conflicting results.[82] Many writers claim that a low-bloom (70–90) pigskin gelatine is most suitable;[83] however, others[84] showed that there was little difference in the grade used between 200 and 50 bloom, and between type A and type B gelatines. Ringland[85] states that high-molecular-weight gelatines should be more effective.
(3) It has been found that casein can react in a similar way to gelatine but requires a greater concentration; the flocs take longer to form and are less dense.

In practice, many wine or juice processes use a single grade of gelatine and adjust it to their daily requirements. There are available several gelatine products tailor-made for the wine industry: concentrated solutions of hydrolysed gelatine containing preservatives are popular and low in price. There are also special 'fining' gelatines, which are made by careful blending to give a wide and consistent molecular weight distribution. These 'fining' gelatines are made by blending together a range of bloom and viscosity values to give a wide molecular weight spectrum.[86]

When fining white wine or juice, it is unlikely that an efficient floc will form, owing to the low levels of tannin present. The use of silica sols in recent years has enabled the gelatine to be used more effectively. Silica sols are added with a slight excess of gelatine to provide bulk to the floc, resulting in a more efficient removal of polyphenols.

7.e Gelatine Derivatives

Important advances in gelatine derivatives for food industry products have been made in the last two decades or so. The commercial exploitation of gelatine with cold water solubility has taken place, although to only a limited degree, and there is an increasing demand for hydrolysed gelatine.

7.e.1 *Cold-Water-Soluble Gelatine*
Despite the large number of patents applicable to cold-water-soluble gelatine manufacture, there are only three types of product available in the European market. There are considerable problems in producing a gelatine with cold water solubility and at an acceptable price. All methods are based on the fact that gelatine solutions, when dried without passing through the gel phase, are of amorphous structure, and do not show any crystalline

character as do gelatines dried from the gel phase. When these dried solutions are rehydrated at a temperature below the gelling point, they will set and form a rigid gel identical in every way to that of a normal gelatine. Unfortunately, gelatine when dehydrated, is extremely hygroscopic and it is difficult to form gels of moderate concentration. In order to overcome this difficulty, mixtures of gelatine and certain carrier substances are available—glucose syrup and starch are the most popular, as they are ingredients in many of the products for which the gelatine is intended. Drum-dried instant gelatine is also available,[87] but still has to be mixed intimately with other ingredients to ensure complete solution.

These products have their relative advantages and disadvantages, but none of them can produce what must be the ultimate cold-water-soluble gelatine product—a clear 'instant' table jelly. As a surface-active agent and foam stabilizer (e.g. in marshmallows) gelatine will aerate when introduced into cold water with mechanical action and so can be found as an ingredient in instant mousses, desserts, cheesecake fillings and whipped-cream stabilizers.

A typical recipe for an instant dessert displaying the non-sticky texture characteristics of gelatine, using the spray-dried gelatine/glucose mixture GP 201, is presented here.

INSTANT CHOCOLATE DESSERT MIX

Finely milled sugar	35 g
GP 201	12 g
Whipping agent (e.g. DP 49, ex DMV)	30 g
De-fatted cocoa powder	8 g
Colour	q.a.
Flavouring/salt	q.a.

Mix thoroughly and whisk into 250 ml cold milk for 2–3 min, pour into individual serving dishes and refrigerate for 15 min.

7.e.2 Hydrolysed Gelatines

Hydrolysed gelatines are available from many manufacturers as a pale-yellow powder, usually spray-dried. They differ from normal gelatines in that they are soluble in cold water, and they possess no gelling power.

The hydrolysis of gelatine is complex; there must be strict control of the reaction to avoid the formation of undesirable products such as 'bitter peptides', etc. Collagen itself is sometimes used, thereby avoiding the gelatine extraction stage. Several manufacturers market a range of

TABLE 9
PROPERTIES OF SOME HYDROLYSED GELATINES

Analysis	#A	#C	#E	#O
Moisture	7·0	7·0	7·0	7·0
Ash	2·0	2·0	2·0	2·0
N (%)	17·0	17·0	17·0	17·0
Viscosity (mpoise), 10%, 25°C	20–25	35–45	35–45	17–19
Average molecular weight (kilodaltons)	3–4	10–12	10–12	1–2
Emulsion properties	Fair	Fair	Excellent	Fair

hydrolysed gelatines, each with controlled molecular weight ranges which can vary from below 1000 to 15 000.

Differences in physical and chemical properties are brought about by changing the hydrolysis techniques, i.e. pH of hydrolysis, temperature profile, and the enzymes used at each stage of the reaction.

Table 9 gives typical values for a range of hydrolysed gelatines available from one manufacturer. Types 'A' and 'C' can be used, for example, in tablet binding and granulation agents, replacing gelatine that is not cold water soluble under normal conditions of use. The hydrolysed gelatines give the tablet a better rate of dissolution and disintegration. Type 'E' is used for its emulsification properties (e.g. in meat/fat emulsion) and also as an encapsulation agent for flavour concentrates, taking advantage of its low carbohydrate content.

Hydrolysed gelatine can be kept as a concentrated solution in a bulk pack after UHT treatment, or more commonly with preservative added. These are often supplied to the wine industry as ready-to-use fining agents. Hydrolysed gelatines still possess many of the useful properties of gelatine, e.g. emulsion stability, colloid protection and flocculation promotion. In addition, because they are both non-gelling and cold-water-soluble, they can be used to good effect in situations that require the properties of a concentrated and totally clear protein solution.

Other examples of use are mainly directed towards the meat industry, where it can be added at proportions of up to 3%. Examples of uses are in reconstituted hams and 'stitch pumping', where it can act as a fat emulsifier and as a water-retention aid. It can also be added to raise the total protein content in these products, in order to meet the regulations in existence in some countries. The market for hydrolysed gelatines remains a small but an expanding one. Much of the production is destined for the cosmetic market, where it is added to shampoos, for example, as 'protein' and to creams and lotions as 'hydrolysed collagen', where use is made of its characteristics of absorption by skin and hair.

The price of food-grade hydrolysed gelatine is roughly that of a high-grade gelatine, partially owing to the cost incurred in spray-drying the processed materials.

7.f Nutritional Aspects

Owing to the fact that gelatine does not contain any tryptophan (Table 2), it cannot be used as a 'complete protein'; however, it has increased proportions of certain amino acids (e.g. lysine). It can therefore be used to supplement other proteins to give a mixture with a higher protein value than each component. When it is mixed with beef protein, the net protein value can rise from 84% to 99%.[88]

The calorific value of gelatine is only 3·5 kcal/g.

REFERENCES

1. Glanville, R. W. & Khun, K., The primary structure of collagen. In *Properties of Fibrous Proteins. Scientific, Industrial and Medical Aspects*, Vol. 1, ed. D. A. D. Parry & L. K. Creamer. Academic Press, London, 1979, pp. 133–50.
2. Bornstein, P. & Traub, W., The chemistry and biology of collagen. In *The Proteins*, Vol. IV. Academic Press, London, 1979, pp. 412–605.
3. Kuhn, K. & Glanville, R. W., Molecular structure and higher organisation of different collagen types. In *Biology of Collagen*, ed. A. Viidik & J. Vuust. Academic Press, London, 1980, pp. 1–14.
4. Asghar, A. & Henrickson, R. L., Chemical, biochemical, functional, and nutritional characteristics of collagen in food systems. In *Advances in Food Research*, Vol. 28. Academic Press, London, 1982, pp. 232–372.
5. Light, N. D. & Bailey, A. J., Covalent crosslinks in collagen. In *Properties of Fibrous Proteins. Scientific, Industrial and Medical Aspects*, Vol. 1, ed. D. A. D. Parry & L. K. Creamer. Academic Press, London, 1979, pp. 151–206.
6. Barnard, K., Light, N., Sims, T. & Bailey, A., Chemistry of the collagen crosslinks. *Biochem. J.*, **244** (1987) 303–9.
7. Hamilton, P. B. & Anderson, R. A., Demonstration of orintuine in gelatine by ion-exchange chromatography. *J. Biol. Chem.*, **211** (1954) 95–102.
8. Davis, P., The guanidino side chains and the protective colloid action of gelatine. In *Recent Advances in Gelatin and Glue Research*, ed. G. Stainsby. Pergamon Press, London, 1957, pp. 225–30.
9. Eastoe, J. E. & Leach, A. A., The chemical constitution of gelatin. In *The Science and Technology of Gelatin*, ed. A. G. Ward & A. Courts. Academic Press, London, 1977, pp. 78, 84, 85, 93.
10. Koeppf, P., Die Anwendung der Gelelektrophorese bei der Herstellung von Gelatine. *Leder*, **31** (1980) 83–9. See also *The Use of Electrophoresis in Gelatin Manufacture*, Fourth Conference on Photographic Gelatin, Oxford, September 1979.

11. Tomka, I., Die makromolekulare Charakterisierung der Gelatine. *Chimia*, **37**(2) (1983) 33–40.
12. Lorry, D. & Verdins, M., Determination of molecular weight distribution of gelatines by HPSEC. Presented at the Fourth IAG Conference, Fribourg, September 1983.
13. Tomka, I., Electrophoresis as a routine tool to investigate photographic gelatines. Presented at the Fourth Conference on Photographic Gelatin, Oxford, September 1979.
14. Bohonek, J., Spühler, A., Ribeaud, M. & Tomka, I., Structure and formation of the gelatin gel. In *Photographic Gelatin II*, ed. R. J. Cox. Academic Press, London, 1976, pp. 37–56.
15. Ohno, T., Ishikawa, T. *et al.*, On the relationship between molecular weight and viscosity or jelly strength of fractionated gelatin. In *Proceedings of Fifth IAG Conference*, ed. H. Amman-Brass & J. Pouradier. IAG, Fribourg, Switzerland, 1989.
16. Tavernier, B. H., Molecular mass distribution of gelatin and physical properties. In *Proceedings of Fifth IAG Conference*, ed. H. Amman-Brass & J. Pouradier. IAG, Fribourg, Switzerland, 1989.
17. Aoyagi, S., Matsumoto, T. & Ishikawa, T., SDS gel electrophoretic characterisation of gelatin polypeptides. In *Photographic Gelatine*, ed. S. Band. RPS, Bath, 1987, pp. 38–41.
18. Aoyagi, S., Electrophoretic characterisation of photographic gelatin. In *Proceedings of Fifth IAG Conference*, ed. H. Amman-Brass & J. Pouradier. IAG, Fribourg, Switzerland, 1989.
19. Johnston-Banks, F. A., From tannery to table. *J. Soc. Leather Technol. Chem.*, **68** (1984) 141–5.
20. Saunders, P. R. & Ward, A. G., Mechanical properties of degraded gelatines. *Nature (London)*, **176**(4470) (1955) 26–7.
21. Hinterwalder, R., The evolution of the gelatin technology. In *Proceedings of Fifth IAG Conference*, ed. H. Amman-Brass & J. Pouradier. IAG, Fribourg, Switzerland, 1989.
22. De Clercq, M., Chemical aspects of photogelatin production. In *Proceedings of Fifth IAG Conference*, ed. H. Amman-Brass & J. Pouradier. IAG, Fribourg, Switzerland, 1989.
23. Cole, C. G. B. & McGill, A. E. J., The properties of gelatines derived by alkali and enzymic conditioning of bovine hide from animals of various ages. *J. Soc. Leather Technol. Chem.*, **72**(5) (1988) 159–64.
24. Ward, A. G., An accelerated alkaline pretreatment process for the preparation of gelatine from collagen, Parts I, II and III. *Gelatine and Glue Research Association Res. Rep.*, C1, C3, C5 (1955).
25. Isdon, B. & Braswell, E., Gelatin. In *Advances in Food Research*, Vol. 7, ed. E. M. Mrak & G. F. Stewart. Academic Press, New York, 1957, pp. 235–338.
26. Eastoe, J. & Ward, A., *Practical Analytical Methods for Connective Proteins*. Spon, London, 1963, pp. 99–100.
27. Li-Juan, C., Determination of iso-electric point and its distribution of photogelatin by iso-electric focusing method. Presented at the Fourth IAG Conference, Fribourg, September 1983.
28. Toda, Y., The relationship between the iso-electric point distribution and

molecular weight distribution of gelatin. In *Photographic Gelatin*, ed. S. Band. Royal Photographic Society, Bath, 1987.
29. Maxey, C. R. & Palmer, M. R., The isoelectric point distribution of gelatine. In *Photographic Gelatin II*, ed. R. J. Cox. Academic Press, London, 1976, pp. 27–36.
30. Tavernier, B. H., Electric charge potential of gelatin. In *Proceedings of Fifth IAG Conference*, ed. H. Amman-Brass & J. Pouradier. IAG, Fribourg, Switzerland, 1989.
31. Eastoe, J. & Ward, A., *Practical Analytical Methods for Connective Proteins*. Spon, London, 1963, pp. 36–41.
32. Stegemann, H. & Stakder, K., Determination of hydroxyproline. *Clin. Chim. Acta*, **18** (1967) 267–73.
33. Analysis of SO_2 in gelatine. BSI 757:1976.
34. Wainewright, F., Physical tests for gelatin and gelatin products. In *The Science and Technology of Gelatin*, ed. A. Ward & A. Courts. Academic Press, London, 1977. (See also BSI 757:1976.)
35. Kemp, P. & Stainsby, G., Mucoids in skin and their significance to the tanner. *J. Soc. Leather Technol. Chem.*, **65**(5) (1981) 85–90. (See also Refs 1–4.)
36. Cunningham, L. W., Ford, J. D. & Sergest, J. P., The isolation of identical hydroxylysyl glycosides from hydrolysates of soluble collagen and human urine. *J. Biol. Chem.*, **242**(10) (1967) 2570–6.
37. Croome, R. J., Variation of viscosity of gelatine solutions with temperature. *J. Appl. Chem.*, **3** (1953) 330.
38. Watanabe, M., Toyokawa, A., Shimada, A. & Arai, S., Proteinaceous surfactants produced from gelatine by enzymic modification. *J. Food Sci.*, **46** (1981) 1467.
39. Arai, S. *et al.*, Physicochemical properties of enzymatically modified gelatin as a proteinaceous surfactant. *Agric. Biol. Chem.*, **48**(7) (1984) 1861–6.
40. Arai, S. *et al.*, Enzymatically modified gelatin as an antifreeze protein. *Agric. Biol. Chem.*, **48**(8) (1984) 2173–5.
41. Eagland, D., Pilling, G. & Wheeler, R. G., Studies of the collagen fold formation and gelation in solutions of a monodisperse α-gelatine. *Faraday Discuss. Chem. Soc.*, **57** (1974) 181–200.
42. Hauschka, P., Harrington, W. & Rao, N., Collagen structure in solution, I–V. *Biochemistry*, **9** (1970) 3714–24, 3725–33, 3734–44, 3745–53, 3754–63.
43. Harrington, W. & Rao, N., Collagen structure in solution. I. Kinetics of helix regeneration in single chain gelatine. *Biochemistry*, **9** (1970) 3714.
44. Finer, E. G., Franks, F., Phillips, M. C. & Sugget, A., Gel formation from solutions of single chain gelatin. *Biopolymers*, **14** (1975) 1995–2005.
45. Ledward, D. A., Gelation of gelatin. In *Functional Properties of Food Macromolecules*, ed. J. R. Mitchell & D. A. Ledward. Elsevier Applied Science Publishers, London, 1986, pp. 171–201.
46. Stainsby, G., Gelatin gels. In *Advances in Meat Research*, Vol. 4, *Collagen as a Food*. Van Nostrand Reinhold, London, 1985, pp. 209–22.
47. Privalov, P. L., Stability of proteins. In *Advances in Protein Chemistry*, Vol. 35. Academic Press, London, 1982, pp. 55–94.
48. Eastoe, J. E. & Leach, A. A., A survey of recent work on the amino acid

composition of vertebrate collagen and gelatin. In *Recent Advances in Gelatine and Glue Research*, ed. G. Stainsby. Pergamon Press, London, 1958, pp. 173–8.
49. Norland, R. E., Fish gelatine. In *Photographic Gelatine*, ed. S. Band. RPS, Bath, 1987, pp. 266–71.
50. Croome, R. J., The kinetics of the development of rigidity in gelatin gels. In *Photographic Gelatin II*, ed. R. J. Cox. Academic Press, London, 1976, pp. 101–20.
51. Veis, A., *The Macromolecular Chemistry of Gelatin*. Academic Press, London, 1964.
52. Tkocz, C. & Kuhn, K., The formation of triple-helical collagen molecules from α1 or α2 chains. *Eur. J. Biochem.*, **7** (1969) 454–62.
53. Hayashi, A. & Oh, S., Gelation of gelatin solution. *Agric. Biol. Chem.*, **47**(8) (1983) 1711–16.
54. Djabourov, M., A review on gelatin gelation: recent experiments and modern concepts. In *Proceedings of Fifth IAG Conference*, ed. H. Amman-Brass & J. Pouradier. IAG, Fribourg, Switzerland, 1989.
55. Heidemann, E., Peng, B., Neiss, H. G. & Moldehn, R., Use of a microgelometer in determination of gel strength of collagen fragments. *Das Leder*, **40**(5) (1989) 87–91.
56. Grand, R. J. A. & Stainsby, G., N-terminal imino acids and gelatin gelation. In *Photographic Gelatin II*, ed. R. J. Cox. Academic Press, London, 1976, pp. 11–26.
57. Grose, S. & Rose, P. I., Non-gelling components from gelatin gels: extraction and identification. In *Photographic Gelatin II*, ed. R. J. Cox. Academic Press, London, 1976, pp. 73–100.
58. Heidemann, E. & Müller, H. T., Isolation and characterisation of peptides from acid processed pigskin gelatin. In *Proceedings of Fifth IAG Conference*, ed. H. Amman-Brass & J. Pouradier. IAG, Fribourg, Switzerland, 1989.
59. Szücs, M., The setting ability and the gelatin conformation. In *Proceedings of Fifth IAG Conference*, ed. H. Amman-Brass & J. Pouradier. IAG, Fribourg, Switzerland, 1989.
60. Rose, P., Computer-assisted modelling of gelatin properties: effects of peptide size and alpha chain sequence. *J. Photographic Sci.*, **34** (1986) 114–17.
61. Ledward, D. A. & Stainsby, G., Gelatine in chemically modified gelatins. *Gelatine and Glue Research Association Research Report*, January 1968, A38.
62. Slade, L. & Levine, H., Polymer-chemical properties of gelatin in foods. In *Advances in Meat Research*, Vol. 4, *Collagen as a Food*. Van Nostrand Rheinhold, New York, 1985, pp. 251–66.
63. Robinson, J., Kellaway, I. & Marriot, C., The effect of blending on the rheological properties of gelatin solutions and gels. *J. Pharm. Pharmac.*, **27** (1975) 818–24.
64. Marrs, W. M. & Weir, G. S. D., The gel properties of gelatin systems containing sorbitol and simple sugars. *British Food Research Association Res. Rep. No. 244*, 1976.
65. Marrs, W. M. & Weir, G. S. D., The effect of starch hydrolysates on the gelling properties of gelatin solutions. *British Food Research Association Tech. Circ. No. 646*, 1977.

66. Gelatin/carbohydrate interactions and their effect on the structure and texture of confectionery gels. *Prog. Food Nutr. Sci.*, **6** (1982) 259–68.
67. Tolstoguzov, V. B., Functional properties of protein–polysaccharide mixtures. In *Functional Properties of Food Macromolecules*, ed. J. R. Mitchell & D. A. Ledward. Elsevier Applied Science Publishers, London, 1986, pp. 385–415.
68. Tolstoguzov, V. B., Physikochemische Aspekte der Herstellung künstlicher Nahrungsmittel. *Die Nahrung*, **18** (1974) 523.
69. Tolstoguzov, V. B. *et al.*, On protein functional properties and the methods of their control, Part 1. *Die Nahrung*, **25**(3) (1981) 231; Part 2, **25**(9) (1981) 817.
70. Tolstoguzov, V. B., Interaction of gelatin with polysaccharides. In *Gums and Stabilisers for the Food Industry 5*. ed. G. O. Phillips, D. J. Wedlock & P. A. Williams, OUP, Oxford, UK, 1990.
71. Mohamed, S. & Stainsby, G., The ability of various proteins to form thermostable gels with propylene glycol alginate. *Food Chem.*, **13** (1984) 241–55.
72. Mohamed, S. & Stainsby, G., The digestibility of gelatin complexed with propylene glycol alginate. *Food Chem.*, **18** (1985) 193–7.
73. Ring, S. & Stainsby, G., Filler reinforcement of gels. *Prog. Food Nutr. Sci.*, **6** (1982) 323–9.
74. Boyar, M. M., Kilcast, D. & Fry, J. C., Use of gel-based model food systems in texture measurement. In *Gums and Stabilisers for the Food Industry 2*, ed. G. O. Philips, D. J. Wedlock & P. A. Williams.
75. Henry *et al.*, Texture of semi-solid foods: sensory and physical correlates. *J. Food Sci.*, **36** (1971) 155–61.
76. Szczeniak, A. S., Textural characterisation of temperature-sensitive foods. *J. Texture Studies*, **6** (1975) 139–56.
77. Muñoz, A. M., Pangbourn, R. M. & Noble, A. C., Sensory and mechanical attributes of gel texture. 1—Effect of gelatine concentration; 2—Gelatin, sodium alginate and kappa carrageenan gels. *J. Texture Studies*, **17** (1986) 1–16, 17–36.
78. Larson-Powers, N. & Pangborne, R. M., Paired comparison and time intensity measurements of the sensory properties of beverages and gelatines containing sucrose or synthetic sweeteners. *J. Food Sci.*, **43** (1978) 41–6.
79. Christenson, O. & Trudoe, J., Effect of other hydrocolloids on the texture of kappa carrageenan gels. *J. Texture Studies*, **11** (1980) 137–47.
80. Eves, A. & Jones, S., Electromyography of confectionery products. *Leatherhead Food Research Association Res. Rep. No. 595*.
81. Heatherbell, D. A., Fruit juice clarification and fining. *Confructa*, **28**(3) (1984) 192–7.
82. Bannach, W., Food gelatine in the beverage industries—an important help in juice and wine fining. *Confructa*, **28**(3) (1984) 198–206.
83. Wucherpfenng, K. P., Possman, P. & Kettern, W., Einfluss der gelatineart auf die schönungswirkung. *Flussiges Obst.*, **9** (1972) 388–406.
84. Hamatscek, J., Entwicklung eines kontinuierlichen weinklärverfahrens mit hilfe von zentrfugalseparatoren zur trubabtrennung unter besonderer berücksichtigung der kontinuierlichen gelatine–kieselsols–schönung maischeerhitzter rotweine. Dissertation, Univ. Hohenheim, 1982.
85. Ringland, C. & Eschenbruch, R., Gelatine for juice and wine fining. *Food Technol. NZ*, **18**(8) (1983) 40–1.

86. Cole, C. G. B., The use of gelatine in wine fining. In *Emulsifiers, Stabilisers and Thickeners in the Food Industry 1*, ed. P. B. Bush, I. R. Clarke, M. J. Kort & M. F. Smith. Natal Technicon Printers, Durban, 1986, pp. 227–35.
87. Kaufmann, R. & Janssens, L., Gelatine as a food ingredient and introduction to an original and pure instant gelatine. In *Emulsifiers, Stabilisers and Thickeners in the Food Industry 1*, ed. P. B. Bush, I. R. Clarke, M. J. Kort & M. F. Smith. Natal Technicon Printers, Durban, 1986.
88. The amino acid composition and nutrition value of proteins. 5. Amino acid requirements as a pattern for protein evaluation. Rama-Rao, Norton, Johnson.

RECOMMENDED READING

Ames, W. (1944). The preparation of gelatine, Parts I, II, III, IV. *J. Soc. Chem. Ind. London*, **63**, 200, 234, 277, 303.
Anon. (1976). The use of gelatin in confectionery products. *Confect. Prod.*, August, 359–60.
Anon. (1983). Gums–jellies–pastilles, Part 1—Preparation of high quality gelatine jellies. *Sweetmaker, Confect. Prod.*, June, 331–2.
Cumper, C. W. N. & Alexander, A. E. (1952). The viscosity and rigidity of gelatin in concentrated solutions I and II. *Australian J. Sci. Res.*, **A5**, 146–59.
Ellis, N. A. (1982). New plant produces gelatin. *Food Processing*, August.
Feddern, H. (1982). Uses and action of gelatine as a whipping agent and stabiliser. *Rev. Choc. Confect. Bakery*, **7**(1), 20–2.
Herramann, P. (1979). Production of gelatine from cattle bones. *Food Eng. Int.*, September.
Hoffmann, P. (1985). The use of gelatine in confectionery. *Susswaren Tech. Wirtsch.*, **29**(1/2), 33–8.
Lower, E. S. (1983). Utilising gelatin: some applications in the coatings, adhesives, and allied industries. *Pigment and Resin Technol.*, July, 9–14.
Mageean, P. & Jones, S. 1989). Low fat spread products. *Food Science and Technology Today*, September, 162–4.
Williams, C. T. (1960). The function of gelatine in sugar confectionery. *Confect. Manufact.*, October, 155–6, 167.

Chapter 8

MIXED POLYMER GELS

EDWIN R. MORRIS
*Department of Food Research and Technology,
Cranfield Institute of Technology,
Silsoe College, Silsoe, Bedford MK45 4DT, UK*

1 INTRODUCTION

Food products very often include in their formulation more than one hydrocolloid to achieve the desired physical structure, perceived 'eating quality' and behaviour during processing. Many such mixed systems behave as would be expected from the known properties of the individual polymers, as detailed in other chapters. In others, however, the properties of the blend are superior to those of either component alone or are qualitatively different. For example, some binary mixtures of gelling hydrocolloids give stronger gels than either of the individual polymers or, equivalently, give the same gel strength at a lower total concentration, with obvious cost advantages. More spectacularly, some polymers that are individually non-gelling will form gels on mixing. Such advantageous, non-additive behaviour is often described as 'synergism', and forms the main topic of this chapter.

Synergistic gel formation can occur in a variety of ways. The most direct mechanism is by formation of covalent bonds between the two polymers. This is not common in food systems, but a few examples are given in the following section. More generally, the behaviour of mixed polymer solutions is often dominated by the enthalpy of segment–segment interactions (in contrast to solutions of small molecules, in which entropy of mixing has a major effect). Attractive interactions (i.e. when the mixing process is exothermic) normally occur only for polymers of opposite charge, and often result in precipitation of a 'complex coacervate' (with practical applications for microencapsulation), but in some cases can lead to gel formation, as described in Section 3 on polyanion–polycation interactions. In most other cases the interactions are unfavourable (endothermic) and can lead to mutual exclusion of each component from the 'polymer domain' of the other, with increase in the effective concentration of both. At sufficiently high concentration the system can separate into two phases, giving a route to production of multitextured gels, as outlined in the section on 'polymer exclusion and biphasic gels'. Another way in which mixed gels can form is by cooperative association of long stretches of the two polymers into mixed 'junction zones', analogous to the junctions in many of the single-component gel systems described in previous chapters. It seems reasonably certain that the alginate–pectin acid gel system described in the final section of this chapter involves mixed junctions of this type. In many mixed-gel systems of practical importance the mechanism of gelation is unknown or is controversial. In particular, the synergistic gelation of galactomannans or related plant polysaccharides with xanthan, or with helix-forming polysaccharides in the agar/carrageenan series, is a topic of considerable current debate. These systems, which have extensive food applications, are discussed in detail in the remaining, largest, section of this review.

It is also possible for mixed systems to give the spurious appearance of polymer synergism when the true interaction is between one of the macromolecules and small molecules or ions present in the other polymer sample. For example, since alginate is very sensitive to calcium ions, small amounts of Ca^{2+} as a contaminant, or counter-ion, in another polymer sample could lead to gel formation without the necessary involvement of the other polymer. Similarly, sugar used to 'standardize' some commercial hydrocolloid samples might promote gelation of other polymers that are particularly sensitive to solvent quality. Since effects of this type really involve only one of the polymeric constituents, they are not considered further in the present chapter, but should be borne in mind in interpreting, and using, any apparent enhancement of properties in mixed polymer systems.

Although the mechanism of gel formation, characterization of physical properties and practical applications in food systems are outlined in each section of this review, coverage is by no means exhaustive. Detailed discussion of specific topics can be found in other reviews.[1-9] The reader is also referred to a particularly lucid and enjoyable overview of the general principles of multicomponent gel formation by Dr V. J. Morris.[10]

2 COVALENT INTERACTIONS

The most obvious and direct way in which two polymers can interact to form a gel network is by formation of covalent bonds between the chains. Perhaps the best-documented example of this behaviour for food hydrocolloids is the interaction of propylene glycol alginate (PGA) with gelatine under mildly alkaline conditions (pH ~ 9·6) to give gels[11,12] that are stable to above 100°C.

Infrared spectroscopy[11] indicates that the cross-links are amide bonds formed between carboxy groups of the alginate and uncharged amino groups from gelatine (Fig. 1). Since there are few histidine residues in gelatine, and the reaction occurs at pH values at which arginine is still fully ionized, the most likely site of reaction is at the ε-NH_2 of lysine and hydroxylysine. Similarly, the requirement[13] for a high degree of esterification implicates the propylene glycol ester groups as the reactive sites on the alginate. PGA samples with a high mannuronate content appear to give stronger gels than those in which guluronate predominates,[11] suggesting that ester groups attached to the flat, ribbon-like polymannuronate sequences are more accessible for reaction than those in the highly buckled polyguluronate regions (Fig. 2). This interpretation is consistent with the observation[11] that propylene glycol pectate, in which

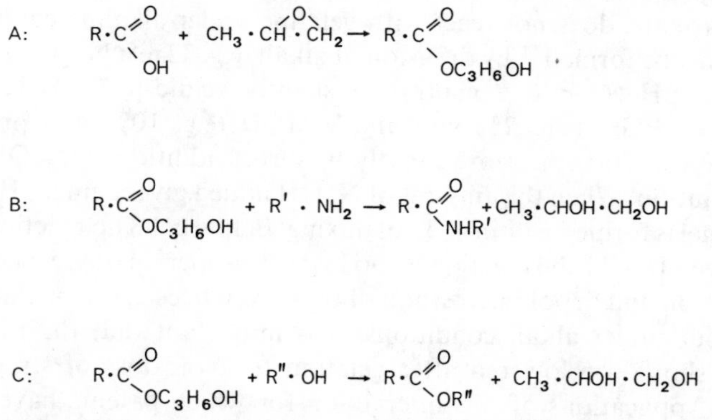

A: $R \cdot C(=O)OH + CH_3 \cdot CH(O)CH_2 \rightarrow R \cdot C(=O)OC_3H_6OH$

B: $R \cdot C(=O)OC_3H_6OH + R' \cdot NH_2 \rightarrow R \cdot C(=O)NHR' + CH_3 \cdot CHOH \cdot CH_2OH$

C: $R \cdot C(=O)OC_3H_6OH + R'' \cdot OH \rightarrow R \cdot C(=O)OR'' + CH_3 \cdot CHOH \cdot CH_2OH$

FIG. 1. A: Production of propylene glycol alginate (PGA) by reaction of partially neutralized alginic acid with propylene oxide. In commercial PGA the degree of esterification is typically 40–85%.[32] B: Chemical cross-linking between PGA and uncharged ε-amino groups of lysine or hydroxylysine residues of gelatine. C: Chemical cross-linking by transesterification of PGA with hydroxyl groups of other polymers, such as starch.

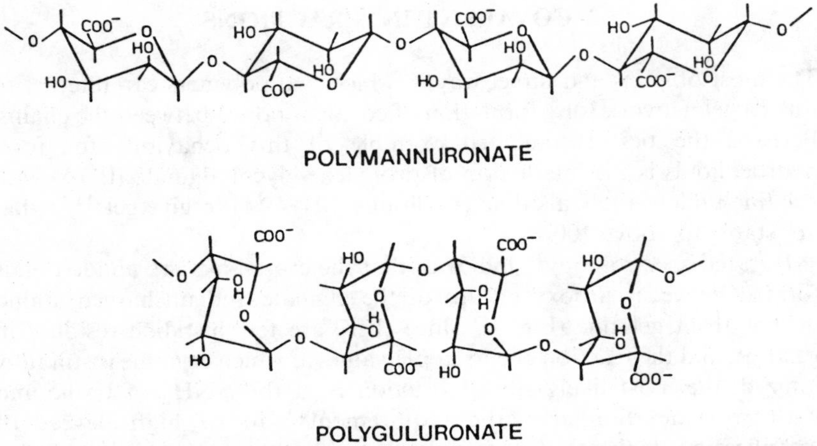

POLYMANNURONATE

POLYGULURONATE

FIG. 2. Comparison of backbone geometry in alginate polymannuronate and polyguluronate sequences. The flat, ribbon-like structure of polymannuronate may make the carboxy groups more accessible than in the highly buckled polyguluronate structure and hence explain the greater chemical reactivity of mannuronate-rich samples in, for example, extrusion of soya or cross-linking of PGA with gelatine.

the polygalacturonate backbone geometry is closely similar to that in polyguluronate, does not react with gelatine under alkaline conditions.

Gels may be formed[11] by diffusion of alkali (e.g. 32 mM Na_2CO_3/104 mM $NaHCO_3$; pH 9·6) into a neutral or slightly acidic (e.g. pH 6·5) mixed solution of PGA (e.g. 2% w/v) and gelatine (e.g. 10% w/v) previously dissolved at $c.$ 50°C, or more rapidly by direct addition of NaOH to the mixed solution. When the amount of NaOH added gives a final pH of $c.$ 9·6 a strong gel is formed within 4–5 s of mixing. Below pH 8 no reaction occurs and above pH ~ 11 the reaction is too fast for proper mixing. Since protein deamination and depolymerization of polyuronate esters by β-elimination both occur under alkali conditions, it is important that the pH of the product should be lowered after gelation to avoid loss of structure on storage. Applications of this interaction for which patents have already been filed include binding of fish protein concentrate into a product resembling fish muscle tissue[14] and formation of sausage skins by interaction of PGA with pastes, slurries or dispersions of collagen.[15]

PGA under mildly alkaline conditions also shows an analogous cross-linking reaction with a number of other polymers that contain hydroxyl groups,[13,16] probably by formation of ester linkages (Fig. 1). Polyvinyl alcohol and starch are particularly effective, and large increases in solution viscosity, or formation of thermally stable mixed gels, can occur at comparatively low concentrations of PGA. For example, on mixing thin-boiling starch (Viscosol 310; 8% w/v) with PGA (80% esterified; 2·0% w/v), raising the pH by addition of sodium carbonate (1 mmol/g PGA; i.e. 1·17% w/v) and acidifying after 15 min, a stiff gel was obtained.[13] As in the case of the amide cross-linked networks discussed above, reduction of pH after gel formation is essential to prevent subsequent alkaline degradation. Other weak alkalis, such as trisodium phosphate or sodium metasilicate, can be used in place of sodium carbonate, but sodium hydroxide is unsuitable,[13] probably owing to excessive depolymerization under local conditions of very high pH during mixing. The PGA–starch product can be broken down by enzymes, and might therefore be acceptable for food use.[13]

There is now reasonable evidence that the effect of alginate on the extrusion of soya[17–21] is due to specific, although as yet incompletely characterized, chemical interaction between protein and polysaccharide. Initial studies[17,18] on a laboratory-scale single-screw extruder showed that while most polysaccharides (including guar gum, locust-bean gum, carboxymethylcellulose, pectin and carrageenan) had little effect on the extrusion behaviour of soya grits, incorporation of alginate at the 1% level gave a marked reduction in extruder torque and product temperature. This effect has subsequently been confirmed on a twin-screw cooking extruder,[19] and pressure measurements at the die[18] show that it is due to a lowering in viscosity of the soya melt.

Subsequent investigations of soya in the presence and absence of alginate by differential scanning calorimetry and vacuum drying[20] strongly indicate that the viscosity decrease is due to the formation of additional water when soya and alginate are heated together. The interaction is specific for alginate in which mannuronate residues predominate ('high-M'). As discussed above, a similar specific reactivity of mannuronate-rich samples is observed in formation of amide bonds between propylene glycol alginate and lysine residues of gelatine, and in both cases may be due to the greater physical accessibility of carboxy groups in polymannuronate rather than polyguluronate (see Fig. 2).

There is also evidence of specific involvement of glutamic-acid residues from the protein. In a recent investigation of chemical changes in heated mixtures of proteins and alginate[21] only soya and gluten (proteins rich in glutamic acid) showed significant formation of additional water in the presence of high-M (but not high-G) alginate. Other proteins tested (pancreatin, ovalbumen, haemoglobin, whole blood albumen, casein and gelatine), all of which have a lower glutamic acid content, were inactive with both high-G and high-M alginate.

Neutral soya isolate heated (185°C; 35 min) in the presence of high-M alginate (Kelco Manucol DM; 67% mannuronate; 2% incorporation) showed,[21] in addition to formation of water, a marked reduction (from $c.19\%$ to $c.8\%$) in glutamic acid (amino-acid side-chain $R = CH_2CH_2COOH$) and a marked increase (from $c.9\%$ to $c.15\%$) in aspartic acid ($R = CH_2COOH$). The detailed chemistry of the interaction has still to be established, but it seems clear that there is a specific requirement for high-M alginate and for proteins with a high content of glutamic acid, and that the observed reduction in temperature and torque during extrusion is due to formation of additional water by chemical reaction between the protein and the polysaccharide.

3 POLYANION–POLYCATION INTERACTIONS

Another obvious way in which two different polymers can associate with one another is by charge–charge attraction, for example between an anionic polysaccharide and a protein below its isoelectric point. The rheological properties of mixed gels formed by gelatine and sodium alginate or low-methoxy pectin, under conditions where the polysaccharide alone does not gel, show evidence of this type of interaction.[3,22,23]

Freshly prepared gels of gelatine (e.g. 5% w/v) with alginate or pectate (e.g. 2·5%) below the isoelectric point of gelatine (e.g. at pH 3·9) melt over the same temperature range as gelatine alone (30–40°C), but on ageing they become thermostable. Thus, as illustrated in Fig. 3, the normal melting transition of gelatine is still observed as a decrease in rigidity, but the gel remains intact to much higher temperature (> 80°C), and the strength of the residual network increases with ageing time (typically over a period of about a week). This behaviour cannot be ascribed to the covalent cross-linking discussed in the previous section, since amide-bond formation occurs under alkaline rather than acid conditions. The gels are also stable in 7M urea, arguing against hydrogen bonding or hydrophobic interactions, but the enhanced thermal stability can be eliminated by high salt (e.g. 0·3M NaCl) or by raising the pH to above the isoelectric point of gelatine, as expected for an ionic network.

Normally, however, charge–charge attraction results in precipitation of the neutral complex, rather than in formation of a hydrated gel, and this behaviour is used commercially in microencapsulation. For example,

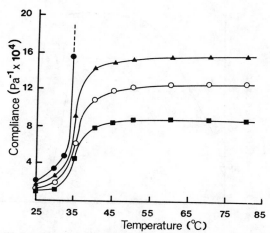

FIG. 3. Thermal stability of complex gels of gelatine (5%) and sodium alginate (2·5%) at pH 3·9 after ageing for 2 h (●), 1 day (▲), 3 days (○) and 5–8 days (■). (From Tolstoguzov,[3] with permission.)

gelatine and gum arabic form a coacervate that will deposit round dispersed oil droplets to produce microencapsulated spheres.[24] More recently the use of gellan gum in combination with gelatine has been explored.[25,26] Coacervation was observed[26] only in the pH range 3·5–5·0, where the gelatine and gellan gum had net positive and negative charge, respectively, and encapsulated droplets or beads were produced by the following preferred procedure. Hot (>60°C) solutions of gelatine (0·02–5·0% w/v) and gellan gum (fully deacylated; K^+ salt form; 0·02–1·5% w/v) were adjusted to pH 6–9 (with 0·1N NaOH) and mixed in equal volumes with the core material (oil droplets or solid particles) at a temperature above the gelation temperature of both polymers. The mixed solution was then adjusted to pH 3·5–5·0 (with 0·1N HCl) and cooled; the resulting coacervate droplets were then washed with water, fixed with glutaraldehyde (2·5–6·0%), washed again, dehydrated with isopropanol (70%) and concentrated to a slurry or dried to a free-flowing powder. In the case of encapsulated oils, however, the capsules often burst on drying to the powder form.

Under less acidic conditions the interaction of gellan gum with gelatine results in a synergistic enhancement of gel strength,[25] which may significantly reduce the total hydrocolloid requirement for specific applications. Native (acylated) gellan gum shows no such synergism, but there is a progressive increase in the strength of interaction as acyl

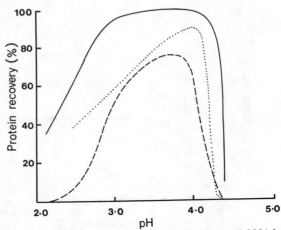

FIG. 4. Efficiency of protein recovery from a model system (0·20% bovine serum albumin; ionic strength 0·20) using 0·05% alginate (——), pectate (···) or carboxymethylcellulose (– – –). (From Imeson,[27] with permission.)

substituents are removed (by hydrolysis under alkaline conditions), with fully deacylated gellan (the normal commercial form) exhibiting the strongest interaction with gelatine. The optimum ratio of the two polymers is about 1:1, but useful effects can be obtained over the range 5:1 to 1:5, with total polymer concentrations of, typically, 0·2–0·4% w/w.[25] Mixed gels may be formed by dispersing the two polymers together in normal tap water (e.g. by stirring for c. 2 h), heating to 80°C for about 10 min and storing at room temperature (e.g. overnight). Possible applications (subject to food approval for gellan gum) might include[25] water- and milk-based desserts, syrups and toppings, fabricated fruits, vegetables, meat and fish, aspics and petfoods.

Another possible commercial use of polyanion–polycation interactions is in recovery of protein from abattoir effluent or from whey.[27] Figure 4 compares the efficiency of alginate, pectate and carboxymethylcellulose (CMC) in precipitating protein from a model system of bovine serum albumin (0·2%) at an ionic strength of 0·2, conditions similar to those in situations of practical importance. In all cases, optimum recovery is achieved under moderately acid conditions (typically pH ~ 4) and drops off sharply at higher pH (as the isoelectric point of the protein is reached) and, more gradually, at lower pH (as the negative charge on the polysaccharide is suppressed). There is also an optimum in protein:polysaccharide ratio, typically in the range 3:1 to 6:1. At higher ratios there is insufficient polysaccharide to precipitate all the protein; conversely, at very low ratios there is insufficient protein to neutralize the charge on the polysaccharide. As illustrated in Fig. 4, alginate is a more effective precipitant than either pectate or CMC, and can be used over a wider pH range. It is also least sensitive to ionic strength.[28]

Another important advantage of alginate is that its gelling properties (see Chapter 2) may be used to structure the recovered protein into a useful product. For example, the following procedure could be adopted for utilization of abattoir effluent. First remove solid debris such as hair, fat and skin (using, for example, a tilted-screen separator), add sufficient sodium alginate solution (2% w/v) to give a protein: polysaccharide ratio in the range 10:1 to 3:1, acidify to pH ~ 4 (using, for example, 10% sulphuric acid), and recover the resulting complex by filtration and pressing. The protein–alginate isolate may then be solubilized by slurrying in water and adding sufficient alkali to raise the pH to well above the isoelectric point of the protein (e.g. to pH ~ 6), mixed with other appropriate ingredients such as meat or fish mince, and structured using calcium alginate gel technology (see Chapter 2 for more details).

For example, gelled chunks suitable for use in canned petfoods may be moulded using a conventional 'internal-set' procedure with a sparingly soluble calcium salt (e.g. $CaSO_4 \cdot 2H_2O$) and sequestering agent (e.g. tetrasodium pyrophosphate); reformed fish fillets may be formed using a diffusion process, by filling a mixture of solubilized protein–alginate isolate and fish mince into moulds, spraying the surface with calcium chloride solution to form a skin, and completing the gelation process by immersing the fillet in a calcium bath; and fibres suitable for preparing meat and fish analogues may be formed by extruding the solubilized protein–alginate isolate through a spinneret into a calcium chloride bath.[29] Alternatively, the precipitated protein–alginate complex may be cross-linked directly using calcium salts, to give material similar to minced meat.[30]

Alginate may also be used[27,31] to recover protein from whey. An advantage of this procedure is that it allows the (charged) protein to be precipitated free of (neutral) lactose, thus avoiding problems from crystallization of lactose in final products, or discoloration of whey protein isolates during heat treatment in the presence of contaminating lactose.

Propyleneglycol alginate (PGA), in which some (typically 40–85%) of the carboxyl groups are esterified, also interacts with proteins below their isoelectric point, and since PGA is less susceptible than underivatized alginates to precipitation under acidic conditions, the resulting complexes may give a hydrated network rather than an insoluble precipitate. One practical application of this behaviour[32] is the use of PGA in stabilization of salad dressings. The strength of the thixotropic gel network formed by interaction of PGA with proteins also present in the formulation decreases with increasing degree of esterification, allowing flow properties to be controlled by choice of a suitable grade of PGA.

4 POLYMER EXCLUSION AND BIPHASIC GELS

4.a Thermodynamic Incompatibility

Since entropy of mixing (which favours single-phase systems) is dependent on the number of molecules present, rather than on their concentration by weight, it becomes largely insignificant for polymers. The behaviour of mixed polymer solutions is therefore determined predominantly by energies of interaction between chains.[2,3,33] This is true even when the energies involved are small, as in the case of transient contacts between disordered chain segments, rather than the much stronger cooperative interactions of ordered sequences typical of the junction zones in most polysaccharide gels.[34] When the net energy of interaction is favourable (i.e. when the mixing process is exothermic), as in the polyanion–polycation systems discussed in the previous section, the two polymers may associate into a single gel-like phase or insoluble precipitate. However, in the far more common situation where the interactions between the two polymers are less favourable than those between like segments of each type, there is a tendency for each to exclude the other from its polymer domain, so that the effective concentration of both is raised.

At sufficiently high concentrations, the system can separate into two liquid phases or, in extreme cases, one component may be driven out of solution by the other. Since the number of (energetically unfavourable) segmental contacts that a given polymer chain can make increases with increasing flexibility (as well as with increasing molecular weight), conformationally mobile synthetic polymers such as polyethylene glycol (PEG) are generally more effective in excluding other polymers than are polysaccharides (since steric clashes between adjacent sugar residues severely limit the conformational freedom of most carbohydrate chains[34]). Dextran, however, is a notable exception (since the 1,6-linkage separates the residues by three covalent bonds rather than two, thus conferring much greater flexibility), and it is therefore often used in experimental studies of polymer exclusion and in preparative techniques for isolation of proteins by precipitation.[2,3,33]

Even in single-phase mixed polymer systems mutual exclusion effects can

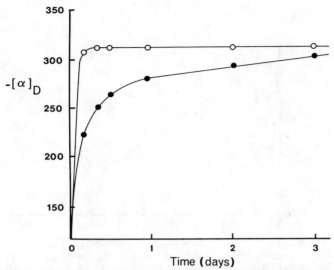

FIG. 5. Rate of triple-helix formation in gelatine (4%; 6°C) in the presence (○) and absence (●) of dextran (molecular weight 65 000; 0·2%) as monitored by the time course of optical rotation change. (From Tolstoguzov et al.,[38] with permission.)

produce large changes in physical properties. In particular, if one of the polymers can exist in two different conformations that occupy different volumes, the relative stability of the more compact form will be enhanced by the presence of an incompatible polymer species. For example, both dextran and PEG have been shown to inhibit separation of the DNA double helix into individual strands,[35] and stabilization of multi-subunit proteins or protein fibres by PEG has also been demonstrated.[36,37]

In the area of food gels the same behaviour is illustrated by the striking effect of polymeric co-solutes on the rate of formation and physical properties of gelatine gels.[38] As shown in Fig. 5, low levels of dextran (0·2%) greatly accelerate conversion from the disordered sol state to the more compact triple helical form responsible for gel formation. The effect on gel strength after a fixed setting time is illustrated in Fig. 6. On addition of increasing amounts of dextran to a fixed concentration of gelatine, the initial effect is to decrease the compliance of the gel (i.e. to increase its rigidity), as expected from increased helix formation. A point is then reached, however, beyond which further increase in dextran concentration raises the compliance once more (i.e. weakens the gel). The concentration of dextran at which this reversal occurs increases with decreasing molecular

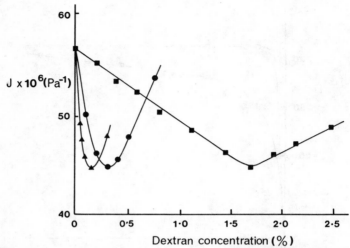

FIG. 6. Effect of dextran on the compliance ($J = 1/G$) of gelatine gels (10%) after a fixed gelation period (30 min). Results are shown for varying concentrations of dextrans of molecular weight 500 000 (▲), 65 000 (●) and 15 000–20 000 (■). (From Tolstoguzov et al.,[38] with permission.)

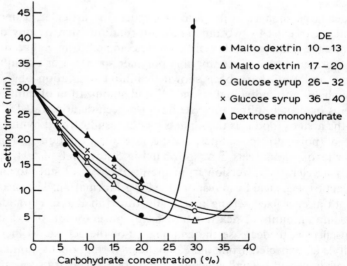

FIG. 7. Effect of starch hydrolysates on the setting times of an acid pigskin gelatine sol (3%) at 25°C; $DE = 100/n$, where n is the degree of polymerization. (From Marrs,[39] with permission.)

weight, as shown, and is associated with the onset of phase separation. Closely similar behaviour has been demonstrated using PEG samples of different molecular weights.[39] Gelatine gelation is also accelerated substantially by maltodextrins, and even by dextrose, although, as illustrated in Fig. 7, the concentrations required to produce useful enhancement are, of course, very much higher than for polymers.[39] Amylopectin has also been reported to show pronounced incompatibility with gelatine,[40,41] and indeed one of the first documented examples of polymer exclusion was between gelatine and soluble starch.[42] Although the effect on gel strength and setting rate has not been studied in detail, it seems likely that mutual exclusion may be an important determinant of structure development and final texture in food products incorporating both starch and gelatine.

Although the most striking incompatibilities are normally observed between different classes of polymer (e.g. polysaccharide–protein), a few examples of polysaccharide–polysaccharide exclusion effects have recently been reported. Figure 8 shows the effect of dextran ($\bar{M}_r \simeq 5 \times 10^5$) on the rigidity of amylose gels (3·2%). A steady increase is observed[43] up to a dextran concentration of $c.$ 3%, where the rigidity is about three times higher than for amylose alone, but at higher concentrations of dextran the gels become weaker. This behaviour closely parallels the initial strengthening and subsequent weakening of gelatine gels with increasing dextran concentration, shown in Fig. 6, and is again due to polymer incompatibility. Concentrated mixtures of amylose and dextran (e.g. 3·5% of both) become translucent immediately after mixing and, if held at elevated temperature (75°C) to prevent gelation, gradually separate into two distinct layers.[43]

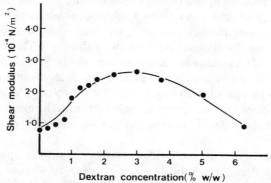

FIG. 8. Effect of dextran ($\bar{M}_r \simeq 5 \times 10^5$) on the gel rigidity of amylose ($\bar{M}_r \simeq 7 \times 10^5$; isolated from pea starch). (From Kalichevsky et al.,[43] with permission.)

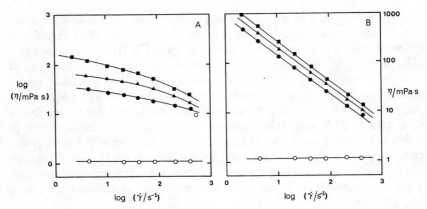

FIG. 9. Effect of wheat-starch solubles on the viscosity (η) of gum solutions over a range of shear rates ($\dot{\gamma}$). A: Guar gum (0·25%) alone (●) and in the presence of 0·4% (▲) and 0·6% (■) starch solubles. B: Xanthan (0·25%) alone (●) and in the presence of 0·6% (▲) and 1·0% (■) starch solubles. The viscosity of a 1·0% solution of starch solubles alone (○) is also shown on both graphs. (From Christianson et al.,[45] by courtesy of the American Association of Cereal Chemists.)

When the same mixtures are allowed to gel at lower temperature, microscopy studies reveal droplets of dextran-rich solution within the amylose network, leading to the observed reduction in gel strength. However, at lower concentrations, before the onset of phase separation, the net effect of the exclusion of amylose from the polymer domain of dextran is to raise its effective concentration and thus promote gelation.

While, for reasons of chain flexibility discussed previously, dextran is particularly effective in promoting exclusion of other polymers, other less-flexible polysaccharides can show analogous behaviour. In particular, a surprising incompatibility has recently been demonstrated[44] between amylose and amylopectin at concentrations above c. 2·5% of both. The full implications of this incompatibility between the two major components of starch have yet to be explored, but it seems likely to influence both the physical properties of starch gels and the time–temperature course of gelatinization, as both polymers are released from the granule. It may also shed light on the influence of other hydrocolloids on the paste viscosity of starch.

A number of polysaccharides, notably xanthan, guar gum and carboxymethylcellulose, show[45] large increases in viscosity when mixed with comparatively small amounts ($>c.$ 1% w/v) of the soluble material released from starch granules during gelatinization, although at the

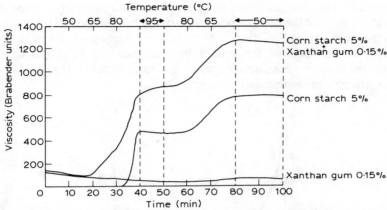

FIG. 10. Brabender amylographs showing the effect of xanthan on paste viscosity during the gelatinization of corn starch. (From Howling,[46] with permission.)

concentrations used the starch solubles alone have very low viscosity (Fig. 9). Similarly, when starch granules are gelatinized in the presence of these polysaccharides, a substantial enhancement of paste viscosity is observed. As illustrated in Fig. 10, the effect is particularly noticeable[7,45,46] during the early stages of granule swelling, when the viscosity increase for starch alone is barely detectable. Although originally interpreted in terms of the formation of starch–polysaccharide complexes, the recent studies outlined above suggest that the observed increases in viscosity may be due to complex mutual exclusion effects between the added polysaccharide and the amylose and amylopectin components of starch.

Whatever the detailed origin of the functional interactions between starch and other food polymers, they are already used in a number of practical applications. For example, when starch is gelatinized in the presence of xanthan, and the resulting mixture is dried under controlled conditions, a product is obtained that has many of the properties of chemically-modified pre-gelatinized starches, including enhanced resistance to acid, heat and shear.[47] The interaction of starch with xanthan can also be used to produce gluten-free bread, giving a matrix that during baking forms a structure similar to that in normal breads[48] and that will tolerate enrichment with added protein such as soy.

4.b Phase Separation

In the systems discussed so far, the useful effects of polymer incompatibility, such as increased gel strength and decreased setting time, have been confined to the concentration range in which both polymers coexist in a single phase, and the onset of phase separation has been associated with an undesirable reversal of these effects (e.g. weakening of gel structure). However, partition of the system into separate liquid phases can in principle be of very considerable practical value, by providing a route to the formation of multi-textured products.[4]

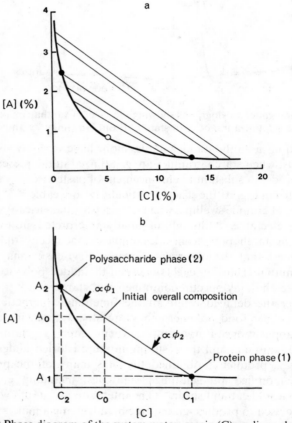

FIG. 11. (a) Phase diagram of the system water–casein (C)–sodium alginate (A) at pH 7·2 and 25°C. The bold line is the binodal or 'cloud-point' curve and the faint lines are 'tie-lines' joining points (●) representing the composition of coexisting protein-rich and polysaccharide-rich phases. With decreasing concentration the tie-lines converge to the critical point (○), below which the system remains as a single homogenous phase. (From Suchkov et al.,[49] with permission.) (b) Relationship between the composition and volume fraction of the two phases.

Figure 11(a) shows a phase diagram for a typical thermodynamically incompatible protein–polysaccharide system, casein alginate.[49] The bold line is the binodal or 'cloud-point' curve. When the concentrations of protein and polysaccharide present fall below this line, the system remains in a single phase, showing the (often useful) exclusion effects outlined previously. However, if we attempt to prepare a solution in which the protein and polysaccharide concentrations lie above the binodal (for example by mixing concentrated solutions of the two polymers), the mixture will separate spontaneously into two liquid phases, one enriched in protein and depleted in polysaccharide, and the other polysaccharide-enriched and protein-depleted. The points defining the composition of each phase lie on the binodal and the 'tie-line' joining them passes through the point corresponding to the initial composition of the mixture (i.e. the overall concentrations of protein and polysaccharide). All mixtures whose initial composition falls on the same tie-line will also separate into phases with these same final concentrations, but the relative volumes of the two phases will vary, as discussed below. For each system the phase separation behaviour can be defined by a family of (empirically derived) tie-lines that, with decreasing concentration, finally degenerate to a single 'critical point' (Fig. 11(a)).

The volume fraction (ϕ) in each phase is inversely proportional to the degree of enrichment of the dominant polymer (i.e. the more concentrated the phase, the less there is of it). Quantitatively, if we denote the volume fraction, polysaccharide (alginate) concentration and protein (casein) concentration in the protein-rich phase (phase 1) as ϕ_1, A_1 and C_1, respectively, and the corresponding parameters for the polysaccharide-rich phase 2 as ϕ_2, A_2 and C_2, then

$$\phi_1 A_1 + \phi_2 A_2 = A_0$$

and

$$\phi_1 C_1 + \phi_2 C_2 = C_0$$

where A_0 and C_0 are the concentrations of polysaccharide and protein, respectively, over the whole system. Since $\phi_1 + \phi_2 = 1$, it can be easily shown that

$$\phi_1 = (A_2 - A_0)/(A_2 - A_1) = (C_0 - C_2)/(C_1 - C_2)$$
$$\phi_2 = (A_0 - A_1)/(A_2 - A_1) = (C_1 - C_0)/(C_1 - C_2)$$

and

$$\phi_1/\phi_2 = (A_2 - A_0)/(A_0 - A_1) = (C_0 - C_2)/(C_1 - C_0)$$

An equivalent, and perhaps simpler, way of expressing these relationships is to say that the length of the tie-line (Fig. 11(b)) between the overall composition of the system and the composition of one of the phases is directly proportional to the volume fraction of *the other* phase.

As illustrated in Fig. 11, there is in general a gross asymmetry in the phase diagrams of protein–polysaccharide systems (note the difference in the axis scales), so that, with comparable concentrations of each polymer in the initial mixture, the final protein concentration in phase 1 is very much higher than the final concentration of polysaccharide in phase 2.[3,4] To a first approximation, this may be regarded as arising from the very much larger volume occupied by an expanded polysaccharide coil than by a compact protein molecule. As might therefore be expected, the degree of asymmetry increases with increasing molecular weight of the polysaccharide.[3]

As a rough rule-of-thumb, phase separation in protein–polysaccharide mixtures will occur only when the total polymer concentration exceeds $c.\,4\%$ w/v,[3] although there are, of course, substantial variations from system to system. For two of the most extensively studied food proteins, soya bean globulin and casein, the effectiveness of charged (anionic) polysaccharides in promoting phase separation decreases through the series pectin > CMC > alginate > gum arabic > dextran sulphate, and with all of these casein is less effective than soya.[3] In all cases the incompatibility of anionic polysaccharides with proteins (i.e. the degree of asymmetry in the phase diagram) increases with increasing salt concentration, but the relative effectiveness of different salts does not correlate with their position in the lyotropic series.

Phase separation is also observed as a general phenomenon in mixed solutions of different globular proteins,[4] or even in mixtures of the native and heat-denatured forms of the same protein, but the separation threshold is normally much higher than for protein–polysaccharide systems (typically at total protein concentrations in the range 12–20%). The only reported exception to this general behaviour is the ovalbumin–BSA (bovine serum albumin) system, which remains as one phase up to a total protein concentration of 50%, irrespective of pH and ionic strength. Unfolded protein structures are normally more effective than native globular structures in promoting exclusion of other polymers (i.e., as expected from space-occupancy considerations, their behaviour is intermediate between that of highly expanded polysaccharide coils and very compact globular proteins). For example, phase separation in mixtures of globular proteins and the denatured ('random-coil') form of gelatine occurs typically at a total protein concentration of $c.\,8\%$.

4.c Structure and Properties of Biphasic Gels

Immediately after phase separation, the system normally exists as an emulsion of the two liquid phases. Since in general the density of the phases will be different (for example, in protein–polysaccharide mixtures the concentrated protein phase will have higher density than the more dilute polysaccharide phase), the system will gradually separate into two discrete layers. In experimental studies of polymer incompatibility this process is often accelerated by centrifugation, for direct determination of phase volume. If, however, conditions are changed to promote gelation of one (or both) of the phases before macroscopic separation into two layers has occurred, then the system may be trapped as a 'gelled emulsion', whose physical properties may be quite different from those of a homogeneous gel of either component. The changes required to produce gelation will, of course, vary from system to system, as discussed in detail in other chapters. For example, gelation of gelatine or 'cold-setting' polysaccharides such as agar and carrageenan would involve cooling a heated premix; formation of cation-specific gels, such as those of alginate or low-methoxy pectin, would require controlled introduction of the appropriate cation; and gelation of globular proteins would normally involve heating.

The microstructure of the resulting mixed gel will be largely dependent on the state of dispersion of the two liquid phases at the time of gelation. In general, if the volume fractions are very different the smaller phase will be present as discrete droplets dispersed in a continuous matrix of the larger phase. If the phase volume of the disperse phase is increased (e.g. by increasing the overall concentration of the dominant polymer in that phase), the emulsion will ultimately invert, with the previously disperse phase becoming the continuous phase, and vice versa.

After gelation, the system may be regarded as a filled network or 'composite' of the type familiar in other polymer areas (e.g. filled rubbers). The mechanical properties of such composites may, to a reasonable approximation, be derived by simple proportion from the properties of the individual phases.[6] When the continuous phase is more rigid than the disperse phase, 'dilution' of the continuous network by easily deformable 'pockets' decreases the force required to produce a given deformation, i.e. the overall rigidity is proportional to the phase volume of the more rigid phase:

$$G = \phi_1 G_1 + \phi_2 G_2$$

where G_1 and G_2 are the rigidity moduli of the stronger and weaker phases, respectively, and G is the overall modulus of the composite. Both constituent phases are deformed by the same amount (i.e. they are subjected to the same strain: *isostrain* conditions). Conversely, when the continuous phase is weaker, the effect of 'dilution' by stronger, less-deformable 'beads' is to decrease the extent of deformation produced by a given force, i.e. the overall *compliance* ($J = 1/G$) is proportional to the phase volume of the more deformable continuous phase:

$$J = \phi_1 J_1 + \phi_2 J_2$$

i.e.

$$1/G = \phi_1/G_1 + \phi_2/G_2$$

Since both phases are now subjected to the same stress (although the more rigid 'beads' are deformed less than the surrounding softer matrix), these are known as *isostress* conditions.

For biphasic mixed gel systems where, as discussed above, phase inversion can occur as the relative volumes of the component phases change, the isostrain and isostress equations define, respectively, upper and lower bounds for the overall rigidity of the system. When the weaker component predominates—and is therefore the continuous phase—the system should follow the lower (isostress) bound; conversely, when the stronger component is the dominant, continuous phase, the system should follow the upper (isostrain) bound. Between these two extremes there may be a concentration range in which both phases are continuous but present in varying proportions, with a corresponding gradual shift of the overall modulus between the two bounds, as outlined schematically in Fig. 12.

The essential features of this comparatively simple treatment have been verified experimentally for two different mixed-gel systems, agar–gelatine[50] and agar–BSA.[51] There are, however, some complications in the behaviour of aqueous mixed gels of biopolymers that do not arise for normal polymer composites. In particular, the above treatment of the dependence of gel properties on phase values implicitly assumes that the relative proportions of the separate phases in the final gel is the same as in the initial mixed solution. As discussed earlier, however, the exclusion properties of polymers depend on their conformation as well as on their primary structure. Since biopolymer gelation almost invariably involves a change in conformation[5,34] there is likely to be an associated change in relative phase volumes (i.e. in the partition of solvent between the two phases—an effect that obviously cannot occur in polymer composites where there is no solvent present).

If, as will normally be the case, the two polymers gel at different rates, one network may already be almost fully formed before gelation of the second

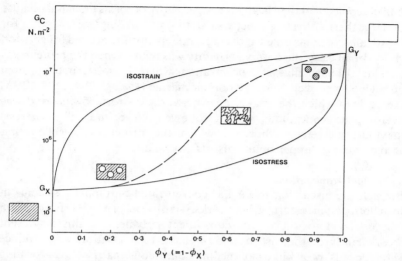

FIG. 12. Mechanical properties of biphasic gels. When the weaker component (X) predominates (i.e. $\phi_X \gg \phi_Y$, where ϕ is phase volume) and forms the continuous phase, the modulus of the composite gel (G_C) follows the 'lower-bound' isostress limit $[1/G_C = \phi_X/G_X + \phi_Y/G_Y]$; with increasing phase volume of the stronger component (Y), the system undergoes phase inversion (passing through a region of intermediate composition where both phases are continuous) and moves to the 'upper-bound' isostrain limit $[G_C = \phi_X G_X + \phi_Y G_Y]$. (From Clark,[6] with permission.)

component begins, and the mechanical properties of the pre-formed network may be largely unaffected by any changes in phase volume resulting from conformational changes in the other polymer as it gels (i.e. the properties of the mixed gel may reflect polymer concentrations in the two phases at the time of gelation, rather than final concentrations). In particular, if the polymer composition of the premix falls below the binodal (Fig. 11)—giving a homogeneous, single-phase solution—then the contribution to the overall properties of the mixed gel from the component that gels first may be close to that expected from its nominal concentration in the whole system. Quantitative analysis of the composition dependence of rigidity in cold-setting mixed gels of agar and gelatine shows evidence of this type of behaviour.[50]

The effect of gelation sequence on final mechanical properties has been demonstrated experimentally for the agar–BSA system.[51] If the mixed solution is first heated to gel the protein, and then cooled, gelation of agar occurs in the presence of the existing BSA network. Conversely, if the premix is cooled and then heated, the agar network forms first and, because of its large thermal hysteresis between setting and melting (see Chapter 1 for more details), remains intact at the higher temperature required for gelation of BSA. In all cases the observed rigidity was higher when the protein was gelled first. In particular, at concentrations where, in isolation, the protein would give firmer gels than the polysaccharide (1% agar; $>c.$ 10% BSA) mixed gels in which the weaker agar network was formed first (and was therefore the continuous phase) followed 'lower bound' (isostress) behaviour; conversely, those in which the protein was gelled first approximated to 'upper bound' (isostrain) rigidity.

4.d Product Implications

Although the discussion so far has concentrated on biphasic systems in which both polymers form gel networks, it is obviously possible for gelation to be confined to one of the constituent polymers. If this is present predominantly in the discontinuous phase, then gelation will produce discrete 'beads' of gel in a non-gelling continuous matrix. A gel-forming continuous phase and non-gelling dispersed phase, however, would yield gels with small pockets of solution dispersed through them. Although these will have the effect of weakening the gel (probably in direct proportion to their phase volume, following 'upper bound' isostrain behaviour for a discontinuous phase with essentially zero modulus) there may, in product situations, be compensating advantages in 'eating quality'.

For example, since hydrocolloid solutions have far better flavour-release characteristics than gels, the presence of non-gelled regions within the overall gel structure should give a higher intensity of perceived flavour for the same objective levels of flavouring, with possible scope for net reduction in ingredient costs. Similarly, the same perceived intensity of taste could be achieved with lower objective concentrations of, for example, sugar or salt, with potential value in formulation of low-calorie or low-salt products. The presence of a dispersed liquid phase within a gelled product might also give a useful way of enhancing perceived 'juiciness'.

The greatest potential value of biphasic gel structures, however, is probably in creation of interesting, non-homogeneous product texture. An obvious crucial issue here is the state of dispersion of the discontinuous phase within the continuous matrix, which will depend on many different factors, including mixing conditions, the difference in density of the two phases, interfacial tension, viscosity of both phases, and the length of time between mixing and gelation (i.e. the time available for small droplets to coalesce). The complex interplay of these factors is not yet fully understood,

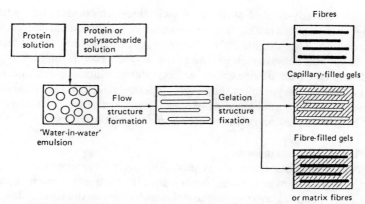

FIG. 13. Schematic illustration of the principle of 'spinneretless spinning' for production of oriented, multi-textured products. (From Tolstoguzov,[3] with permission.)

and individual systems must be tackled on a largely empirical basis. None the less, some encouraging progress has already been made in practical applications of food biopolymers in biphasic systems.

In particular, the use of flow to stretch and orientate the dispersed phase prior to gelation offers a useful route to gelled products with the anisotropic, fibrillar character of natural materials such as meat.[4] The general principle of this technique of 'spinneretless spinning' is outlined schematically in Fig. 13. The product obtained after rapid gelation (induced, as discussed earlier, by any appropriate method, such as heating, cooling or use of a 'setting-bath') may be fibres (if only the dispersed phase gels), a gel filled with liquid channels or capillaries (if only the continuous phase gels), or a homogeneous matrix filled with oriented fibres (if both phases gel). As well as the obvious scope for direct application in development of novel products and processes, the concept of 'spinneretless spinning' may provide useful understanding of existing processes (in particular extrusion cooking) where complex product structure is generated under flow.

5 GALACTOMANNAN SYNERGISM

The most familiar and widely exploited mixed gels of food hydrocolloids are probably those involving galactomannans, such as locust-bean gum, (or other structurally related plant polysaccharides), in combination with κ-carrageenan, furcellaran, agar or xanthan, all of which adopt rigid, ordered structures.[34] The double-helix-forming polysaccharides (carrageenan, furcellaran and agar) are all gelling agents in their own right, and as such are discussed in detail elsewhere in this volume. Xanthan and the plant polysaccharide cosynergists, however, do not gel in isolation, and their structures, sources and physical properties will therefore be summarized here, before discussion of their role in mixed-gel formation.

5.a Plant Galactomannans

The two galactomannans of greatest commercial significance for food use[52] are guar gum and locust-bean gum (alternatively known as carob gum). Both occur as energy-reserve polysaccharides in the seed endosperm of plants in the Leguminoseae family, whose members range from full-sized trees to small herbaceous plants such as clover.[53] Leguminoseae seedpods have a long history of use as animal fodder and, particularly in times of hardship, as human food. The name guar is derived from Sanskrit 'gau-ahar', meaning 'cow-food',[54] while locust-bean pods[54] are probably the 'husks that the swine did eat' with which the biblical Prodigal Son 'would fain have filled his belly', as well as sustaining John the Baptist in the wilderness: 'His meat was locusts and wild honey'.

The carob or locust-bean tree (*Ceratonia siliqua*), which grows on rocky soil in Mediterranean climates, begins to bear fruit after about 5 years but does not attain its full height (of about 8 m) or maximum yield for at least another 20 years. Carob trees can be cultivated but usually grow wild, and are harvested by hand. The guar plant (*Cyamopsis tetragonolobus*), by contrast, can be grown as an annual (nitrogen-fixing) crop, reaching a maximum height of about 60 cm, and can be harvested mechanically. Because of these obvious agronomic advantages, guar gum can be produced at lower cost than locust-bean gum (LBG), with greater flexibility of supply. Although guar gum was first produced in commercial quantities (as a substitute for LBG) only 35 years ago, it now far outstrips LBG as an industrial hydrocolloid (by at least a factor of 20). In both cases the commercial polysaccharide is obtained by milling the seed endosperm after removal of the husk and germ, giving products with a galactomannan content[54,55] of $c.$ 80%, the other main constituents being water ($c.$ 12%), protein ($c.$ 5%), lipid ($c.$ 0·5%) and inorganic material ($c.$ 1%).

The primary structure of Leguminoseae galactomannans[53] is based on a 1,4-linked β-D-mannan backbone substituted to different extents by single sugar side-chains of 1,6-linked α-D-galactose (Fig. 14(b)). Their overall

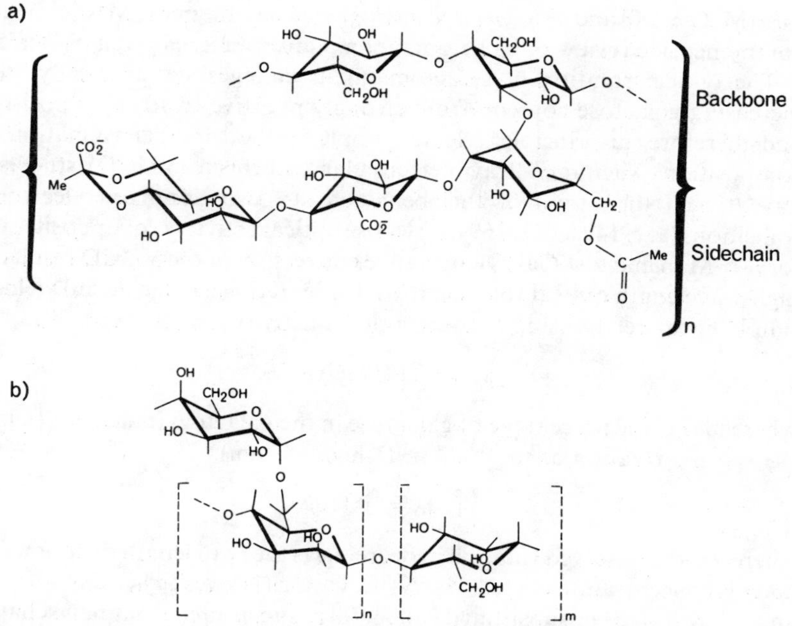

FIG. 14. Primary structure of (a) xanthan and (b) plant galactomannans. The acetate and pyruvate–ketal substituents of xanthan are present in varying proportions in different samples, and there may also be occasional missing side-chains. For the galactomannans, the lengths of substituted and unsubstituted blocks (n and m, respectively) can vary widely within and between samples, but are not now believed to have the large values originally proposed.[87,88]

composition may therefore be characterized by the relative proportions of mannose and galactose (M:G ratio), although, as discussed later, the pattern of substitution is also an important determinant of synergistic gel formation. In general, the ease of solubility increases with increasing galactose content, the parent mannan being totally insoluble in water (although it can be dissolved in strong alkali). Guar gum, which typically has an M:G ratio of around 1·55, dissolves completely on stirring at, or just above, ambient temperature (25–40°C), whereas LBG (M:G ≃ 3·5) is only partially soluble in cold water and indeed can be separated into cold-water-soluble and hot-water-soluble fractions with lower and higher M:G ratios, respectively. Other Leguminoseae of potential value as sources of food galactomannans, or of interest in studies of the mechanism of synergistic gel formation, include *Caesalpinia spinosa* (tara; M:G ≃ 2·7–3·0), *Cassia tora* (M:G ≃ 3·0) and *Trigonella foenum-graecum* (fenugreek; M:G ≃ 1·1). A comprehensive review of galactomannan plant sources is given in Ref. 53.

The optical rotation of galactomannans changes systematically with increasing galactose content, from strongly negative to strongly positive, and therefore, in principle, offers a simple method for determination of composition. Vacuum ultraviolet circular dichroism (VUCD) studies[56] have traced the origin of this behaviour to two dominant electronic transitions (at *c.* 149 and *c.* 169 nm) whose optical activity is of opposite sign for β-D-Man and α-D-Gal. Quantitative conversion of the VUCD results to high-wavelength optical rotation (by Kronig–Kramers transform) yields a simple linear relationship between optical activity and composition:

$$[\alpha]_D = 2\cdot 85(\%G) - 50 \tag{1}$$

where $\%G$ is the percentage of galactose in the galactomannan and $[\alpha]_D$ is the specific rotation at the sodium D-line (589 nm):

$$[\alpha]_D = 100\alpha/lc \tag{2}$$

where α is the observed rotation in degrees, l is the path length in decimetres and c is concentration in g/dl (% w/v). Equation (1) gives $[\alpha]_D$ values of -50 and $+92\cdot 5$ for unsubstituted and fully substituted mannan chains, respectively, in good agreement with measured values close to these extremes.[53] Reported deviations from strict linearity[57] may reflect the influence of differences in fine structure between samples of the same overall composition, but are perhaps more likely to be due to experimental error in chemical analyses. Equation (1) may be recast for direct determination of M:G ratio:

$$M:G = \frac{235 - [\alpha]_D}{50 + [\alpha]_D} \tag{3}$$

In the solid state, the galactomannan backbone adopts[58] an ordered, two-fold conformation with a repeat distance of *c.* 1·04 nm (i.e. 0·52 nm per residue, as expected for an almost fully extended 1,4-diequatorially linked chain). The individual chains pack together into flat sheets, with a spacing of *c.* 0·9 nm between chains. In the parent mannan the sheets are separated by about 0·72 nm, but the presence of galactose side-chains forces them apart, to a separation of about 3 nm or more, the exact spacing being sensitive to humidity but showing no systematic dependence on M:G ratio. Thus, the first few galactose 'spacers' are sufficient to abolish close packing of mannan sheets, and higher degrees of substitution can then be accommodated without further increase in spacing (the 'voids' where side-chains are missing from the lattice being occupied by water).

On dissolving, the galactomannans lose their ordered structure and go into solution as conformationally-disordered, fluctuating 'random coils'.[59,60] There is, however, evidence of some gradual 'renaturation' of intermolecular structure, particularly in the less-soluble galactomannans (i.e. those of low galactose content). For example, concentrated solutions of hot-water-soluble LBG (M:G $\simeq$ 4·5) develop a tenuous, gel-like structure when left to stand for several days at ambient temperature. The process is accelerated at low temperature (e.g. in a refrigerator) and particularly by freezing and thawing, when a rubbery, self-supporting gel structure can be obtained.[61] These gels are very prone to syneresis, and can lose as much as half their water content after a second freeze–thaw cycle. At concentrations below those required for gel formation, freezing and thawing galactomannan solutions can lead to selective precipitation of the less highly substituted chains.

The rheological properties of freshly prepared galactomannan solutions also show evidence of departures from normal 'random coil' behaviour. For all disordered polymers there is an abrupt change in the concentration dependence of solution viscosity at a critical concentration (c^*) at which the individual coils begin to interpenetrate one another.[62] Below c^*, where the chains are free to move independently, viscosity is approximately proportional to $c^{1.3}$ (i.e. doubling concentration c increases viscosity by a factor of c. 2·5). Above c^*, however, chains can move only by 'wriggling' through their neighbours, and viscosity (measured, as discussed below, under 'zero-shear' conditions, before the onset of 'shear thinning') then has a much higher concentration dependence (typically $c^{3.3}$, so that doubling concentration gives a 10-fold increase in viscosity).

The onset of coil overlap and entanglement obviously depends both on the number of coils present (proportional to concentration) and on the volume that each occupies (proportional to intrinsic viscosity $[\eta]$), and can therefore be characterized by the (dimensionless) 'coil-overlap parameter' $c[\eta]$. For normal 'random-coil' polymers (including most polysaccharide thickeners), plots of viscosity against $c[\eta]$ are closely superimposable (Fig. 15), irrespective of polymer type, molecular weight or solvent conditions, and the 'c^* transition' occurs at a value of $c[\eta] \simeq 4$ (where the solution viscosity is about 10 times the viscosity of water). For LBG and guar gum, however, the c^* transition occurs at a lower degree of space occupancy ($c[\eta] \simeq 2.5$), and the subsequent concentration dependence of viscosity is greater (about $c^{4.0}$, so that doubling concentration increases viscosity by a

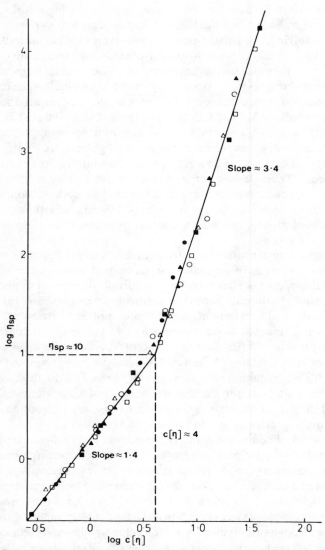

FIG. 15. Concentration dependence of 'zero-shear' viscosity in solutions of conformationally disordered ('random-coil') polysaccharides. The different symbols show results obtained for different polysaccharides. (From Morris et al.,[62] with permission.)

factor of c. 16, rather than c. 10 for most polysaccharides), although at concentrations below c^* their behaviour is the same as that of normal 'random coils'. The most likely interpretation of this behaviour is that at concentrations at which the galactomannan chains are forced into intimate contact with one another normal polymer entanglement is augmented by chain–chain association, as in the solid state.

The apparent viscosity of entangled 'random-coil' polysaccharide solutions (i.e. at concentrations above c^*) is strongly dependent on shear rate ($\dot{\gamma}$). At low shear rates (where there is sufficient time for entanglements pulled apart by the flow of the solution to be replaced by new entanglements between different chain partners, with no net change in the 'cross-link density' of the entangled network), viscosity remains constant at the maximum 'zero-shear' value (η_0). With increasing shear rate, however, a point is eventually reached beyond which re-entanglement can no longer keep up with the rate of destruction of existing entanglements, so that the overall entanglement density is reduced, and viscosity falls.[63]

The absolute values of η_0 and the shear rates at which 'shear thinning' begins can, of course, vary enormously from sample to sample. However, if measured viscosities are expressed as a fraction of the maximum 'zero-shear' viscosity (i.e. as η/η_0) and shear rates are similarly expressed relative to the shear rate, $\dot{\gamma}_f$, required to reduce viscosity to a fixed fraction (f) of η_0, then, as shown in Fig. 16, individual shear-thinning curves for all 'random polysaccharides', including LBG and guar gum, converge to a single 'master curve', irrespective of polysaccharide type, concentration (above c^*), molecular weight and solvent conditions.[12]

This generalized form of shear thinning can be fitted,[64] to well within experimental error, by a simple equation:

$$\eta = \frac{\eta_0}{1 + m(\dot{\gamma}/\dot{\gamma}_f)^p} \quad (4)$$

where m is an adjustable constant that depends on the value chosen for f. Since, by definition, $\eta = f\eta_0$ when $\dot{\gamma} = \dot{\gamma}_f$ then $f\eta_0 = \eta_0/[1 + m(\dot{\gamma}_f/\dot{\gamma}_f)^p]$, i.e. $f = 1/[1 + m]$.

The shear-thinning equation can therefore be further simplified by choosing a reference value of $f = 0.5$ (i.e. expressing all shear rates as a fraction of the shear rate required to halve the viscosity from its maximum 'zero-shear' value), since when $f = 0.5$, $m = 1$. Thus

$$\eta = \frac{\eta_0}{1 + (\dot{\gamma}/\dot{\gamma}_{1/2})^p} \quad (5)$$

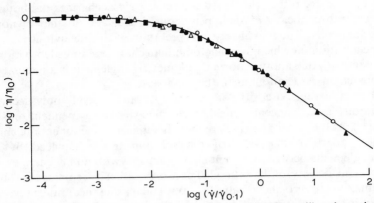

FIG. 16. Generalized form of shear thinning in 'random-coil' polysaccharide solutions. The different symbols show results obtained for different polysaccharides. Measured viscosities (η) are scaled to the maximum ('zero-shear') viscosity at low shear rate (η_0) and shear rates ($\dot{\gamma}$) are scaled to the shear rate ($\dot{\gamma}_f$) required to reduce viscosity to $f\eta_0$. In this case a value of $f = 0.1$ has been used, but the same superposition would be obtained with other values of f, the only difference being in the position of the curve relative to the horizontal axis. (From Morris et al.,[62] with permission.)

At very low shear rates, where $\dot{\gamma}/\dot{\gamma}_{1/2}$ is much less than 1, this approximates to $\eta = \eta_0$ (corresponding to the shear-independent 'plateau' region), while at high shear rates, where $\dot{\gamma}/\dot{\gamma}_{1/2}$ is much greater than 1, the equation reduces to

$$\eta = \frac{\eta_0}{(\dot{\gamma}/\dot{\gamma}_{1/2})^p} \quad \text{i.e. } \log \eta = \log \eta_0 - p \log(\dot{\gamma}/\dot{\gamma}_{1/2})$$

Thus $-p$ is the terminal slope of $\log \eta$ vs. $\log \dot{\gamma}$ at high shear rate, and has a constant (empirical) value of $p = 0.76$.

From eqn (5)

$$\eta_0 = \eta + (\eta \dot{\gamma}^p)/\dot{\gamma}_{1/2}^p \quad \text{i.e. } \eta = \eta_0 - (\eta \dot{\gamma}^p)/\dot{\gamma}_{1/2}^p$$

Thus, for any 'random-coil' polysaccharide solution, a plot of η vs. $\eta \dot{\gamma}^{0.76}$ should give a straight line of intercept η_0 and slope $-1/\dot{\gamma}_{1/2}^{0.76}$ (i.e. $\dot{\gamma}_{1/2} = (-1/\text{slope})^{1.316}$). Figure 17 shows the standard of linearity observed for a typical solution of guar gum.[65] Perhaps surprisingly, plots constructed from viscosity measurements made under ill-defined shear regimes (e.g. on a Brookfield viscometer, substituting spindle speed for shear rate) also show good linearity and give η_0 values very close to those obtained under well-defined conditions (e.g. using cone-and-plate geometry).

The advantage of this procedure (which can be carried out in a few

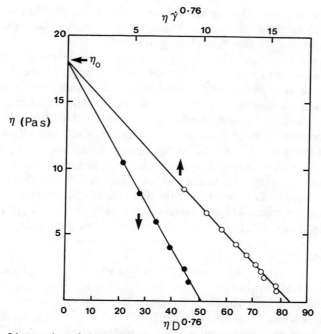

FIG. 17. Linear plot of shear thinning, illustrated for a typical solution of guar gum (Meyprogat 150, Meyhall, 1% w/v). Results obtained under an ill-defined shear regime (●), as a function of spindle speed (D) on a Brookfield viscometer, extrapolate to a zero-shear viscosity value (η_0) very close to that obtained from measurements at well-defined shear rate ($\dot{\gamma}$) on cone-and-plate geometry (○).

minutes using a pocket calculator) is that it provides a simple route to two parameters, η_0 and $\dot{\gamma}_{1/2}$, which, unlike observed viscosities, are independent of measuring conditions, and which, using eqn (5), define completely the viscosity of the sample at all shear rates. In particular, as shown in Fig. 15, η_0 at a known polymer concentration is directly related to intrinsic viscosity, which in turn is related to molecular weight (M) by the Mark–Houwink equation: $[\eta] = kM^\alpha$. Thus, in principle, it should be possible to calculate molecular weight from a few simple viscosity measurements. This approach has recently been calibrated for guar gum[65] using the known[9] Mark–Houwink parameters of $k = 3.8 \times 10^{-4}$ dl/g and $\alpha = 0.723$, yielding the relationship

$$\log M = 5.28 - 1.2 \log c + 0.26 \log \eta_0 \qquad (6)$$

where c is the concentration of guar gum in dl/g (% w/v) and η_0 is expressed in mPa s (centipoise). The method was calibrated using Brookfield viscosity readings plotted as shown in Fig. 17 to obtain η_0, and can be applied using any experimentally convenient concentration(s).

Investigations using rigorously purified enzymes to remove galactose side-chains without cleavage of the mannan backbone[66] have shown that the intrinsic viscosity of galactomannans is dependent solely on chain length, not on the degree of substitution. Thus, eqn (6) can also be applied to derive the molecular weight of other galactomannans if allowance is made for the contribution of galactose side-chains to the overall molecular weight. Taking the M:G ratio for guar gum as 1·55 (i.e. 61% mannan backbone) it can readily be shown[65] that the appropriate correction is to multiply the apparent molecular weight from eqn (6) by $0.61(1+R)/R$, where R is the M:G ratio of the galactomannan. The correction factor is small, ranging from 0·61 for unsubstituted mannan to 1·22 for a fully substituted chain (M:G = 1), and is of course 1 for guar gum.

5.b Xanthan

Xanthan, the extracellular bacterial polysaccharide from *Xanthomonas campestris*, was first isolated during an intensive screening programme at the USDA Northern Regional Research Laboratory in Peoria, and was originally known as polysaccharide B1459. Currently it is produced by large-scale aerobic fermentation, and is available commercially[67] in both food-grade and industrial-grade forms under trade names such as Keltrol and Kelzan (Kelco Inc.) and Rhodopol and Rhodigel (Rhône-Poulenc). Its primary structure[68,69] is based on a β-1,4-D-glucan (cellulose) backbone, substituted on O-3 of alternate backbone residues by charged trisaccharide side-chains of β-D-Man*p*-1,4-β-D-GlcA*p*-1,2-α-D-Man*p* to give a branched pentasaccharide repeating unit (Fig. 14(a)). The main source of structural variation is the presence or absence of 4,6-linked pyruvate ketal substituents on terminal mannose residues and/or acetate substituents on O-6 of the inner mannose residues. In normal commercial xanthans the degree of substitution is usually around 30–40% for pyruvate and 60–70% for acetate. However, there can be substantial variations in the extent of substitution[70] within and between chains, and in particular it is possible to separate xanthan from a single fermentation batch into high- and low-pyruvate fractions.[71] There is also evidence that the proportion of side-chains may fall short of the idealized repeating structure (Fig. 14(a)) by up to *c.* 5%,[72] but it is not yet known whether this undersubstitution occurs randomly or is concentrated in regions of unsubstituted cellulose backbone.

In contrast to most polysaccharide thickeners, xanthan normally exists in solution as a rigid, conformationally ordered structure, rather than as a fluctuating 'random coil'.[73] The ordered form is stabilized by salt[73-76] but at sufficiently low ionic strength it can be 'melted out' on heating. The order–disorder transition has been monitored by many physical techniques,[73-78] including optical rotation, circular dichroism, solution viscosity, NMR, differential scanning calorimetry and light scattering, and shows the sharp, sigmoidal form characteristic of a cooperative conformational change, with no detectable thermal hysteresis between formation and melting of the ordered structure.

X-ray fibre diffraction studies of xanthan in the solid state show a 5-fold helix with a pitch of 4·7 nm.[79,80] Since the backbone repeating unit is a disaccharide, each turn of the helix corresponds to 10 glucose residues, giving a rise per residue of 0·47 nm, rather than 0·52 nm in cellulose (or mannan). The quality of the X-ray data is not yet sufficient to decide between two possible candidate structures: a single helix stabilized by packing of side-chains along the polymer backbone[79] and a coaxial double helix[80] in which both strands have a conformation close to that in the single-helix model.

Solution-state evidence is conflicting. The disorder–order transition obeys first-order kinetics[75] and, under non-aggregating conditions, has no associated change in molecular weight,[74-76] indicating single-helix formation, but the mass-per-unit-length from light scattering is twice that expected for a single xanthan chain.[78] One proposal that might accommodate most of the current evidence is of a double helix that, on thermal denaturation over the experimentally accessible temperature range, does not fully disorder, so that the two strands remain associated and therefore renature as a single species (i.e. with first-order kinetics and no weight change).

Whatever the detailed nature of the ordered structure, it has profound effects on solution rheology.[64] Firstly, xanthan solutions show quite different shear-thinning behaviour from the generalized shear-rate dependence observed for disordered 'random-coil' polysaccharides (Fig. 16). As shown in Fig. 18, the double-logarithmic plot of viscosity vs. shear rate for xanthan under most salt conditions of practical importance is linear, with no indication of a 'Newtonian plateau' at low shear rates, and a substantially higher slope than the maximum value of -0.76 observed for random coils.

When the physical properties of xanthan solutions are measured under small-deformation conditions (e.g. low-amplitude oscillation), solid-like

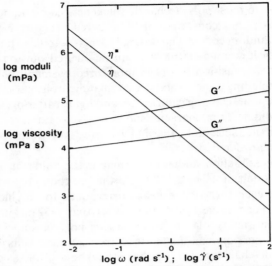

FIG. 18. 'Weak-gel' properties of xanthan (2% w/v). Storage and loss moduli (G' and G'', respectively) show little variation with frequency (ω), and G' exceeds G'' at all frequencies, as in a true gel. Dynamic viscosity (η^*) is substantially higher than rotational viscosity (η) at equivalent values of frequency and shear rate ($\dot\gamma$), in contrast to normal polysaccharide solutions, where the two coincide.

properties (characterized by the storage modulus, G') predominate over liquid-like response (loss modulus, G''), and both moduli show only slight variation with frequency (ω), behaviour typical of a gel rather than a normal solution.[64] The viscosity of xanthan solutions, measured under small-deformation oscillatory conditions (dynamic viscosity, $\eta^* = (G'^2 + G''^2)^{1/2}/\omega$), is substantially higher (Fig. 18) than the normal viscosity (η) obtained from conventional large-deformation measurements (e.g. in a rotational viscometer), in contrast to 'random-coil' polysaccharides in which both types of measurements give the same values.[62,81] Thus, xanthan solutions have a gel-like structure, but one that can readily be broken down to give a freely flowing solution. This combination of solid-like properties at rest and liquid-like properties under applied stress ('weak gel' behaviour) is of considerable practical value,[7] giving solutions that can be pumped or poured but are capable of holding particles in suspension or stabilizing emulsions over long times (weeks or months).

The solution properties of xanthan are remarkably stable to temperature and ionic strength, as would be expected for a rigid molecular structure. In particular, while the viscosity of 'random-coil' polyelectrolytes decreases with increasing salt concentration (owing to progressive suppression of intramolecular charge–charge repulsion and consequent contraction of coil dimensions[82]), xanthan solutions at concentrations above $c.$ 0·25% w/v show[70] a noticeable increase in viscosity (and enhancement of 'weak gel' properties) with increasing salt concentration (which can be attributed to reduction in intermolecular repulsion promoting network formation through helix–helix association). The effect is particularly pronounced for calcium ions,[83] and decreases through the series $Ca^{2+} > K^+ > Na^+$.

5.c Influence of Fine Structure on Galactomannan Interactions

As mentioned above, the tendency of galactomannans to self-association (e.g. on freezing and thawing) depends, to a good first approximation, on the degree of substitution, with materials of low galactose content interacting more strongly than more fully substituted samples. The same is also true in mixed synergistic interactions of galactomannans with other polysaccharides such as xanthan and agarose.[53] However, galactomannans with nearly identical overall composition can show appreciable differences in both self-association and synergistic interactions. For example, the galactomannan from *Leucaena leucocephala*, which has the same galactose content as guar gum ($c.$ 40%), forms weak, but cohesive, gels with xanthan at concentrations of both polymers as low as 0·1%, whereas the interaction of guar gum with xanthan is limited to enhancement of viscosity.[84,85] Similarly, the galactomannan from *Sophora japonica* has about the same galactose content as the hot-water-soluble fraction of LBG ($c.$ 18%), but shows appreciably weaker interactions both in self-association (freeze–thaw precipitation from dilute solution) and in the strength (yield stress) of mixed gels with xanthan.[85,86] These differences must obviously arise from differences in the distribution of galactose side-chains along the mannan backbone (i.e. differences in fine structure).

Early investigations of galactomannan fine structure by analysis of the products from digestion with α-galactosidase and β-mannanase enzymes were interpreted in terms of a block-like distribution of side-chains, with stretches of unsubstituted mannan backbone alternating with fully substituted regions.[87,88] This was an attractive model, since it appeared to offer a simple interpretation of the ability of galactomannans, particularly those of low galactose content, to form gels under forcing conditions (e.g. after freezing and thawing at high concentration), by association of unsubstituted mannan (as in the solid state), with the fully substituted interconnecting regions solubilizing the network.[61] Long stretches of

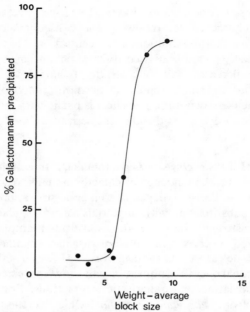

Fig. 19. Dependence of galactomannan self-association (as judged by the proportion precipitated from dilute solution after freezing and thawing) on the average length of totally unsubstituted regions of the mannan backbone. (From Dea et al.,[86] with permission.)

unsubstituted mannan also seemed likely sites of interaction with other polysaccharides in synergistic gel formation.[61,73] Subsequent studies using very carefully purified enzymes,[84] however, suggested a much more random structure, but it was pointed out in the same paper that regions of regular alternating structure, present in significant proportions in some of the galactomannans studied, might act as interaction sites since, in the preferred 2-fold conformation of the mannan chain,[58] these sequences would have one side totally unsubstituted (with the other side fully substituted). The same argument can be extended to sequences that are not strictly alternating, but which, in the 2-fold chain conformation, have one unsubstituted side (with an irregular pattern of side-chain substitution on the other side).

Characterization of galactomannan fine structure by enzymic digestion with β-mannanase rigorously purified from contaminating α-galactosidase activity has been developed using computer simulation of enzymic attack to identify the pattern of substitution that is most consistent with the oligosaccharide fragments obtained experimentally.[85,86] In particular, this approach gives a way of exploring the length distribution of potential binding sites (unsubstituted sequences or unsubstituted sides) for comparison with the degree of self-association and strength of synergistic interaction observed experimentally.

One of the most striking conclusions from these studies is that the extent of self-association, characterized as the proportion of the galactomannan precipitated from dilute solution (0·1% w/v) after freezing and thawing, is directly related to the average length of totally unsubstituted regions of mannan backbone, with almost no precipitation occurring below a threshold block length of about six consecutive unsubstituted mannose residues and almost complete precipitation above it (Fig. 19). Above this threshold the strength of mixed gels formed with xanthan increases with increasing average length of totally unsubstituted backbone regions, but below the threshold, where the unsubstituted regions are too short to associate with one another on freezing and thawing, they also seem incapable of participating in formation of mixed gels, since the order of synergistic gel strength then follows the average length of unsubstituted *sides* (which for all the galactomannans studied was greater than six residues[86]).

5.d Mechanism of Gelation

The nature of the synergistic gels formed by galactomannans and related plant polysaccharides is the subject of considerable controversy. Figure 20 shows a schematic model first proposed in 1972 for the formation of mixed gels by association of unsubstituted regions of the galactomannan chain to the double helix of κ-carrageenan, agar or furcellaran.[89–91] In terms of this model, the subsequent evidence on the effect of fine structure on galactomannan interactions, outlined above, would imply that unsubstituted sides can form analogous, but substantially weaker, mixed junctions, with a minimum sequence-length requirement of about six mannose residues in both cases. However, the concept of mixed junction zones in these systems has been questioned in a series of recent publications,[92–95] and it has been proposed instead that gelation results from polymer incompatibility, as discussed in the previous section, with the galactomannan driving helix–helix association in the algal polysaccharide. Microscopy studies at a resolution of $c.\ 1\ \mu m$ show no evidence of phase separation,[94] but this obviously does not preclude microheterogeneity at an even shorter length scale, or exclusion effects within a single phase.

The main evidence for the 'polymer exclusion' model comes from X-ray

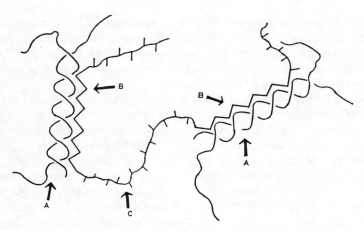

FIG. 20. Junction-zone model for synergistic gelation of galactomannans with helix-forming polysaccharides in the agar–carrageenan series. This proposal invokes direct association between the double helix of the algal polysaccharide (A) and unsubstituted regions of the galactomannan (B), with substituted regions (C) cross-linking and solubilizing the gel network. (From Rees,[90] with permission.)

fibre-diffraction studies of dried gels incorporating κ-carrageenan or furcellaran in combination with LBG, gum tara or konjac mannan (which, as discussed in more detail later, has the same β-1,4-diequatorial linkage geometry as the galactomannan backbone, and shows similar synergistic gelation). In all cases the diffraction patterns are similar to those of the algal polysaccharides alone, with no evidence of additional diffractions from mixed junctions. It could also be argued that, since exclusion effects are a general, expected property of mixed polymer systems, they provide the simplest 'minimum interpretation' of synergistic gelation, and should therefore be accepted as the correct model unless convincingly disproved (the 'Occam's razor' principle).

The main evidence in favour of the 'mixed junction' proposal is the striking dependence of synergistic gel formation on galactomannan structure (overall galactose content and side-chain distribution), although this could perhaps also be interpreted in terms of polymer incompatibility, with the algal polysaccharides driving self-association of mannan chains. However, there are two additional lines of evidence that are more difficult to reconcile with simple exclusion.

(1) Algal polysaccharides in the agar and carrageenan families include in their primary sequence occasional 'rogue residues' (1,4-linked

residues in the open-chair form rather than the anhydride form) which cannot be incorporated in the ordered, helical structure. Selective cleavage of the polymer chain at these residues[89] gives short segments which form helices but do not gel, no matter how high the concentration. They do, however, form firm, rubbery gels with LBG and gum tara,[89] which is consistent with their participation in mixed junctions, cementing together galactomannan chains, but difficult to interpret in terms of polymer incompatibility, since such low-molecular-weight species are unlikely to exert a major exclusion effect at the concentrations used (2% w/v), and, since the segments themselves do not gel even at very high concentration, the converse effect of galactomannan driving segment–segment association would also not be expected to lead to gel formation.

(2) Galactomannan of low molecular weight and low galactose content (20%), produced by enzymic modification of guar gum,[9] *decreases* gel strength when mixed with 0·2% agarose (in contrast to the marked enhancement observed with galactomannans of similar, or higher, galactose content, but higher molecular weight). At even lower concentration of agarose (0·1%), the short galactomannan chains caused rapid precipitation. This is again very hard to explain in terms of polymer incompatibility, but consistent with short, sparingly substituted chains participating in formation of mixed junctions with agarose helices, to promote aggregation without contributing significantly to development of a supporting gel network.

On balance, therefore, the available evidence probably favours the 'mixed junction' proposal, but with the qualification from X-ray studies[94] that the junctions may involve substantial helix aggregates rather than individual helices, a possibility accepted when the 'mixed junction' model was first proposed.[89] A requirement for large 'bundles' of helices would also be consistent with the direct correlation between synergistic gel strength and extent of aggregation (as judged by criteria such as minimum gelling concentration, turbidity and thermal hysteresis between gel formation and melting) across the agar–carrageenan series.[90] Thus, at one extreme agarose shows very pronounced aggregation behaviour and is also very effective in synergistic gelation, whereas, at the other extreme, ι-carrageenan has little tendency to aggregate and does not form synergistic gels, while κ-carrageenan, furcellaran and substituted agars show intermediate behaviour in both respects.

The structure of the synergistic gels formed by xanthan is also controversial, but in this case the nature of the controversy is quite different. Unlike polysaccharides in the agar–carrageenan series, xanthan does not gel, although, as discussed previously, it does show very tenuous 'weak-gel' properties at low deformations. The mechanical properties of the synergistic gels formed between xanthan and galactomannans or related plant polysaccharides, however, lie beyond any realistic estimates of 'upper-bound' behaviour (Fig. 12) for the individual polymers and therefore cannot reasonably be explained in terms of polymer incompatibility. There is also direct evidence of specific binding between galactomannan and xanthan from gel-permeation chromatography.[96] When LBG of reduced molecular weight (to prevent macroscopic gel formation) is mixed with xanthan, it elutes from the GPC column in two separate bands: a fraction of high galactose content with the same retention time as the galactomannan alone, and a fraction of low galactose content that co-elutes with the xanthan.

As for the algal polysaccharides, a 'mixed-junction' model has been proposed,[61,73] analogous to that outlined in Fig. 20 but with the 5-fold helix of xanthan replacing the double helix of agar, carrageenan or furcellaran. In this case, X-ray diffraction patterns of dried gels are not dominated by the helical polysaccharide, but show additional diffraction intensities indicative of specific interactions between the two components.[94,97,98] These, however, have been interpreted as binding of the plant polysaccharides (LBG,[97] gum tara[94] and konjac mannan[98]) to regions of the xanthan chain in the 2-fold conformation typical of cellulose (and mannan), rather than to the 5-fold xanthan helix. Since techniques such as optical rotation show no detectable difference in the extent of conformational reordering when xanthan is cooled in the presence or absence of interacting galactomannan,[61] any such 2-fold junctions must involve only a small fraction of the xanthan chains.[94] In this case, therefore, the 'Occam's razor' argument favours the type of interaction shown in Fig. 20: if the 5-fold structure of xanthan is the only form detectable in synergistic gels, it seems reasonable to assume that it is the form involved in junction formation, rather than invoking an experimentally undetectable alternative structure, unless there is compelling evidence to the contrary.

The main evidence for the '2-fold junction' proposal is that when the polysaccharides are mixed under conditions in which the xanthan helix is already present (low temperature and/or sufficient concentration of appropriate counter-ions) the mixture remains fluid, and a cohesive gel is formed only after heating to a temperature at which xanthan is converted to the disordered form, and then re-cooling. This, however, seems far from conclusive, since the process of mixing could destroy the gel network as it forms (as happens in many other gelling systems; for example, direct mixing of solutions of calcium chloride and alginate or pectin gives a microgel slurry, or a precipitate, rather than a recognizable gel). Although the mechanical properties of 'cold-mixed' xanthan–galactomannan systems are far weaker than those obtained after heating and cooling when measured under large-deformation conditions (e.g. by creep compliance), the elastic response to low-amplitude oscillation at high frequency (G' at $c.$ 50–100 rad/s) is closely similar (to within a factor of 2), and very much greater than that of either polysaccharide alone at the same total polymer concentration. This is exactly the behaviour expected of a 'smashed gel' in comparison with the corresponding intact network, and has been observed independently for mixtures of xanthan with LBG[99] and with konjac mannan,[98] although in the latter case the results were interpreted as showing no interaction under 'cold-mix' conditions.

5.e Upgrading Galactomannans by Enzymic De-branching

As discussed previously, guar gum is substantially less expensive than LBG and the supply is more secure and better able to respond to changes in demand. However, it interacts far less strongly with galactomannans, forming cohesive gels only with agarose, but not with more commercially important materials such as xanthan and κ-carrageenan.[53,89] Indeed, addition of guar gum to gelling mixtures of xanthan–LBG can substantially *decrease* gel strength,[100] presumably by competition for binding sites on the xanthan chain without significant contribution to network formation. Since the overriding structural difference between the two polymers is the higher content of galactose in guar gum, it would seem an attractive commercial proposition to improve the synergistic properties of guar up to, or perhaps beyond, those of LBG, by partial removal of galactose side-chains.

Selective cleavage of galactosyl linkages without accompanying cleavage of mannosyl linkages in the polymer backbone cannot be achieved by known chemical routes, but should, in principle, be possible using α-galactosidase enzymes. The technical feasibility of doing so has been demonstrated experimentally.[66,84] Again, a major problem is to avoid accompanying cleavage of the polymer backbone, since the most obvious sources of appropriate de-branching enzymes (germinating guar seeds, or microorganisms that grow on guar) also produce enzymes with mannanase or mannosidase activity.[101] However, careful purification by affinity

TABLE 1
PHYSICAL PROPERTIES OF GALACTOMANNANS PRODUCED BY ENZYMIC DE-BRANCHING OF GUAR GUM AND THEIR EFFECTIVENESS IN SYNERGISTIC GELATION WITH XANTHAN

Sample	Galactose (%)		Intrinsic viscosity, $[\eta]^a$ in units of dl/g polymer or dl/g mannose		Mixed-gel properties[b]		
	By analysis[c]	From OR^d			G' (Pa)	Yield stress $(N)^e$	MP $(°C)^f$
A (guar)	38·7	40·0	15·1	24·6	57	g	g
B	33·2	34·0	16·9	25·3	145	g	g
C	27·6	28·0	17·2	23·8	263	g	36
D	23·2	22·5	18·8	24·5	425	2·6	38
E	19·0	18·7	20·7	25·5	586	4·4	41
F	15·3	13·7	20·5	24·2	655	10·9	45
G	12·8	12·2	20·4	23·4	674	12·9	46
H	9·9	h	h	h	579	9·8	48
LBG	22·8	23·0	12·2	15·8	280	3·4	39

From McCleary et al.,[102] with permission.
[a] Average of values from Contraves Low Shear 30 and Ubbelohde suspended-level viscometers.
[b] Galactomannan (1% w/v) and xanthan (0·5% w/v) heated together for 5 min at 120°C.
[c] Average of values from GLC and enzymic analysis.
[d] Calculated from optical rotation values by eqn (3).
[e] Measured by compression of cylindrical samples of 12·8 mm diameter and 12·3 mm depth, aged for 24 h.
[f] Gel melting point, estimated from the point of inflection of plots of G' vs. temperature.
[g] Either gels did not form or were too weak to allow measurement of properties.
[h] Not determined owing to polymer solubility problems.

chromatography can yield α-galactosidase free of interfering enzyme activities.[66,101,102]

Table 1 shows the physical properties of a range of galactomannan samples prepared[102] from a single batch of guar gum, using purified α-galactosidase to remove different proportions of the galactose side-chains. Galactose content determined by chemical and enzymic assays is in good agreement with values obtained from optical rotation using eqn (3). Intrinsic viscosity shows a slight *increase* with progressive de-branching, rather than the decrease that would be expected if there were any significant cleavage of the polymer backbone. This increase is associated with the progressive reduction in molecular weight per backbone residue: when intrinsic viscosity is expressed relative to the concentration of the mannan backbone alone (as shown in Table 1) it remains constant, to within experimental error, at the same value as obtained for the unmodified sample of guar gum, indicating that (a) there is no detectable chain cleavage and (b) the hydrodynamic volume of galactomannans is determined solely

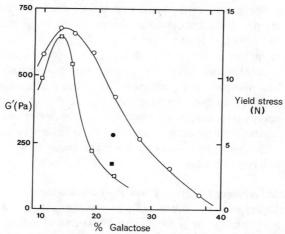

FIG. 21. Yield stress (□, ■) and storage modulus G' (○, ●) for mixed gels of xanthan (0·5%) with galactomannans (1·0%) of varying galactose content. Open symbols show results for enzymically de-branched guar gum (the sample of highest galactose content being the unmodified starting material) and filled symbols show results for a typical sample of LBG. (From McCleary et al.,[102] with permission.)

by the length of the mannan backbone, and is unaffected by the degree of branching.[84]

Figure 21 and Table 1 show the physical properties of the mixed gels formed by these enzymically modified materials in combination with xanthan, in comparison with a typical sample of LBG. Gel strength, measured under both large-deformation conditions (yield stress) and small-deformation conditions (G'), increases with progressive de-branching down to a galactose content of $c.\,13\%$ (M:G $\simeq$ 7). At even lower galactose content, the measured values decrease, almost certainly owing to the onset of insolubility in the galactomannan. Gel melting points increase steadily with extent of de-branching for all samples, including the sparingly soluble material of lowest galactose content.

The physical properties of the mixed gels formed by LBG are similar to those of the enzymically modified samples at comparable overall galactose content. The slight differences shown in Fig. 21 are probably due to differences in fine structure (length distribution of unsubstituted regions and unsubstituted sides) and to the somewhat lower molecular weight of the LBG sample used (Table 1). At lower galactose content (down to $c.\,13\%$) the samples prepared by enzymic de-branching of guar gum outperform LBG in synergistic gel formation. Thus, if an economically viable process for production of such materials could be devised, they would not only provide an acceptable alternative to LBG but could also extend the range of mixed-gel properties.

One of the major commercial obstacles to simply scaling-up the laboratory process for enzymic de-branching is the cost of recovery of the modified galactomannan from comparatively dilute solution, since any feasible route, either by direct drying or by solvent precipitation, would involve high energy consumption. However, it has been reported in a recent patent[103] that the de-branching enzyme will function in concentrated pastes of guar gum (up to $c.\,40\%$ w/v), which could point the way to a practical commercial process.

5.f Synergistic Gelation of Other Plant Polysaccharides

The β-1,4-linked mannan backbone of galactomannans (Fig. 14(b)) can be regarded as a variant of the β-1,4-glucan backbone of cellulose (differing only in the configuration of the hydroxyl group at C-2). Other natural variants that share the same 1,4-diequatorial linkage geometry include xylans (in which the C-6 hydroxymethyl group of cellulose is replaced by hydrogen) and their substituted derivatives such as arabinoxylans, glucomannans (in which both mannose and glucose residues occur in the polymer backbone), and substituted cellulose derivatives, such as tamarind-seed polysaccharide. Several of these show synergistic interactions analogous to those of the galactomannans,[9] but the strength of interaction is normally low (similar to that of galactomannans with a very high galactose content), and formation of true gels normally occurs only with agarose (if at all).

A notable exception, however, is the glucomannan from the tubers of *Amorphophallus konjac* (konjac mannan), whose interactions are comparable, or superior, to those of LBG.[9,61] The glucomannan backbone contains $c.\,40\%$ glucose and $c.\,60\%$ mannose, with both residues distributed fairly evenly along the chain rather than in blocks, and is solubilized by acetylation of about one residue in six.[9,53] Acetylation does not, however, prevent ordered packing of the glucomannan chains in the solid state.[98]

Konjac mannan interacts with non-gelling concentrations of both agarose and κ-carrageenan to give gels similar in mechanical properties and melting point to those obtained using LBG.[9] With xanthan, however, konjac mannan forms gels at unusually low concentration (as low as $c.\,0.02\%$ total polysaccharide) and with unusually high melting points ($c.\,63°C$ for a 1:1 mixture with xanthan at a total polysaccharide concentration of 0·5%, in comparison with c. 41°C for a corresponding xanthan/LBG gel). Furthermore, removal of the acetate groups from xanthan gives a slight increase in both self-association ('weak-gel' properties) and synergistic interaction with galactomannans such as LBG, but substantially destabilizes the interaction with konjac mannan, in particular reducing the gel melting and setting temperatures by c. 20°C (to give values similar to those obtained with galactomannans). Thus, the interaction of konjac glucomannan with unmodified xanthan appears to show a strength and specificity that is not shared by the galactomannans.[61] Commercial utilization of this interaction, either directly or as part of a ternary mixed-gel system with galactomannan, seems an attractive possibility in many product areas. Although konjac mannan is not a permitted food additive in the West, it has a long history of food use in Japan, which might facilitate USDA and EC approval.

Another plant polysaccharide of potential value in synergistic gel formation is the seed xyloglucan from tamarind (*Tamarindus indica*). Like konjac, tamarind has a long history of food use, particularly in India, and has been used in the West, although it is not currently a permitted additive owing to lack of exploitation. The seed amyloid has a β-1,4-linked glucan (cellulose) backbone in which about two-thirds of the residues are substituted at O-6 by α-D-xylose side-chains, about half of which are themselves substituted at O-2 by β-D-galactose,[9,104] giving both monosaccharide and disaccharide side-chains (with all residues in the pyranose ring form). The native polysaccharide shows only weak synergistic activity, interacting with agarose to about the same extent as very highly substituted galactomannans such as gum fenugreek.[9] However, an enzymic debranching procedure, analogous to that outlined above for galactomannans, has recently been reported,[104] and this may offer a route to production of material with commercial potential for mixed-gel formation.

5.g Mechanical Properties of Synergistic Gels

Figure 22 shows the effect[9] of addition of low concentrations of galactomannan (0·1%) on the rate of gelation and ultimate gel strength of agarose (0·2%). Neutral, unsubstituted agarose interacts more strongly with galactomannans than any other known co-synergist, showing a strong gelling interaction with guar gum and even with gum fenugreek, in which the mannan backbone is almost fully substituted (M:G $\simeq$ 1:1). At the concentrations used,[9] LBG increases the final rigidity (G') of the gel by a factor of about 3.

Natural agars frequently carry substituents, the most common of which

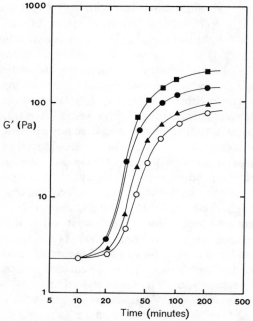

FIG. 22. Development of rigidity (storage modulus G') at 25°C after cooling solutions of agarose (0·2%) alone (○) and in the presence of 0·1% gum fenugreek (▲), guar gum (●) or LBG (■). (From Dea and Rees,[9] with permission.)

are O-methyl groups. As shown in Fig. 23, interaction with LBG (quantified[47] by the magnitude of the increase in optical rotation accompanying gel formation) decreases systematically with increasing methoxyl content for a wide range of agar samples. Two samples (with O-methyl content of 1·91% and 2·37%) that lie off the line were also anomalous in their gelation temperature in the absence of galactomannan,[105] suggesting that in some instances the position and/or distribution of O-methyl groups along the agarose backbone, as well as their overall concentration, may have a significant effect on both self-association and synergistic interaction with galactomannans.[9]

In mechanical terms, the presence of O-methyl substituents is reflected in both a decrease in rigidity of synergistic gels and in the requirement for higher concentrations of galactomannan to induce gel formation at non-gelling concentrations of agar.[9] For example, a 0·05% solution of unsubstituted agarose is gelled by addition of 0·05% LBG, whereas 0·6% LBG is needed to gel an equivalent 0·05% solution of agar with a methoxyl

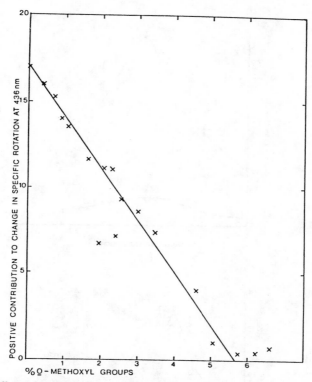

FIG. 23. Effect of O-methyl substituents on the interaction of agar (0·1%) with LBG (0·3%) as characterized by the magnitude of the positive optical rotation change accompanying gelation. (From Dea and Rees,[9] with permission.)

content of 2·26% (in the middle of the range of samples shown in Fig. 23). This extent of substitution is roughly equivalent to methylation of $c.$ 25% of the disaccharide residues in the polymer chain. When more than half the disaccharides are substituted (e.g. in the most highly methylated samples in Fig. 23, with an O-methyl content above $c.$ 5%), interaction is undetectable by the criterion of optical rotation change and rheological measurements show only very slight enhancement of gel strength.[9]

Charged substituents can have an even greater effect. For example, the agar from *Gracilaria compressa*, which has a pyruvate ketal content of $c.$ 3%, shows less interaction with LBG than agars with a corresponding content of neutral O-methyl substituents.[9] More importantly, the degree of sulphation appears to be the major determinant of differences in the

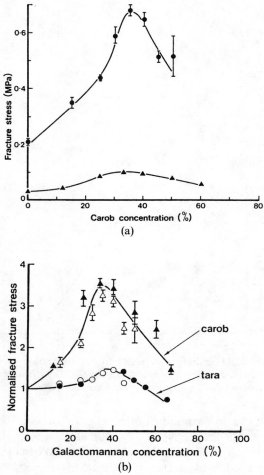

FIG. 24. (a) Variation in the strength (resistance to fracture under compression) of κ-carrageenan–LBG mixed gels on varying the proportion of LBG (carob) at fixed total polymer concentrations of 2·0 (●) and 1·5 (▲) % w/v. (b) Composition dependence of fracture stress in κ-carrageenan–galactomannan mixed gels, normalized to the value obtained for carrageenan alone. Results are shown for mixed gels with LBG at total polymer concentrations of 2·0 (△) and 1·5 (▲) % w/v (from (a)), and with gum tara at 3% (○) and 2% (●) total polymer concentration. (From Cairns et al.,[106] with permission.)

strength of synergistic interaction across the agar–carrageenan series. Furcellaran (which has the same backbone repeating structure as carrageenan) has a sulphate substituent on only c. 50% of the constituent disaccharides and interacts far more strongly than κ-carrageenan, in which each disaccharide is sulphated, while ι-carrageenan, which has two sulphate groups per disaccharide, shows no detectable synergism with galactomannans or related plant polysaccharides. Although some agars have sulphate substituents, the degree of sulphation is generally far lower than in carrageenans. However, *Gloiopeltis furcata* and *Gloiopeltis cervicornis* yield agars sulphated at O-6 of every D-galactose residue and at O-2 of c. 15% of the 3,6-anhydro-L-galactose residues, giving an overall sulphate content somewhat higher than in κ-carrageenan.[53] As might therefore be expected, these agarose sulphate polysaccharides interact with galactomannans to a comparable, or slightly lesser, extent than κ-carrageenan. For example, a non-gelling 1% solution of agarose sulphate from *G. cervicornis*[9] is gelled by addition of 1% LBG, with elevation of the melting temperature of the agar helix by about 10°C.

As in the case of single-polymer gels, the strength of binary mixed gels is strongly dependent on total polymer concentration, but is also substantially affected by the relative proportions of the two polymers. Figure 24(a) shows the effect of composition on the yield stress of a series of mixed gels of κ-carrageenan and LBG, prepared by holding total polysaccharide concentration constant but varying the concentration of LBG.[106] Results obtained at two different total polysaccharide levels (2% and 1·5%) show the same overall trend: an initial substantial increase in yield stress with increasing proportion of LBG up to a carrageenan:galactomannan ratio of about 2:1, with a subsequent sharp decrease at higher galactomannan concentrations (since, of course, LBG alone is non-gelling under normal conditions). The extent of deformation required to rupture the gel (breaking strain) also shows a corresponding initial increase and subsequent decrease with increasing concentration of LBG.[92]

Although the measured yield stress values at the higher total polysaccharide concentration are much larger than those obtained at the lower concentration, the two sets of data converge to a single line when results for the mixed gels are expressed relative to the yield stress of κ-carrageenan alone at the same overall polymer concentration. As shown in Fig. 24(b), a similar common curve can be obtained for mixed gels of κ-carrageenan with gum tara,[106] but in this case the enhancement of yield stress by the galactomannan is much smaller, and the optimum relative concentration of galactomannan is somewhat higher (c. 40%).

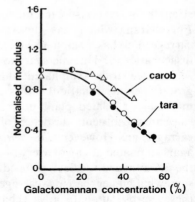

FIG. 25. Composition dependence of normalized compression modulus for mixed gels of κ-carrageenan with LBG (carob) or gum tara. Symbols as in Fig. 24(b). (From Cairns et al.,[106] with permission.)

When the same synergistic gels are measured under small deformation (from the initial slope of stress–strain compression curves), a similar normalization can be achieved (Fig. 25), but the moduli decrease with increasing concentration of galactomannan over the composition range where maximum yield stress is observed (Fig. 24), with gum tara causing a greater reduction in rigidity than LBG.[104] Thus, the overall effect of substituting κ-carrageenan by galactomannan, up to a replacement level of c. 30–40%, is to reduce the 'brittleness' of the gels (i.e. to decrease gel rigidity and increase the force, and deformation, required for rupture).

As in the case of carrageenan, progressive replacement of xanthan by galactomannan also causes an initial large increase in gel strength with a

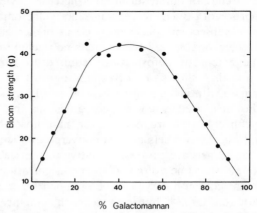

FIG. 26. Composition dependence of gel strength[107] in xanthan–LBG gels at a fixed total polymer concentration of 1% w/v.

subsequent decrease at high levels of replacement. Figure 26 presents results from an early patent[107] showing the composition dependence of gel strength (measured as 'Bloom strength' from a simple penetrometer test) for a series of xanthan–LBG gels at fixed total polymer concentration, but varying the relative proportions of the two polysaccharides. The resistance to deformation remains fairly constant for LBG:xanthan ratios in the range 0·25–1·5, but drops off steeply at higher or lower proportions of galactomannan.

The measurements shown in Fig. 26 were made on gels aged for 17 h after cooling the mixed solutions to room temperature from $c.$ 80°C. Recent work, however, indicates complex changes in the structure and rheology of xanthan–LBG gels during this ageing period.[108] Figure 27 shows mechanical spectra of a mixed solution of xanthan–LBG at high temperature (65°C), and of the resulting gel, measured shortly after cooling to room temperature (25°C). The high-temperature measurements have the form expected[64] for a polymer solution: G' (characterizing solid-like behaviour) falls below G'' (characterizing viscous flow), except at high frequencies, and both moduli increase steeply with increasing frequency. Similarly, the values obtained after 30 min at room temperature are typical

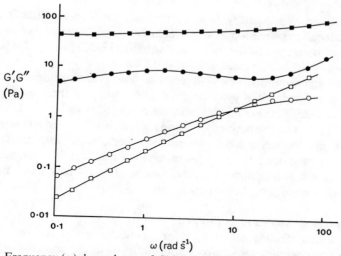

FIG. 27. Frequency (ω) dependence of G' (■, □) and G'' (●, ○) for a xanthan–LBG mixed gel system (0·5% total polysaccharide concentration; 0·1 M NaCl; xanthan:LBG = 70:30) in the sol state at 65°C (open symbols) and in the gel state after 30 min at 25°C (filled symbols). (From Cuvelier and Launay,[108] with permission.)

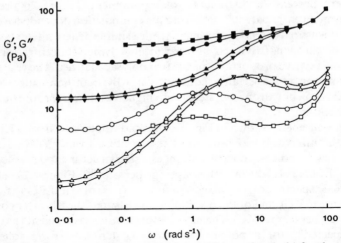

FIG. 28. Variation in G' (filled symbols) and G'' (open symbols) for the xanthan–LBG gel system shown in Fig. 27, after ageing for 30 (■, □), 190 (●, ○), 1400 (▲, △) or 2350 (▼, ▽) min at 25°C. (From Cuvelier and Launay,[108] with permission.)

of a polysaccharide gel: G' is much higher than G'', with little frequency dependence in either modulus. On longer storage, however, the measured values of both moduli decrease substantially (Fig. 28) at low frequencies, reaching final stable values after about 24 h.

This behaviour obviously suggests a slow rearrangement of the gel network formed initially on cooling. The nature of the rearrangement, however, is not yet known. It should also be noted that the results shown in Figs 27 and 28 were obtained[108] using a xanthan sample in an almost pure sodium salt form. Since the presence of other counter-ions, in particular Ca^{2+}, can produce large changes in xanthan rheology,[83] the same ageing effects may not occur in more typical commercial xanthans, which normally contain a significant proportion of calcium counter-ions.

5.h Practical Applications in Food

As well as increasing gel strength (or giving the same strength at a lower polymer concentration), incorporation of galactomannans or related polysaccharides in agar or carrageenan gels confers a less brittle, more elastic, texture and also reduces syneresis. Both of these effects are probably due to the plant-polysaccharide chains restricting helix–helix aggregation in the algal polysaccharide. The optimum ratio of the two components in the mixed-gel system depends on whether it is more important to optimize gel strength or to control syneresis. For example, maximum gel strength in κ-carrageenan/LBG gels is attained at a carrageenan:LBG ratio of about 2:1, but a ratio of about 1:4 is required to minimize syneresis.[8] Thus, addition of appropriate amounts of galactomannan offers a general, versatile method for modifying texture in the agar gel and carrageenan systems discussed in detail elsewhere in this volume (see Chapters 1 and 3).

Xanthan–galactomannan mixes have a very wide range of food applications.[8] Guar gum does not gel with xanthan, but gives a substantial enhancement of solution viscosity. One of the main attractions of this interaction is the potential for cost saving, since guar gum is far less expensive than xanthan. However, in products where the characteristic 'weak-gel' properties of xanthan (such as very rapid recovery of particle-suspending ability after shear) are of prime importance, use of guar gum as an 'extender' can result in an inferior product.[7]

For gelling systems, a galactomannan of lower galactose content is required. This is almost invariably LBG, but some formulations using gum tara have been patented.[109] A major disadvantage of the thermoreversible xanthan/LBG gel system for food applications is the very cohesive, elastic texture,[7] which would be unacceptable for direct use in simple products such as table jellies. Mixed gels of xanthan with the less-interactive gum tara have a less cohesive structure, and a more acceptable 'short' texture.[109] Gum tara has a Codex A2 food status, and is thus temporarily accepted as a food additive with a provisional maximum ADI (acceptable daily intake) of 12·5 mg/kg according to the FAO/WHO Codex Alimentarius Commission; it does not have an EEC E number, presumably owing to lack of exploitation, but is permitted for food use in Finland and Switzerland.[106] With increasing pressure on supplies of LBG, gum tara may become an attractive alternative in synergistic gel systems.

The undesirable textural attributes of xanthan–LBG can be modified to a form suitable for use in a wide range of food products by inclusion of other hydrocolloids such as starch, CMC, carrageenan, guar gum or microcrystalline cellulose,[7,8,110–114] or by the presence of high concentrations of other food ingredients such as fat[115] or meat.[116] Product applications[111–116] include bakery and pie fillings, puddings, dips and spreads, acidified milk gels, icing and confectionery products, tomato aspic, and meat products such as paté, luncheon meat and canned pet food. In contrast to the very 'long', elastic texture of xanthan–LBG, gellan gels have the opposite problem, of a somewhat 'short', brittle texture. The two gelling systems can therefore be used in combination[117] to provide a useful range and diversity of gel texture. Further variation can be achieved by using gum tara, konjac mannan or cassia galactomannans in place of LBG.

As well as its convenience as a thermoreversible gel system, melting and setting at moderate temperature (typically 40–45°C), xanthan–LBG can confer additional specific advantages in particular products. For example, in canned products incorporating large pieces to be set in the gel matrix (e.g. meat chunks), the 'weak-gel' properties of xanthan can retard sedimentation prior to formation of the true, synergistic gel network and so prevent separation of the product into a high-solids layer at the bottom of the can with a jelly layer above.[116] Similarly, the particle-suspending properties of xanthan allow milk and fruit juice to be blended in stable, gelled products, without the problems of curdling normally encountered in dairy systems at low pH.[112,113] Other specific advantages include decreased setting time in high-solids starch-jelly confectionery[114] and improved thermal and freeze–thaw stability in icings for commercial products.[115]

In applications where a firm, cohesive gel structure is required, it is essential to cool the two polysaccharides together from high temperature (typically c. 80°C). However, as discussed previously, cold mixing of xanthan with LBG gives semi-gelled products with much higher viscosity than xanthan alone. Detailed discussion of such fluid systems is outside the scope of the present review, but they have many practical applications in pourable products.[107]

6 ALGINATE–PECTIN MIXED GELS

As detailed elsewhere in this volume, alginates (particularly those with a high content of poly-L-guluronate), form firm gels in the presence of calcium ions but precipitate at low pH, and pectins (particularly those with a high content of methyl ester substituents) will gel at low pH, but only under conditions of low water activity (e.g. 60% sucrose). Mixtures of the two, however, give firm, cuttable gels at low pH,[118–122] irrespective of water activity, and the presence of calcium ions in the initial solutions is antagonistic to gelation. In contrast to normal alginate or pectin gels, the mixed gels are usually thermoreversible, but the melting point increases with decreasing pH[122] and, under sufficiently acidic conditions, gel structure may be retained at 100°C.[119]

The interaction is reported[122] to have been discovered during a search for thermoreversible gelling systems that could be used in low-sugar, low-calorie jams and jellies, and can involve pectin either naturally present in

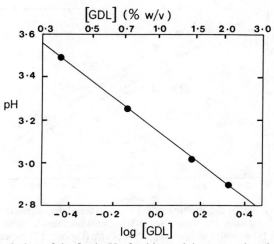

FIG. 29. Variation of the final pH of cold-set alginate–pectin mixed gels with the concentration of glucono-δ-lactone (GDL) used in their preparation. The results shown are for a total polysaccharide concentration of 1·2% w/v, using a 1:1 mixture of 'high-G' alginate (Protanal LF 250; 70% guluronate) and high-methoxy pectin (Grinsted Mexpectin RS 400; 70% methyl ester), and are taken from values reported in Fig. 3 of Ref. 122. (Toft et al., with permission.)

the fruit or added as a separate ingredient. In the initial studies,[118–120] gels were formed by cooling hot, acidified mixed systems, but cold-setting gels can also be prepared[121,122] using the dissociation of glucono-δ-lactone (GDL) to lower pH *in situ*. When polysaccharide concentration is held constant, the final pH decreases linearly with the logarithm of GDL concentration (Fig. 29), as might be expected, since $pH = -\log_{10}[H^+]$ and, to a first approximation, the concentration of hydrogen ions generated should be directly dependent on the amount of GDL present. As illustrated in Fig. 30, the melting point of the resulting gels shows an approximately linear increase with decreasing pH[122] and extrapolates to room temperature at pH ~ 3·8, the highest pH at which gel formation can be detected.[119,122] Gel rigidity and breaking stress (Fig. 31) also increase with decreasing pH.[121,122]

The stability of the gels, whether characterized by rigidity modulus, breaking stress or melting temperature, is greater[122] for alginates of high guluronate content. For example, in combination with a normal 'high-methoxy' pectin (c. 70% methyl esterified), the gels formed by the cold-set procedure using a typical commercial 'high-G' alginate (c. 70% guluronate) are about 2–3 times stronger, in terms of both rigidity and break point (Fig.

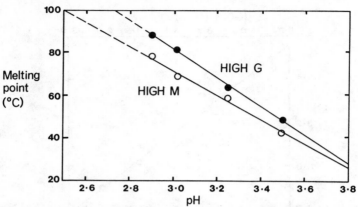

FIG. 30. Dependence of the melting point of cold-set alginate–pectin mixed gels on their final pH. Results are shown for a total polysaccharide concentration of 1·2% w/v, using a 1:1 mixture of high-methoxy pectin (Grinsted Mexpectin RS 400; 70% methyl ester) in combination with either 'high-G' alginate (Protanal LF 250; 70% guluronate; ●) or 'high-M' alginate (Protanal LF 200 MA; 60% mannuronate; ○). (From Toft et al.,[122] with permission, using the relationship shown in Fig. 29 to estimate final pH from GDL concentration.)

31) than those formed at equivalent pH by a typical 'high-M' sample (c. 60% mannuronate). Gel melting points for the same samples (Fig. 30) also differ substantially (e.g. by c. 10°C at pH 3) and extrapolate to 100°C at pH ~ 2·75 and pH ~ 2·5 for 'high-G' and 'high-M', respectively. In earlier studies using gels formed by cooling hot, acidified solutions[119] gel fractions resistant to thermal melting up to 100°C were reported for less-acidic conditions (pH ~ 3·6 for 'high-G' alginate and pH ~ 3·1 for 'high-M'). A possible explanation of this difference is that gels produced by direct acidification at high temperature are less homogeneous, owing to local inhomogeneity of pH during mixing, and include 'islands' of very stable structure.

Although the hydrolysis of GDL in isolation is rapid, alginate and pectin exert a buffering action[121] which delays the onset of gelation (Fig. 32) by increasing the time taken to reach the 'threshold' pH (Fig. 31) for interaction. When polymer concentration is held constant, the 'lag period' can be reduced[121] by increasing the concentration of GDL, but even at very high levels (e.g. 6% GDL for 2% total polymer concentration) there is a substantial delay (c. 20 min) in the onset of gelation. When the concentration of GDL is held constant,[121] the lag period increases (Fig. 33) with increasing polymer concentration (owing to the greater buffering

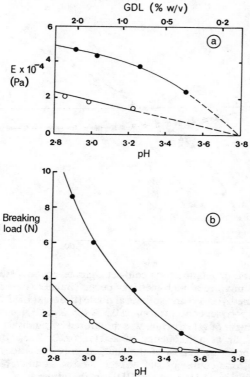

FIG. 31. Dependence of the strength of cold-set alginate–pectin gels on their final pH. Samples and symbols as in Fig. 30. (a) Rigidity modulus (E) measured[122] under compression, in the linear region of the stress–strain curve, on a Stevens LFRA Texture Analyser, using cylindrical pellets of gel with diameter 14 mm and height 15 mm. (b) Breaking load, measured using the same instrument and sample geometry as in (a). (From Toft et al.,[122] with permission, using the relationship shown in Fig. 29 to estimate final pH from GDL concentration.)

capacity) although, of course, the subsequent rise in gel strength is more rapid.

The nature of the interaction has still to be established unequivocally (e.g. by X-ray diffraction), but there are strong indications of a genuine heterologous association between specific chain sequences of the two polymers: alginate poly-L-guluronate 'blocks' and pectin poly-D-galacturonate sequences of low charge density (e.g. sequences with a high content of methyl ester). At the simplest level of experimental evidence, the

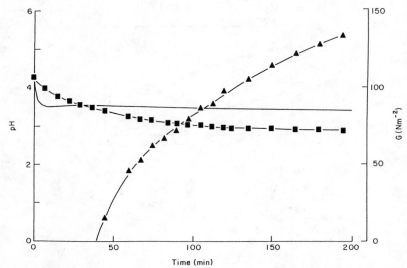

FIG. 32. Time course of gelation for a cold-set alginate–pectin mixed gel. Results are shown for a 1:1 mixture of high-methoxy pectin (Bulmers type 105 'rapid-set' citrus pectin; 70% methyl ester) and sodium alginate (Protanal LF 250 DL-A; 50% guluronate) at total polymer concentration of 0·5% w/w and GDL concentration of 1·5% w/w. Development of gel structure was monitored by rigidity modulus G (▲) measured under shear at a frequency of 200 Hz on a Rank Brothers Pulse Shearometer. The buffering action of the charged polysaccharides is shown by comparison of the time course of pH change for the sample (■) and for an equivalent concentration (1·5% w/w) of GDL in the absence of polymer (——). (From Morris and Chilvers,[121] with permission.)

interaction is enhanced as the proportion of these sequences is increased:[118–120,122] the greater stability of mixed gels involving 'high-G' alginate rather than 'high-M' is illustrated in Figs 30 and 31; conversely, although low-methoxy pectin (or even pectic acid) will form mixed gels with high-G alginate,[120] a much lower pH is required (e.g. 2·7 for typical commercial low-methoxy pectin).

Poly-L-guluronate and poly-D-galacturonate are exact mirror images except at O-3, which is axial in guluronate and equatorial in galacturonate. There is strong evidence[123,124] that in calcium-induced gelation both adopt the 2-fold 'zig-zag' structure characterized in the solid state for polyguluronate,[125] with long arrays of calcium ions sandwiched between pairs of chains in the cavities formed by the buckled chain contour. Adoption of the same ordered, 2-fold structure by sequences of both types

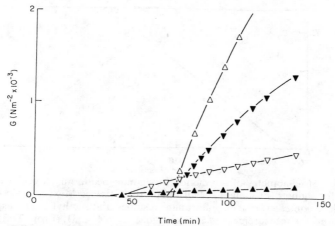

FIG. 33. Effect of total polymer concentration on the time course of gelation in cold-set alginate–pectin gels. Samples and rigidity measurements as in Fig. 32. The alginate:pectin ratio was held constant at 1:1 and a fixed concentration (1·5% w/w) of GDL was used throughout, at total polymer concentrations (% w/w) of 0·5 (▲), 1·0 (▽), 1·5 (▼) and 2·0 (△). (From Morris and Chilvers,[121] with permission.)

would offer a route to formation of sterically regular heterologous junctions in mixed gels.[120]

Inspection of space-filling molecular models indicates that poly-L-guluronate and (esterified) poly-D-galacturonate are capable of packing together in a parallel, 2-fold crystalline array.[120] Although the conformation of the individual chains is the same as in homotypic calcium-mediated junctions, the geometry of the interaction is quite different, analogous to the difference between shaking hands (right-hand, right-hand) and clasping hands (right-hand, left-hand), and instead of leaving cavities capable of accommodating metal ions, the near mirror image chains can form a close-packed, nested structure[120] with opportunities for favourable non-covalent interactions (e.g. between methyl ester groups of pectin and H-1 and H-2 of polyguluronate). In the absence of counter-ions to balance the charge on the polyuronate chains, an assembly of this type would be electrostatically unstable unless charge–density was substantially decreased by esterification (consistent with the preferential interaction of pectins of high methyl-ester content) or by protonation (consistent with the requirement for progressively lower pH with increasing content of unesterified carboxyl groups).

Although this model must still be regarded as highly speculative, it does

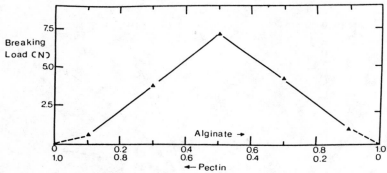

FIG. 34. Stoichiometry of interaction in alginate–pectin mixed gels. Results are shown for alginate (Protanal LF 250) and pectin (Grinsted Mexpectin RS 400) with equivalent content (70%) of guluronate residues and methyl ester, respectively. Conditions for measurement of breaking load as in Fig. 31. (From Toft et al.,[122] with permission.)

offer a unified interpretation of the available experimental evidence. For example, unlike polyguluronate, the polymannuronate (and heteropolymeric) sequences in alginate have quite different geometry from the polygalacturonate backbone of pectin, consistent with the lower reactivity of alginates in which these sequences predominate.[118-120] More directly, it has been shown recently[122] that the rigidity of mixed gels incorporating the same pectin increases systematically with increasing content of long polyguluronate sequences in the alginate.

Evidence that the gelling interaction of pectin with high-G alginates is not due to non-specific polymer incompatibility or exclusion effects includes the observation that other negatively charged polysaccharides, such as carboxymethylcellulose, or uncharged polysaccharides, such as galactomannans, have no such effect,[120] and changes in circular dichroism on lowering pH to induce gelation, which are substantially different from those observed for either polysaccharide in isolation.[120] Moreover, for mixed gels incorporating alginate and pectin with equivalent content of, respectively, guluronate and methylgalacturonate residues (e.g. 70% guluronate, 70% methyl ester), the optimum ratio[122] is 1:1 (Fig. 34), consistent with a stoichiometric, heterotypic interaction as proposed.[120]

Existing or potential applications of the interaction include the preparation of cold-setting fruit gels or flans, stabilization of acidic emulsions such as salad cream or mayonnaise, and preparation of novel multi-textured products, for example using the pectin content of real fruit or alginate present in re-formed fruit (see Chapter 2) to promote binding into a gel matrix.[121] It has also been suggested[121] that the interaction could have medical or pharmaceutical applications, for example by using the acidic environment of the stomach to set up a gelled or semi-gelled structure *in situ*.

A typical procedure for production of a low-solids, low-calorie jam, using the pectin naturally present in the fruit, has been reported by Toft.[119] Apple purée (600 g; solids content 78 g), sucrose (185 g) and soft tap water (100 g; 1·50 dH) were given a short boil; citric acid solution (2 g in 4 ml) and potassium sorbate solution (1 g in 10 ml) were stirred in, the mixture was removed from heat and alginate solution (3 g in 150 ml; M:G = 0·9) was added with vigorous stirring. The temperature was then decreased to 80°C, and the product was filled into 500-ml containers and cooled to room temperature, giving a firm, thermally reversible gel, with a solids content (allowing for loss of 100 g water by evaporation) of $c.\ 27\%$ w/w. On omitting the alginate, the product did not gel.

A typical base formulation for a hot-mix gel[119] might use high-G alginate (3 g/kg product) and high-methoxy pectin (3 g/kg) dissolved together in cold, soft water and brought to the boil, with subsequent addition of citric acid (3 g/kg) and sodium citrate (0·5 g/kg). In the cold-set procedure,[121,122] GDL (typically 10–15 g/kg) would be added as an aqueous dispersion to the cold, mixed solution of alginate and pectin and stirred in rapidly. It is essential that the polymers are fully dissolved and that the GDL is completely dispersed.[121]

Another essential requirement for both procedures (and a major drawback in commercial utilization) is that the polymers must be dissolved in soft (preferably de-ionized) water. For example, Toft[119] has reported that decreased gel strength is noticeable for water hardness above $c.\ 30$ dH, and by 40 dH gelation is eliminated. Some improvement, however, is possible using a suitable sequestering agent, and Calgon (e.g. at 0·5 g/kg) has been recommended.[119] A likely reason for this behaviour is that, under conditions where both are possible, calcium-induced association of alginate polyguluronate sequences occurs in preference to the heterologous interaction with pectin. Once the mixed-gel network is established, however, subsequent addition of calcium has no such deleterious effect, and indeed can significantly enhance the strength of the gel.[120,122] In terms of the proposed model for the interaction,[120] this might indicate that the outer faces of the participating polyguluronate sequences (i.e. the side of the 2-fold chain that is not bound to polygalacturonate) may be capable of normal 'egg-box' binding of calcium,[123] with consequent consolidation of the gel network.

By analogy with the role of methyl ester groups in reducing the charge on pectin and so facilitating interaction with alginate by reducing electrostatic repulsion between chains, we might expect esterification of the carboxyl groups on alginate to have a similar effect. This is indeed the case, and the formation of mixed gels of propylene glycol alginate and low-methoxy pectin has been reported.[119] The gels are soft and thixotropic, but are very resistant to thermal melting (as might be expected, since the net charge density of the system will be even lower than in mixtures of underivatized alginate with high-methoxy pectin) and appear less sensitive to calcium (which might also be expected, since the calcium-binding interaction of low-methoxy pectin—in this case the charged species—is weaker than that of alginate).

ACKNOWLEDGEMENTS

Many of the authors whose work is cited in this review have kindly explained and amplified their published findings in personal conversations, Dr A. H. Clark, who has patiently answered many lengthy enquiries, and particular thanks are due to Dr I. C. M. Dea, Dr I. T. Norton, Dr J. R. Mitchell, Dr V. J. Morris, Dr S. B. Ross-Murphy, Dr I. W. Sutherland and Professor V. B. Tolstoguzov, for helpful discussions.

REFERENCES

1. Stainsby, G., Proteinaceous gelling systems and their complexes with polysaccharides. *Food Chem.*, **6** (1980) 3.
2. Laurent, T. C., Preston, B. N., Comper, W. D. & Sundelöf, L.-O., The interactions in concentrated polysaccharide solutions. *Prog. Food Nutr. Sci.*, **6** (1982) 69.
3. Tolstoguzov, V. B., Functional properties of protein–polysaccharide mixtures. In *Functional Properties of Food Macromolecules*, ed. J. R. Mitchell & D. A. Ledward. Elsevier Applied Science Publishers, London, 1986, p. 385.
4. Tolstoguzov, V. B., Some physico-chemical aspects of protein processing into foodstuffs. *Food Hydrocolloids*, **2** (1988) 339.
5. Clark, A. H. & Ross-Murphy, S. B., Structural and mechanical properties of biopolymer gels. *Adv. Polymer Sci.*, **83** (1987) 57.
6. Clark, A. H., The application of network theory to food systems. In *Food Structure and Behaviour*, ed. J. M. V. Blanshard & P. Lillford. Academic Press, London, 1987, p. 13.
7. Sanderson, G. R., The interactions of xanthan gums in food systems. *Prog. Food Nutr. Sci.*, **6** (1982) 77.

8. Igoe, R. S., Hydrocolloid interactions useful in food systems. *Food Technol.* (April 1982) 72.
9. Dea, I. C. M. & Rees, D. A., Affinity interactions between agarose and β-1,4-glycans: a model for polysaccharide association in algal cell walls. *Carbohydrate Polymers*, **7** (1987) 183.
10. Morris, V. J., Multicomponent gels. In *Gums and Stabilisers for the Food Industry—3*, ed. G. O. Phillips, D. J. Wedlock & P. A. Williams. Elsevier, London, 1986, p. 87.
11. McKay, J. E., Stainsby, G. & Wilson, E. L., A comparison of the reactivity of alginate and pectate esters with gelatin. *Carbohydrate Polymers*, **5** (1985) 223.
12. Agfa, A. G., Water-insoluble film forming nitrogenous products. British Patent 9 624 83 (1964).
13. McDowell, R. H., New reactions of propylene glycol alginate. *J. Soc. Cosmetic Chem.*, **21** (1970) 441.
14. Unilever Ltd, British Patent 1 443 513 (1976).
15. Collagen Products Ltd, British Patent 1 450 687 (1976).
16. Alginate Industries Ltd, A method for modifying alkylene glycol alginates. British Patent 1 135 856 (1968).
17. Smith, J., Mitchell, J. R. & Ledward, D. A., Effect of the inclusion of polysaccharides on soya extrusion. *Prog. Food Nutr. Sci.*, **6** (1982) 139.
18. Berrington, D., Imeson, A., Ledward, D. A., Mitchell, J. R. & Smith, J., The effect of alginate inclusion on the extrusion behaviour of soya. *Carbohydrate Polymers*, **4** (1984) 443.
19. Imeson, A. P., Richmond, P. & Smith, A. C., The extrusion of soya with alginate using a twin-screw cooker extruder. *Carbohydrate Polymers*, **5** (1985) 329.
20. Oates, C. G., Ledward, D. A., Mitchell, J. R. & Hodgson, I., Physical and chemical changes resulting from heat treatment of soya and soya alginate mixtures. *Carbohydrate Polymers*, **7** (1987) 17.
21. Oates, C. G., Ledward, D. A., Mitchell, J. R. & Hodgson, I., Glutamic acid reactivity in heated protein and protein–alginate mixtures. *Int. J. Food Sci. Technol.*, **22** (1987) 477.
22. Muchlin, M. A., Wajnermann, E. S. & Tolstoguzov, V. B., Complex gels of proteins and acid polysaccharides. *Nahrung*, **20** (1976) 313.
23. Tschmak, G. Ja., Wajnermann, E. S. & Tolstoguzov, V. B., The structure and properties of complex gels of gelatin and pectin. *Nahrung*, **20** (1976) 321.
24. Glicksman, M., *Gum Technology in the Food Industry*. Academic Press, New York, 1969.
25. Shim, J. L., Gellan gum/gelatin blends. US Patent 4 517 216 (1985).
26. Chilvers, G. R. & Morris, V. J., Coacervation of gelatin–gellan gum mixtures and their use in microencapsulation. *Carbohydrate Polymers*, **7** (1987) 111.
27. Imeson, A. P., Recovery and utilisation of proteins using alginate. In *Gums and Stabilisers for the Food Industry 2—Applications of Hydrocolloids*, ed. G. O. Phillips, D. J. Wedlock & P. A. Williams. Pergamon Press, Oxford, 1984, p. 189.
28. Imeson, A. P., Watson, P. R., Mitchell, J. R. & Ledward, D. A., Protein recovery from blood plasma by precipitation with polyuronates. *J. Food Technol.*, **13** (1978) 329.

29. Imeson, A. P., Ledward, D. A. & Mitchell, J. R., The effect of calcium and pH on spun fibres produced from plasma–alginate mixtures. *Meat Sci.*, **3** (1979) 287.
30. Tolstoguzov, V. B., Izjumov, D. B., Grinberg, V. Ya., Marusova, A. N. & Chekhovskaya, V. T., US Patent 3 829 587 (1974).
31. Shank, J. L., US Patent 3 404 142 (1968).
32. McDowell, R. H., New developments in the chemistry of alginates and their use in food. *Chemistry and Industry* (1975) 391.
33. Albertsson, P.-A., *Partition of Cell Particles and Macromolecules*, 2nd edn. Wiley-Interscience, New York, 1971.
34. Rees, D. A., Morris, E. R., Thom, D. & Madden, J. K., Shapes and interactions of carbohydrate chains. In *The Polysaccharides*, Vol. 1, ed. G. O. Aspinall. Academic Press, New York, 1982, p. 195.
35. Laurent, T. C., Preston, B. N. & Carlsson, B., Interaction between polysaccharides and other macromolecules. Conformational transitions of polynucleotides in polymeric media. *Eur. J. Biochem.*, **43** (1974) 231.
36. Reinhart, G. D., Influence of polyethylene glycols on the kinetics of rat liver phosphofructokinase. *J. Biol. Chem.*, **255** (1980) 10 576.
37. Lee, J. C. & Lee, L. L. Y., Interaction of calf brain tubulin with poly(ethylene glycols). *Biochemistry*, **18** (1979) 5518.
38. Tolstoguzov, V. B., Belkina, V. P., Gulov, V. Ja., Titova, E. F. & Belavzeva, E. M., State of phase, structure and mechanical properties of the gelatinous system water–gelatin–dextran. *Die Stärke*, **26** (1974) 130.
39. Marrs, W. M., Gelatin/carbohydrate interactions and their effect on the structure and texture of confectionery gels. *Prog. Food Nutr. Sci.*, **6** (1982) 259.
40. Doi, K., Formation of amylopectin granules in gelatin gels as a model of starch precipitation in plant plastids. *Biochim. Biophys. Acta*, **94** (1965) 557.
41. Grinberg, V. Ya. & Tolstoguzov, V. B., Thermodynamic compatibility of gelatin with some D-glucans in aqueous media. *Carbohydrate Res.*, **25** (1972) 313.
42. Beijerinck, M. W., Über Emulsionsbildung bei der Vermischung Wasseriger Lösungen gewisser Gelatinierender Kolloide. *Z. Chem. Ind. Kolloid*, **7** (1910) 16.
43. Kalichevsky, M. T., Oford, P. D. & Ring, S. G., The incompatibility of concentrated aqueous solutions of dextran and amylose and its effect on amylose gelation. *Carbohydrate Polymers*, **6** (1986) 145.
44. Kalichevsky, M. T. & Ring, S. G., Incompatibility of amylose and amylopectin in aqueous solution. *Carbohydrate Res.*, **162** (1987) 323.
45. Christianson, D. D., Hodge, J. E., Osborne, D. & Detroy, R. W., Gelatinization of wheat starch as modified by xanthan gum, guar gum, and cellulose gum. *Cereal Chem.*, **58** (1981) 513.
46. Howling, D., The influence of the structure of starch on its rheological properties. *Food Chem.*, **6** (1980) 51.
47. Cheng, H. & Wintersdorff, P., Xanthan gum-modified starches. US Patent 4 298 729 (1981).
48. Christianson, D. D., Gardner, H. W., Warner, K., Boundy, B. K. & Inglett, G. E., Xanthan gum in protein-fortified starch bread. *Food Technol.* (June 1974) 23.

49. Suchkov, V. V., Grinberg, V. Ya. & Tolstoguzov, V. B., Steady-state viscosity of the liquid two-phase disperse system water–casein–sodium alginate. *Carbohydrate Polymers*, **1** (1981) 39.
50. Clark, A. H., Richardson, R. K., Ross-Murphy, S. B. & Stubbs, J. M., Structural and mechanical properties of agar/gelatin co-gels. Small-deformation studies. *Macromolecules*, **16** (1983) 1367.
51. Clark, A. H., Richardson, R. K., Robinson, G., Ross-Murphy, S. B. & Weaver, A. C., Structure and mechanical properties of agar/BSA co-gels. *Prog. Food Nutr. Sci.*, **6** (1982) 149.
52. Whistler, R. L. & BeMiller, J. N., *Industrial Gums*. Academic Press, New York, 1973.
53. Dea, I. C. M. & Morrison, A., Chemistry and interactions of seed galactomannans. *Adv. Carbohydrate Chem. Biochem.*, **31** (1975) 241.
54. Herald, C. T., Guar gum. In *Food Hydrocolloids*, Vol. 3, ed. M. Glicksman. CRC Press, Boca Raton, Florida, 1986, p. 171.
55. Herald, C. T., Locust/carob bean gum. In *Food Hydrocolloids*, Vol. 3, ed. M. Glicksman. CRC Press, Boca Raton, Florida, 1986, p. 161.
56. Buffington, L. A., Stevens, E. S., Morris, E. R. & Rees, D. A., Vacuum ultraviolet circular dichroism of galactomannans. *Int. J. Biol. Macromol.*, **2** (1980) 199.
57. Leschziner, C. & Cerezo, A. S., Correlation of chemical composition and optical rotation of water-soluble galactomannans. *Carbohydrate Res.*, **11** (1969) 113.
58. Winter, W. T., Song, B. K. & Bouckris, H., Structural studies of galactomannans and their complexes. *Food Hydrocolloids*, **1** (1987) 581.
59. Robinson, G., Ross-Murphy, S. B. & Morris, E. R., Viscosity–molecular weight relationships, intrinsic chain flexibility and dynamic solution properties of guar galactomannan. *Carbohydrate Res.*, **107** (1982) 17.
60. Launay, B., Doublier, J. L. & Cuvelier, G., Flow properties of aqueous solutions and dispersions of polysaccharides. In *Functional Properties of Food Macromolecules*, ed. J. R. Mitchell & D. A. Ledward. Elsevier, Barking, Essex, 1986, p. 1.
61. Dea, I. C. M., Morris, E. R., Rees, D. A., Welsh, E. J., Barnes, H. A. & Price, J., Associations of like and unlike polysaccharides: mechanism and specificity in galactomannans, interacting bacterial polysaccharides and related systems. *Carbohydrate Res.*, **57** (1977) 249.
62. Morris, E. R., Cutler, A. N., Ross-Murphy, S. B., Rees, D. A. & Price, J., Concentration and shear rate dependence of viscosity in random coil polysaccharide solutions. *Carbohydrate Polymers*, **1** (1981) 5.
63. Graessley, W. W., The entanglement concept in polymer rheology. In *Advances in Polymer Science*, Vol. 16. Springer-Verlag, Berlin, 1974.
64. Morris, E. R., Rheology of hydrocolloids. In *Gums and Stabilisers for the Food Industry—2*, ed. G. O. Phillips, D. J. Wedlock & P. A. Williams. Pergamon Press, Oxford, 1984, p. 57.
65. Morris, E. R., in preparation.
66. McCleary, B. V., Amado, R., Waibel, R. & Neukom, H., Effect of galactose content on the solution and interaction properties of guar and carob galactomannans. *Carbohydrate Res.*, **92** (1981) 269.

67. McNeeley, W. H. & Kang, K. S., Xanthan and some other biosynthetic gums. In *Industrial Gums*, ed. R. L. Whistler & J. N. BeMiller. Academic Press, New York, 1973, p. 473.
68. Jansson, P.-E., Kenne, L. & Lindberg, B., Structure of the extracellular polysaccharide from *Xanthomonas campestris*. *Carbohydrate Res.*, **45** (1975) 275.
69. Melton, L. D., Mindt, L., Rees, D. A. & Sanderson, G. R., Covalent structure of the extracellular polysaccharide from *Xanthomonas campestris*: evidence from partial hydrolysis studies. *Carbohydrate Res.*, **46** (1976) 245.
70. Smith, I. H., Symes, K. C., Lawson, C. J. & Morris, E. R., Influence of the pyruvate content of xanthan on macromolecular association in solution. *Int. J. Biol. Macromol.*, **3** (1981) 129.
71. Sandford, P. A., Watson, P. R. & Knutson, C. A., Separation of xanthan gums of differing pyruvate content by fractional precipitation with alcohol. *Carbohydrate Res.*, **63** (1978) 253.
72. Sutherland, I. W., Bacterial surface polysaccharides: structure and function. *Int. Rev. Cytology*, **113** (1988) 187.
73. Morris, E. R., Rees, D. A., Young, G., Walkinshaw, M. D. & Darke, A., Order–disorder transition for a bacterial polysaccharide in solution. *J. Mol. Biol.*, **110** (1977) 1.
74. Milas, M. & Rinaudo, M., Conformational investigation of the bacterial polysaccharide xanthan. *Carbohydrate Res.*, **76** (1979) 189.
75. Norton, I. T., Goodall, D. M., Frangou, S. A., Morris, E. R. & Rees, D. A., Mechanism and dynamics of conformational ordering in xanthan polysaccharide. *J. Mol. Biol.*, **175** (1984) 371.
76. Muller, G., Anrhourrache, M., Lecourtier, J. & Chaveteau, G., Salt dependence of the conformation of a single-stranded xanthan. *Int. J. Biol. Macromol.*, **8** (1986) 167.
77. Milas, M. & Rinaudo, M., Properties of xanthan gum in aqueous solutions: role in the conformational transition. *Carbohydrate Res.*, **158** (1986) 191.
78. Liu, W., Sato, T., Norisuye, T. & Fujita, H., Thermally-induced conformational change of xanthan in 0·01M sodium chloride. *Carbohydrate Res.*, **160** (1987) 267.
79. Moorhouse, R., Walkinshaw, M. D. & Arnott, S., Xanthan gum—molecular conformation and interactions. In *Extracellular Microbial Polysaccharides*. ACS Symposium Series, **45** (1977) 90.
80. Okuyama, K., Arnott, S., Moorhouse, R., Walkinshaw, M. D., Atkins, E. D. T. & Wolf-Ullish, Ch., Fibre diffraction studies of bacterial polysaccharides. In *Fibre Diffraction Methods*. ACS Symposium Series, **141** (1980) 411.
81. Cox, W. P. & Merz, E. H., Correlation of dynamic and steady-flow viscosities. *J. Polymer Sci.*, **28** (1958) 619.
82. Smidsrød, O. & Haug, A., Estimation of the relative stiffness of the molecular chain in polyelectrolytes from measurements of viscosity at different ionic strengths. *Biopolymers*, **10** (1971) 1213.
83. Ross-Murphy, S. B., Morris, V. J. & Morris, E. R., Molecular viscoelasticity of xanthan polysaccharide. *Faraday Symp. Chem. Soc.*, **18** (1983) 115.
84. McCleary, B. V., Enzymic hydrolysis, fine structure, and gelling interaction of legume-seed D-galacto-D-mannans. *Carbohydrate Res.*, **71** (1979) 205.

85. Dea, I. C. M., Clark, A. H. & McCleary, B. V., Effect of galactose-substitution-patterns on the interaction properties of galactomannans. *Carbohydrate Res.*, **147** (1986) 275.
86. Dea, I. C. M., Clark, A. H. & McCleary, B. V., Effect of molecular fine structure of galactomannans on their interaction properties—the role of unsubstituted sides. *Food Hydrocolloids*, **1** (1986) 129.
87. Courtois, J. E. & Le Dizet, P., Action de l'α-galactosidase du café sur quelques galactomannanes. *Carbohydrate Res.*, **3** (1966) 141.
88. Courtois, J. E. & Le Dizet, P., Recherches sur les galactomannanes VI—Action de quelques mannanases sur diverses galactomannanes. *Bull. Soc. Chim. Biol.*, **52** (1970) 15.
89. Dea, I. C. M., McKinnon, A. A. & Rees, D. A., Tertiary and quaternary structure in aqueous polysaccharide systems which model cell wall cohesion: reversible changes in conformation and association of agarose, carrageenan and galactomannans. *J. Mol. Biol.*, **68** (1972) 153.
90. Rees, D. A., Polysaccharide gels—a molecular view. *Chemistry and Industry* (1972) 630.
91. Rees, D. A., Shapely polysaccharides. *Biochem. J.*, **126** (1972) 257.
92. Miles, M. J., Morris, V. J. & Carroll, V., Carob gum–κ-carrageenan mixed gels: mechanical properties and X-ray fibre diffraction studies. *Macromolecules*, **17** (1984) 2443.
93. Cairns, P., Miles, M. J. & Morris, V. J., X-ray diffraction studies of kappa carrageenan–tara gum mixed gels. *Int. J. Biol. Macromol.*, **8** (1986) 124.
94. Cairns, P., Miles, M. J., Morris, V. J. & Brownsey, G. J., X-ray fibre-diffraction studies of synergistic, binary polysaccharide gels. *Carbohydrate Res.*, **160** (1987) 411.
95. Cairns, P., Miles, M. J. & Morris, V. J., X-ray diffraction studies on konjac mannan–kappa carrageenan mixed gels. *Carbohydrate Polymers*, **8** (1988) 99.
96. Cheetham, N. W. H., McCleary, B. V., Teng, G., Lum, F. & Maryanto, Gel-permeation studies on xanthan–galactomannan interactions. *Carbohydrate Polymers*, **6** (1986) 257.
97. Cairns, P., Miles, M. J. & Morris, V. J., Intermolecular binding of xanthan gum and carob gum. *Nature (London)*, **322** (1986) 89.
98. Brownsey, G. J., Cairns, P., Miles, M. J. & Morris, V. J., Evidence for intermolecular binding between xanthan and the glucomannan konjac mannan. *Carbohydrate Res.*, **176** (1988) 329.
99. Robinson, G., Functional interactions of industrial polysaccharides. PhD Thesis, Cranfield Institute of Technology, Bedford, UK, 1990.
100. Morris, E. R., Rees, D. A., Robinson, G. & Young, G. A., Competitive inhibition of interchain interactions in polysaccharide systems. *J. Mol. Biol.*, **138** (1980) 363.
101. McCleary, B. V. & Neukom, H., Effect of enzymic modification on the solution and interaction properties of galactomannans. *Prog. Food Nutr. Sci.*, **6** (1982) 109.
102. McCleary, B. V., Dea, I. C. M., Windust, J. & Cooke, D., Interaction properties of D-galactose-depleted guar galactomannan samples. *Carbohydrate Polymers*, **4** (1984) 253.
103. McCleary, B. V., Critchley, P. & Bulpin, P. V., Reducing galactose content of

galactomannan by treating concentrated aqueous compositions with alpha galactosidase. European Patent 121 960, Unilever NV, 1984.
104. Reid, J. S. G., Edwards, M. & Dea, I. C. M., Enzymatic modification of natural seed gums. In *Gums and Stabilisers for the Food Industry—4*, ed. G. O. Phillips, D. J. Wedlock & P. A. Williams. IRL Press, Oxford, 1988, p. 391.
105. Guiseley, K. B., The relationship between methoxyl content and gelling temperature of agarose. *Carbohydrate Res.*, **13** (1970) 247.
106. Cairns, P., Morris, V. J. & Brownsey, G. J., Comparative studies of the mechanical properties of mixed gels formed by kappa carrageenan and tara gum or carob gum. *Food Hydrocolloids*, **1** (1986) 89.
107. Schuppner, H. R. Jr, Heat reversible gel and method. Australian Patent 401 434 (1967).
108. Cuvelier, G. & Launay, B., Viscoelastic properties of xanthan–carob mixed gels. In *Gums and Stabilisers for the Food Industry—3*, ed. G. O. Phillips, D. J. Wedlock & P. A. Williams. Elsevier, London, 1986, p. 147.
109. General Foods Corporation, Gelling agents for desserts and cold dishes, containing xanthan gum and tara gum or starch. US Patent 3 721 571 (1973).
110. Cottrell, I. W. & Kang, K. S., Xanthan gum, a unique bacterial polysaccharide for food applications. *Develop. Ind. Microbiol.*, **19** (1978) 117.
111. Kelco Division of Merck & Co. Inc., San Diego, USA, *Concept Bulletins*. K # 4, Instant chocolate pudding; K # 8, Canned Danish pudding; K # 15, Canned chip dip; K # 16, Canned imitation sour cream; K # 18, Canned tomato aspic; K # 20, Canned lemon cream pie filling; K # 22, Bakery fillings; K # 26, Canned chicken liver paté; K # 28, Lemon pie fillings.
112. Igoe, R. S., Gelling composition containing xanthan—carob gums, carboxymethylcellulose—for use in food compositions containing fruit juice and milk or cream. UK Patent 1 485 112, Merck & Co. Inc., 1977.
113. Igoe, R. S., Palatable milk and fruit juice composition incorporating a thickener comprising xanthan gum, locust-bean gum and carboxymethylcellulose. US Patent 4 046 925, Merck & Co. Inc., 1977.
114. Cheng, H., Rapid setting starch gel confectionery containing mixture of xanthan gum and locust-bean gum. US Patent 4 219 582, Merck & Co. Inc., 1980.
115. Cheng, H., Foodstuff icing composition containing fat and gelatinization system of xanthan gum and carob-bean gum as stabilizer for baked and fried foodstuffs. US Patent 4 135 005, Merck & Co. Inc., 1979.
116. Schuppner, H. R. Jr, Homogenous gelled meat product for use as luncheon meat or dog food. US Patent 3 519 434, Kelco Company, 1970.
117. Sanderson, G. R., Clark, R. S., Clare, K. & Pettitt, D. J., Low-acetyl gellan gum blends. US Patent 4 647 470, Merck & Co. Inc., 1987.
118. Steinnes, A., Alginate in Lebensmitteln. *Giordian* (1975) 228.
119. Toft, K., Interactions between pectins and alginates. *Prog. Food Nutr. Sci.*, **6** (1982) 89.
120. Thom, D., Dea, I. C. M., Morris, E. R. & Powell, D. A., Interchain associations of alginate and pectins. *Prog. Food Nutr. Sci.*, **6** (1982) 97.
121. Morris, V. J. & Chilvers, G. R., Cold-setting alginate–pectin mixed gels. *J. Sci. Food Agric.*, **35** (1984) 1370.
122. Toft, K., Grasdalen, H. & Smidsrød, O., Synergistic gelation of alginates and

pectins. In *Chemistry and Function of Pectins*. ACS Symposium Series, **310** (1986) 117.
123. Morris, E. R., Rees, D. A., Thom, D. & Boyd, J., Chiroptical and stoichiometric evidence of a specific primary dimerisation in alginate gelation. *Carbohydrate Res.*, **66** (1978) 145.
124. Morris, E. R., Powell, D. A., Gidley, M. J. & Rees, D. A., Conformations and interactions of pectins. 1. Polymorphism between gel and solid states of calcium polygalacturonate. *J. Mol. Biol.*, **155** (1982) 507.
125. Atkins, E. D. T., Neiduszynski, I. A., Mackie, W., Parker, K. D. & Smolko, E. E., Structural components of alginic acid. II. The crystalline structure of poly-α-L-guluronic acid. Results of X-ray diffraction and polarised infrared studies. *Biopolymers*, **12** (1973) 1879.

Chapter 9

MUSCLE PROTEINS

GRAHAM W. RODGER

*Imperial Chemical Industries plc, Biological Products Business,
PO Box 1, Billingham, Cleveland TS23 1LB, UK*

and

PETER WILDING

*Unilever Research Laboratory, Colworth House,
Sharnbrook, Bedford MK44 1LQ, UK*

1 INTRODUCTION

Meat and fish comprise an important part of the diet of people in many countries of the world, and there are hundreds, if not thousands, of products that bear testimony to the culinary skills that have been developed to utilize these raw materials as food. To attempt a description within a single chapter of how meat and fish can be manipulated to provide such a wide range of products would seem foolhardy but, if we stop and consider for a moment, this is not necessarily so. These raw materials are, in the native state, highly organized systems of proteins which are subject to the same physicochemical laws as any other proteins. It is also true that this highly organized state is responsible for the unique qualities we ascribe to muscle tissue foods, i.e. the sequential breakdown pattern of the food on chewing, and its perception, which we know as texture. And yet muscle did not evolve to provide humans with an interesting eating sensation, nor are all the products based on muscle tissue a testimony to the skill of product developers. Rather, several types of product arose from the need to preserve tissue from times of plenty through to scarcity, via processes that effect changes in the physicochemistry of the system, thus causing a different organoleptic response. Later, products developed through an industry need to upgrade material not capable of use or unacceptable in its natural form. Thus, in our approach to writing this chapter, we decided that a simple review of the chemistry of muscle proteins in relation to their food use would only be an addition to what is an already excellent series[1-6] of articles that have gone before. For this reason we have adopted a slightly different approach and will attempt to describe in fairly general terms how the physicochemical properties of muscle proteins have allowed the evolution, whether by accident or design, of the large range of foods derived totally or in part from muscle tissue. To simplify this task we first classify the foods into three main groups.

(i) Products in which the muscle tissue is consumed intact (e.g. beef joints, fish fillets).
(ii) Comminuted products in which the comminution process is deemed to free the muscle proteins from their in-vivo structural organization to create novel textures (e.g. kamaboko, frankfurter) or reduce the connective tissue contribution to texture (e.g. re-assembled meat products).
(iii) Preserved products in which the preservation systems have effected physicochemical changes that have created own-right products (e.g. marinated herring, smoked salmon, kippers, beef jerky, ham).

In this classification we accept that products do exist that cross over our groupings and an example of this is fermented sausage, where the preservation effects are superimposed on the effects of comminution and protein manipulation. In addition, while we recognize that freezing and

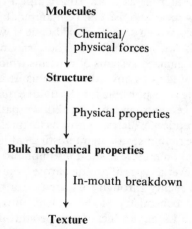

FIG. 1. The dependence of perceived texture upon physical properties and structure according to the scheme proposed by Stanley.[5]

frozen storage as a preservation system may result in gross textural changes (particularly in fish muscle), we would still group frozen products within category (i), since in whole-tissue systems we are unaware of freezing *per se* creating own-right products.

Although our stated objective is to explain how the physicochemistry of muscle tissue affects and allows manipulation of its behaviour as a food, this cannot be achieved without considering the structural organization of both the component proteins and the final product. We believe that the approach first outlined by Stanley[5] and shown in Fig. 1 is the only one capable of allowing a full understanding of the way in which molecular properties are manifest as sensory attributes. We have built this chapter around this central premise; that is, by considering food texture to be determined by the structure of that food (and the mechanical properties of the material comprising that structure), the effects of various processes can be explained systematically. In the sections that follow we only highlight fish and mammalian differences where we know that real effects occur. If none is mentioned we assume the response is, to our knowledge, identical.

2 THE STRUCTURAL ORGANIZATION AND MOLECULAR PROPERTIES OF MUSCLE PROTEINS

In this section we will cover the organization of the major constituent proteins of skeletal muscle and consider, from a food viewpoint, their physical and chemical properties.

2.a The Hierarchical Organization of Muscle Fibres Within a Muscle
2.a.1 Mammalian Skeletal Muscle
There exists in muscle a hierarchy of connective-tissue structures that encompasses fibres, bundles of fibres and the whole muscle. Fibres, each surrounded by their endomysium, are organized in groups or bundles of fibres that are encompassed by a collagenous connective-tissue sheath known as the perimysium. Bundles of fibres are in turn grouped together to form the whole muscle, which is surrounded by another connective-tissue sheath, the epimysium.[7] Figure 2 depicts this hierarchical organization of fibres and connective tissues. Whilst the connective tissues have different names and may contain different collagen types and different degrees of collagen cross-linking, the various mysia are not distinct. In fact the connective-tissue types merge into each other at their junctions, and are better considered as a continuous system. Muscle can thus be regarded as a

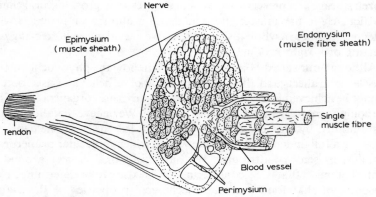

FIG. 2. Diagram depicting the organization of connective tissue and muscle fibres of mammalian muscle. (From Bailey,[1] reproduced with permission.)

complex composite of aligned fibres in a continuous matrix of connective tissue.

2.a.2 Fish Muscle

Fish tissue differs from mammalian tissue at the fibre level and above, although at these higher levels of organization there appears to be similarities in terms of the connective-tissue hierarchy. The skeletal musculature of fish is segmented into a number of muscle blocks known as myotomes, which are separated from each other by encompassing sheaths of connective tissue or myocommata.[8] Each myotome is composed of a large number of muscle fibres, generally shorter than those of mammalian muscles, that run obliquely to the main axis of the muscle system. Figure 3 shows the arrangement of the myotomes within a fillet (or along one side of a fish). Within each myotome the fibres appear to be arranged into bundles

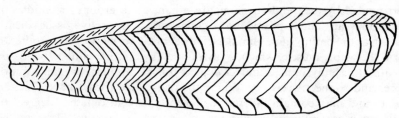

FIG. 3. Diagrammatic representation of the organization of myotomes within a fillet of fish. The internal black lines represent the connective tissues surrounding and separating the myotomes.

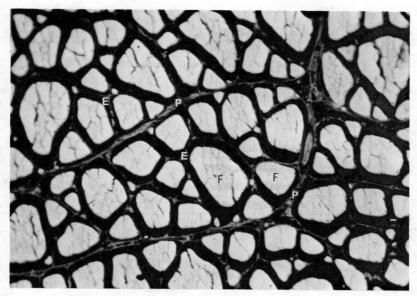

FIG. 4. A micrograph of a thin section of salmon muscle showing the organization of fibres and connective tissue within a myotome. F = muscle fibre; E = endomysium; P = perimysium.

(see Fig. 4) surrounded by connective-tissue sheaths thicker than the endomysia that surrounded each fibre. It is tempting to equate this with the hierarchy that has been identified in mammalian muscle; in this assumption the myotome would be the equivalent of the mammalian muscle. This view is supported by Bremner and Hallett.[9]

2.b The Muscle Cell

The basic unit of muscle is the muscle cell or fibre (see Fig. 5), which can be considered (from the food ingredient viewpoint) to consist of three complex components:

- the contractile apparatus or myofibrils;
- the proteins dissolved in the cell cytoplasm, or sarcoplasmic proteins, which bathe the myofibrils;
- the connective tissue forming the cell envelope.

In meat the fibres are typically up to 30 cm long and 10–100 μm in diameter,[3] whereas in fish the values are of a lower order of magnitude in length but similar diameter.[10]

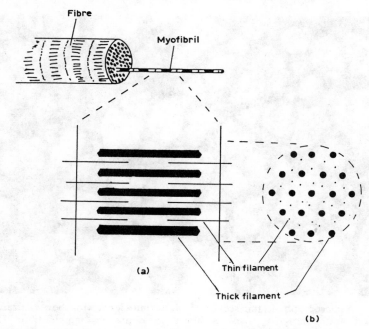

FIG. 5. Schematic representation of the structure of a muscle fibre: (a) organization of thick (myosin) and thin (actin) filaments in the myofibril; (b) transverse section through (a) at point of overlap of thick and thin filaments.

2.b.1 The Contractile Apparatus

About 80% of the muscle fibre volume[11] is occupied by the myofibrils. These are filamentous structures that have a diameter of about 1 μm and run the length of the cell. The myofibrils are composed of some 20 or more proteins,[4] two of which, myosin and actin, account for about 65% of the myofibrillar protein complement and form the basic filament structures (the myofilaments) of the contractile apparatus. Figure 5 represents diagrammatically the organization of the proteins myosin and actin in the myofibril, which form respectively two sets of repeating filaments—the thick and thin filaments—aligned parallel to the long axis of the myofibril.[12] Other proteins that are essential for function *in vivo*, such as troponin and actinin, are considered by us to contribute little to the properties of meat and fish as food when compared to the major protein components actin and myosin. In addition, we will not include the so-called intermediate filament proteins (i.e. desmin, connectin) even though their role in muscle tissue texture is currently under debate. For a fuller description of this class of muscle proteins, see Asghar.[4]

2.b.2 The Sarcoplasmic Proteins

The sarcoplasmic protein pool consists of a large number of soluble proteins, most of which are enzymes and individually do not warrant attention in a food context. However, one of the sarcoplasmic proteins—myoglobin—is important for its contribution to the red colour in muscle tissue. The content of myoglobin varies with muscle type within species (e.g. chicken leg/breast), according to muscle function and across species.[13]

2.b.3 Connective Tissues

The term connective tissue covers a wide variety of structures, such as tendon and ligaments, which exist in mammals but not fish. For the purposes of this chapter we are concerned only with connective tissue at the epi-, peri- and endomysial levels, i.e. the connective tissues that ensheath an individual muscle, bundles of fibres and individual fibres, respectively. Connective tissue consists of cellular and extracellular components. The predominant cell types are mast cells, macrophages and fibroblasts, which are surrounded by an extracellular matrix of modified carbohydrates such as mucopolysaccharides, chondroitin sulphates and hyaluronic acid. Dispersed throughout this matrix is a framework of elastin and collagen fibres.[14] In the connective tissue of muscle the dominant fibre type is collagen.

2.c The Muscle Proteins

The proteins of muscle can be considered as three classes (cf. muscle structure): the myofibrillar proteins, the sarcoplasmic proteins and the connective-tissue/stromal proteins. The major proteins of each class, and therefore the ones we shall concentrate on, are:

myofibrillar—myosin, actin and actomyosin;
sarcoplasmic—considered as a whole;
connective tissue—collagen.

2.c.1 Myosin

Myosin is a protein consisting of a globular head region attached to a rod-like tail.[15] The molecule consists of two main subunits, or polypeptide chains, that form the rod-like tail and the majority of the globular head, the remainder of the head region being made up of three smaller polypeptide chains.[15,16] Myosin contains about 2000 amino-acid residues and has a molecular weight of about 4.8×10^5 daltons. It has an isoelectric point (the pH at which the net charge on the molecule is neutral and the charge density is highest) of 5–5.5.[3] The rod-like tail region is designed to interact with the tails of other myosin molecules and in muscle is assembled into the thick filaments mentioned previously (see Fig. 5). The head regions have sites that interact with actin and therefore can link the thick filaments to the thin filaments.[15] This interaction is modified by adenosine triphosphate (ATP), which also provides the energy for muscle contraction. This in-vivo effect of a natural phosphate is relevant to our understanding of the role of 'active phosphates' used in the processing of muscle foods (see Section 6.d).

2.c.2 Actin
Actin is a globular protein of molecular weight about 40×10^3 daltons. In muscle and most food applications, it is present as a double-stranded helical polymer of globular actin known as F-actin.[17]

2.c.3 Actomyosin
In the absence of ATP or ATP analogues, myosin interacts with actin via its globular head region to form actomyosin. Myosin and actin are more or less in this form in rigor muscle (see Section 5.f.1(iii)) or meat, where the cellular complement of ATP has been depleted. We say 'more or less' because the number of myosin molecules interacting with the actin filaments depends upon the degree of interdigitation of the thick and thin filaments, and thus on the contractile state of the muscle at rigor mortis. Extracted actomyosin is a loosely defined molecule as it may contain variable ratios of actin and myosin. Whilst in meat actomyosin exists in a cross-linked filamentous form, the normal conformation of actomyosin in solution/suspension is that of an arrowhead structure. Although it is possible to shift the conformation to a filamentous form, the equilibrium state is the arrowhead form. In this form individual myosin molecules are attached to the globular molecules of F-actin.

2.c.4 The Sarcoplasmic Proteins
Because of our intent to deal only with those individual proteins or protein systems that are major contributors to tissue food properties, we will not consider any individual proteins in this group. This does not imply that certain proteins in this group (i.e. proteases) do not play an important role in modifying structure and texture, e.g. conditioning in meats (see Section 5.g.1).

2.c.5 Collagen

Collagen refers to a group of fibrous proteins that have a characteristic amino-acid content. The most notable feature is a high content of alanine and pyrrolidine, which together comprise about half the total amino-acid complement.[7] Collagen contains a high level of hydroxyproline, the detection of which is often made use of when quantifying collagen. Collagen also contains hydroxylysine, which plays an important role in the formation of cross-links that stabilize the collagen fibre.[7] Collagen fibres are constructed of rod-like molecules about 280 nm long and 1·4 nm diameter,[7] and each molecule consists of three polypeptide chains intertwined to form a triple helix. (For further information on collagen structure and chemistry refer to Chapter 7.)

3 SOURCE AND PRODUCTION

It is beyond the scope of this chapter to go into details of animal husbandry and fishing methods. In general, the reader can assume that the properties of the muscle proteins described are independent of the source unless specifically stated.

4 COMMERCIAL AVAILABILITY

Again it is beyond the scope of this chapter to cover commercial aspects of meat and fish. There is, however, one source of partially purified muscle myofibrillar protein that warrants a mention here. This material, called surimi, originates from Japan and is becoming available increasingly in Europe in the form of frozen blocks. The properties and uses of surimi will be covered in Section 6.f.6.

5 PROPERTIES OF THE MUSCLE PROTEINS

We have combined the solution and gel properties of the muscle proteins under one section, unlike the method of other chapters, because we believe this approach will make it easier to understand the properties of the individual proteins. We have, however, used the prefixes S for solution properties and G for gel properties.

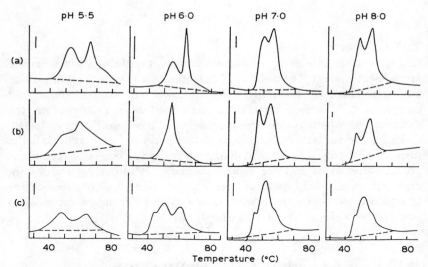

FIG. 6. The effect of pH and ionic strength (I) on the thermal denaturation of rabbit myosin. (a) 0·012 M KCl, 0·5 mM dithiothreitol (DTT), 0·05 M potassium phosphate buffer ($I = 0.05$); (b) 0·212 M KCl, 0·5 mM DTT, 0·05 M potassium phosphate buffer ($I = 0.25$); (c) 0·962 M KCl, 0·5 mM DTT, 0·05 M potassium phosphate buffer ($I = 1$). The vertical bar represents 0·02 mcal/s. The thermograms were obtained by heating 10–15 mg protein in sealed Perkin–Elmer volatile sample pans at a heating rate of 10°C/min using a Perkin–Elmer DSC II.[20]

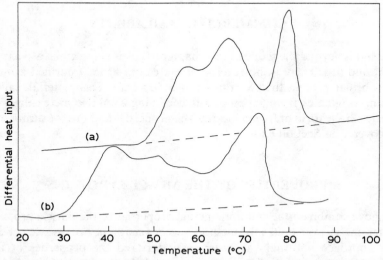

FIG. 7. A comparison of the thermal denaturation as determined by DSC of (a) beef and (b) cod muscle (post-rigor state). (Authors' unpublished data.)

5.a Myosin

S.5.a.1 Solubility

Myosin is usually extracted from comminuted pre-rigor meat (see Section 5.f.1(iii)) with solutions of salt (about 0·5M), and phosphate at about pH 7. As mentioned above, the rod-like tail of myosin has evolved to interact with the tails of other myosin molecules under physiological conditions, and thus over the pH range usually employed in foodstuffs it is essentially insoluble in water and salt solutions of low ionic strength. Therefore, in order to solubilize myosin, the interactions between the rod-like tails need to be minimized. This can be achieved by:

(i) using high sodium chloride concentrations (i.e. >0·3M) and other ions that have a synergistic effect with salt (e.g. the phosphate ion has been shown to weaken the interaction between myosin molecules[18]), and/or
(ii) using high salt concentrations (i.e. >0·3M) in conjunction with elevation of the environmental pH away from the isoelectric pH.

If the pH of a myosin solution (i.e. myosin dissolved in >0·3M NaCl) is decreased below the isoelectric point, the solubility of myosin is decreased.

S.5.a.2 Thermal Behaviour—Heating

When solutions or suspensions of myosin are heated, domains of the molecule undergo conformational change that exposes previously hidden amino-acid side-chains to the solvent, resulting in increased protein–protein interaction.[19] At concentrations of myosin above about 1%, the increased protein–protein interaction may result in the formation of a gel. The technique of differential scanning calorimetry (DSC) has been of particular use for studying the response of proteins to heat under varying environmental conditions. For instance, when myosin at pH 6 and $I=1$ is heated in the DSC calorimeter, three main cooperative thermal transitions are observed at 44, 50 and 59°C.[20] These three transitions have been related to domains of the molecule, namely the head, tail and a flexible region in the tail, often referred to as the hinge.[16] At pH 6 and $I=0.05$, only two transitions, at 56 and 65°C, are observed when myosin is heated in the calorimeter (see Fig. 6). Thus, the temperature at which myosin is thermally denatured depends upon the salt concentration and pH. The response of myosin to heat also depends upon the species from which it originates;[21] for instance, compare the thermal denaturation of beef and cod shown in Fig. 7.

S.5.a.3 Thermal Behaviour—Cooling

(i) *Undercooling.* Under certain conditions, myosin solutions can exist at temperatures below 0°C, when ice would normally exist. This state can be achieved by either judicious cooling of small droplets or the use of cryoprotectants (e.g. glycerol). There is no evidence (to our knowledge) that cooling *per se* leads to permanent modification of protein properties.

(ii) *Freezing.* Freezing of myosin solutions and suspensions invariably results in loss of solubility, and at high concentrations can result in gel

FIG. 8. The structure of heat-gelled myosin as visualized by scanning electron microscopy (SEM). The sample was prepared by heating a solution of myosin (15 mg protein per ml in 0·962M KCl, 0·05M potassium phosphate, pH 6) for 10 min at 70°C. The sample was further prepared for SEM by critical-point drying and coating with gold. (Authors' unpublished data.)

formation.[22] This phenomenon and the response of myosin when in its natural state in muscle will be discussed later (see Section 5.g.2).

G.5.a.4 Gelation—Thermally Induced

As mentioned above, when solutions or suspensions of myosin above 1% are heated to or above denaturation temperatures, the molecules aggregate to produce a gel. As for solubility, the properties of the gel depend upon salt concentration and pH. Figure 8 shows the network structure, as seen by scanning electron microscopy, of a gel made by heating a 1·5% solution of myosin in 0·962M KCl, 0·05M potassium phosphate, pH 6 to 70°C. The development of the gel during heating can be followed by small-deformation oscillatory testing. A typical result is shown in Fig. 9. By comparing the increase in gel properties with the DSC thermogram (Fig. 9 inset), we can see that a gel can form when the myosin head domain is denatured. Subsequent heating then increases the gel properties. For further information on the role of domains of myosin in gel formation, the

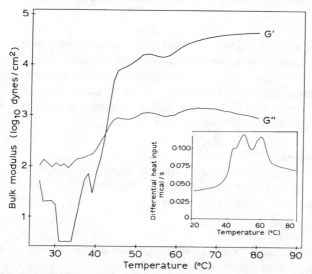

FIG. 9. The thermal gelation of rabbit myosin as followed by small-deformation oscillatory testing. G' = storage modulus; G'' = loss modulus. A solution of myosin (34 mg protein per ml) in 0·962M KCl, 0·05M potassium phosphate, pH 6, was heated at 1°C/min. The measurements were made using a Rheometrics Mechanical Spectrometer at 10% strain and a frequency of 10 rad/s. The inset depicts the thermal denaturation of myosin under similar conditions of pH and I as followed by DSC (heating rate 10°C/min). (Authors' unpublished data.)

reader should consult Samejima *et al.*[23,24] When myosin solutions at pH values near to the isoelectric point are heated, the resulting gels are opaque and often show signs of shrinkage and syneresis.[25]

G.5.a.5 Gelation—Chemically Induced
If myosin solutions at pH values greater than the isoelectric point (e.g. myosin dissolved in 0·6M sodium chloride buffered to pH 6) are titrated below the isoelectric point, two effects can occur depending on the rate of acidification.

(i) Fast acidification results in protein precipitation/flocculation.
(ii) Slow acidification results in the formation of a continuous gel.

G.5.a.6 Gelation—Time Dependent
Above a critical concentration of myosin in solution, the solution rheology gradually changes with time from a viscous liquid to a gel-like state. In the authors' experience, this gel formation can be reversed by shearing.

5.b Actin
In most food gel situations, actin will not be present independent of myosin and thus warrants little attention in this section. Actin is reported to have poor gel-forming properties when compared to myosin or actomyosin.[26] In food systems, therefore, we consider that its main effect is to alter the behaviour of myosin.[27] There is little evidence to suggest that there are species-dependent properties of actin.

5.c Actomyosin
In many respects the properties of actomyosin parallel those of myosin. Therefore, in the following sections we will only highlight those areas where we are aware that differences occur.

S.5.c.1 Solubility
A higher concentration of salt is required to dissolve actomyosin than to dissolve myosin at pH values above the isoelectric point (e.g. at pH 6, 0·3M salt for myosin and 0·6M for actomyosin). Figure 10 depicts the effect of pH and salt concentration on the solubility of extracted actomyosin.[28] If the so-called active phosphates, particularly those containing pyro- and tripolyphosphate, are used in conjunction with salt, then the concentration of the latter necessary to dissolve actomyosin may be reduced. This occurs because pyro- and tripolyphosphate interact with the head region of the

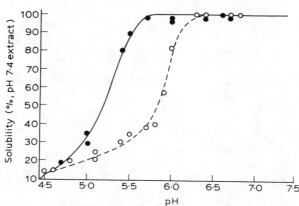

FIG. 10. The effect of pH and salt concentration on the solubility of actomyosin. (●) = 0·67M NaCl; (○) = 0·33M NaCl. (Adapted from Klement et al.,[28] used with permission.)

myosin molecule in a manner similar to that of the interaction of the natural high-energy phosphate ATP.[18] This results in a partial dissociation of the actomyosin to myosin and actin, both of which are soluble in lower salt concentrations.

S.5.c.2 Thermal Properties
The response of actomyosin to heat as evidenced by DSC shows that the behaviour of myosin is modified and that a transition exists that can be assigned to actin (see Fig. 11).

G.5.c.3 Gelation
For similar protein concentrations, heated under identical conditions, rheometric evidence shows that myosin gives a more elastic gel on heating than actomyosin.[27] This effect is not explained solely by a dilution effect of actin on myosin. For more details, the reader should refer to Samejima et al.,[27,29] where the effects of actin:myosin ratios on gel properties are discussed.

5.d Sarcoplasmic Proteins
As mentioned and described previously, the sarcoplasmic proteins will be considered as a whole.

S.5.d.1 Solubility
These proteins as a class are generally soluble in water and at most salt

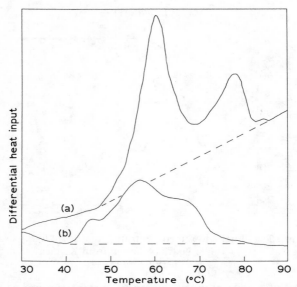

FIG. 11. The thermal denaturation as followed by DSC of extracted actomyosin. (a) 0·012 M KCl, 0·5 mM DTT, 0·05 M potassium phosphate buffer, pH 6 ($I = 0.05$); (b) 0·962 M KCl, 0·5 mM DTT, 0·05 M potassium phosphate buffer, pH 6 ($I = 1$). (Authors' unpublished data.)

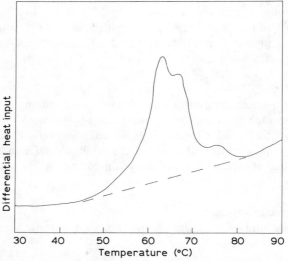

FIG. 12. The thermal denaturation as followed by DSC of extracted rabbit muscle sarcoplasmic proteins. The sample, after concentration by ultrafiltration, was heated at 10°C/min in a Perkin–Elmer DSC II. (Authors' unpublished data.)

concentrations, which means that differential extraction of muscle tissue can give rise to either sarcoplasmic or myofibrillar protein-free fractions.

S.5.d.2 Thermal Properties

(i) *Heating.* Investigations using DSC (see Fig. 12) indicate that the majority of these proteins undergo thermal denaturation at temperatures between those of myosin and actin.[30]

(ii) *Cooling.* The authors are unaware of any major effects of cooling, irrespective of ice formation, on the molecular properties of this group of proteins.

S.5.d.3 Gelation—Thermal and Chemical

Compared to the gel-forming ability of myosin/actomyosin, that of the sarcoplasmic proteins is poor.[26] In addition, at the concentration (about 6%) in muscle they tend to form precipitates rather than gels.

5.e Collagen

S.5.e.1 Solubility

For most practical purposes collagen in connective tissue is insoluble. However, extremes of pH (below 3, above 9) can result in extensive swelling or even dissolution of the collagen. This dissolution/swelling at pH 3 is reduced at high salt concentrations.[31]

S.5.e.2 Thermal—Heating

When collagen is heated, DSC studies show that it undergoes a characteristic transition in molecular conformation that results in visible shrinkage (up to 60%) of the bulk material.[32] In mammalian systems (e.g. bovine), collagen associated with peri- and endomysia show this transition at temperatures of about 66°C, whereas epimysial collagen requires a temperature of only about 60°C (see Fig. 13).

In fish, no distinction between connective-tissue types, either fibre or myotome ensheathing, has been made in terms of their response to heat. The temperature at which the collagen in fish myocommata denatures is lower than that for mammalian muscle collagen; for example, the collagen in salmon myocommata denatures at about 45°C.[33] As heating continues, mammalian collagen is progressively converted to gelatine; the rate and extent depends upon the nature and number of intermolecular cross-links, temperature and time. Thus, collagen from young and immature animals forms gelatine more readily than that from old or mature animals.

Generally, the collagen of fish is far less cross-linked than that of

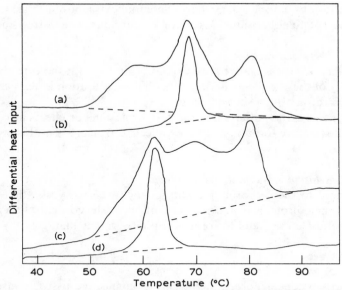

FIG. 13. The difference in thermal denaturation of bovine connective-tissue collagen as determined by tissue type. (a) and (c) DSC thermograms of whole beef muscle; (b) thermogram of excised perimysium; (d) thermogram of excised epimysium. (Authors' unpublished data.)

mammalian tissue and often a significant proportion may be converted to gelatine at the shrinkage temperature.

S.5.e.3 Thermal—Cooling

There is no known effect of low temperatures (undercooling or freezing) on the properties of native collagens. However, if the collagen has been converted to gelatine and subsequently cooled, a gel may form.

G.5.e.4 Gelation

Gelatine is the only form of collagen that will form a gel. In contrast to the thermal gelation of most other proteins, gelatine forms thermally reversible gels on cooling. There is considerable information available on gelatine and the reader is referred to Chapter 7.

5.f Factors Affecting Muscle Tissue as a Raw Material for Whole Muscle Foods

In the previous section, we have described, in fairly general terms, those molecular properties of muscle proteins that we consider are important in

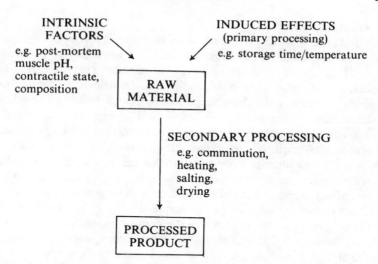

FIG. 14. A simplified scheme showing the interrelationship of the influences that effect the final product properties.

influencing the raw material properties of tissue foods. In this section we will attempt to establish the relationships (cause and effect) that exist between raw material/process interactions and texture. To do this we propose that the outline shown in Fig. 14 suffices to cover the wide range of influences that have been implicated in affecting final sensory quality. As shown, we have categorized these influences into four main groups.

5.f.1 Intrinsic Effects

By this term we refer to properties of the system over which the food processor has little control other than the ability to refuse to purchase, e.g. state of rigor, gender, nutritional state, seasonality, catching ground (for fish). All of these are claimed to affect the properties of tissue foods. Why is this so?

To classify muscle tissue quality by descriptors such as age, seasonality, etc., does not describe the underlying cause of observed effects. Our experience shows that variability can be explained in terms of variations in

- connective tissue;
- pH;
- contractile state;
- protein, lipid and water content.

(i) ANIMAL AGE

Age effects are generally manifest as an increase in sensory toughness in both meat and fish, but the causes of the effect are different. In meat the ageing of the animal results in a progressive increase in the number of heat-stable intermolecular cross-links within the connective tissue. For more detailed information on these changes, see Sims and Bailey.[14] The net effect of this change in cross-link stability is that (i) heat-induced shrinkage increases and (ii) the rate of formation of gelatine decreases. This results in the extension required to break the connective tissue increasing significantly, which probably explains the perceived increase in toughness of such material compared to meat from a young source. Consequently, the observation that meat from old animals requires longer cooking times than that from young animals to achieve acceptable sensory texture is explained via the properties of the collagen in the connective tissue.

In fish, however, effects are probably very different, as age is reflected by the size of the fish, and large fish generally carry a larger glycogen store than small fish. Therefore, at post-mortem, the ultimate pH achieved by large fish is lower than that of small ones. For reasons to be explained under (iii) below, this lower pH is responsible via its effects on the myofibrillar proteins for the increase in sensory toughness. This increase in toughness need not (and in some cases does not) detract from the acceptability of the fish as a food.

(ii) ANIMAL GENDER

An example of gender affecting raw material quality is illustrated by the use of bulls rather than steers in meat production. It appears[34] that certain bulls have the ability (when stressed in a certain manner) to convert their entire muscle glycogen reserve to high-energy phosphate compounds (e.g. ATP, CTP). After slaughter, the production of lactic acid via glycolysis (see (iii) below) cannot proceed because of the lack of glycogen. The ultimate muscle pH is therefore high (about 7), which, via the effects on the myofibrillar proteins in the muscle, results in a meat described as dark, firm and dry (DFD beef).

The darkness relates to the effect of pH on the haem pigment of myoglobin (darker red at high pH); the firmness and dryness relate to the location of water within the tissue, not to the water content.

(iii) RIGOR MORTIS

For ease of understanding some of the above-mentioned effects, in particular those relating to the modification of myofibrillar proteins by pH (and because it comprises the largest visual change accompanying the conversion of living tissue to what we know as tissue foods), we will describe the intrinsic process known as *rigor mortis*.

In vivo, muscle is maintained in a relaxed state, i.e. reduced interaction between the myosin and actin myofilaments, by the presence of ATP and a low free Ca^{2+} level. The ATP is maintained by the aerobic energy pathways of respiration. After slaughter, when the circulation of blood ceases, the muscle metabolism is changed to anaerobic conditions, ATP production from the muscle glycogen store continues. However, because of the lack of oxidative potential, lactic acid is produced, resulting in a lowering of the muscle pH. At a pH of about 6·2 it is believed the organelle responsible for controlling the muscle cell free Ca^{2+} level, the sarcoplasmic reticulum, ceases to scavenge Ca^{2+}, resulting in a contraction of the muscle.[35] Subsequently, lactic acid production continues, reducing the muscle pH until either the muscle glycogen store is exhausted or the enzymes responsible are inhibited by the low pH. The result of these physiological changes is manifest by

- a stiffening of the bulk material, and
- an increase in the apparent wetness/loss of liquid (drip) from the muscle.

The stiffening (or development of *rigor mortis*) occurs as a result of the interactions that link the myosin filaments to the actin filaments in the absence of ATP. We presume that this change in material properties arises because of increased protein density within the fibre effected by the greater degree of filament overlap within the myofibrils, and the enhanced degree of cross-linking between the various filaments that accompanies this increased overlap.

To understand the increase in apparent wetness/drip we need to consider the location of water within a muscle fibre. Typically, a muscle fibre contains about 80% water. As the major portion of the fibre volume is occupied by the contractile elements (the myofibrils), most of the water in the cell must reside within the myofibrils. There are two techniques that have aided particularly the elucidation of the *rigor mortis*-induced changes that occur in muscle structure and the resultant changes in the location of water. These techniques are low-angle X-ray diffraction and proton T_2 nuclear magnetic resonance (NMR). Work by April[36] in 1970 demonstrated the dependence of the size of the myofilament lattice on pH. The distance between the myofilaments decreases as the pH is decreased from 7 to 5. The explanation of this relates to the isoelectric point of the proteins constituting the filaments. At pH 7 (near the in-vivo pH) the proteins, and therefore the myofilaments, have a net negative charge that causes the myofilaments to repel each other. As the net negative charge increases as the pH is increased above the isoelectric point, the repulsive forces between the myofilaments, and therefore the distance between them, increases. Conversely, as the pH decreases towards the isoelectric point (as happens during *rigor mortis*), the net negative charge decreases and the filaments reside closer together. This results in the expulsion of water from within the myofibrils into the intermyofibril or interfibre space, thereby giving rise to drip and a wet appearance. The shift in the distribution of water when a muscle enters *rigor mortis* can be followed by proton T_2 NMR.[37] The physiological state of the muscle before slaughter will affect its properties as meat. For instance, a low level of intramuscular stored glycogen will result in a high ultimate muscle pH. This is generally the situation with fish, which store little muscle glycogen, and when a live animal has been stressed (cf. bulls as mentioned previously) or starved prior to slaughter.

5.g Induced Factors and Their Effect on the Bulk Properties of Muscle Tissue

Another descriptor for 'induced effects' on raw material properties can be post-harvest handling effects. We consider that the major influences in terms of effects on sensory texture arise via (i) electrical stimulation and hanging of meat, and (ii) the freezing and subsequent frozen storage of fish. Effects that arise via bad handling practice, e.g. allowing 'freezer-burn' to develop during frozen storage or microbially induced changes during wet storage, are not considered here.

5.g.1 Electrical Stimulation and Conditioning (Hanging)
In Section 5.f.1(iii) a detailed description of the rigor process was given. The time taken for rigor mortis to develop varies, in the authors' experience, from 1 h for immature fish to several hours, typically around 14 h, for beef animals. By applying a voltage to the carcass it is possible to stimulate the muscles to contract, thus using their energy reserves. The net result is that the rate at which *rigor mortis* develops is increased dramatically. There is considerable literature on the benefits and disadvantages of electrical stimulation.[38,39] In the authors' view, no serious disadvantages have been identified.

The conditioning or hanging of carcass meats in order to enhance meat tenderness has been carried out by butchers for many years. Whilst its effect on the bulk properties of meat has been recognized for many years, the underlying changes in the proteins that constitute the tissue structure have not been positively identified. There has been considerable work on enzymes (e.g. cathepsins, calcium-activated factors) present within meat tissue that could possibly modify its structure during conditioning.[40,41] One important point to note here is that because of these changes meat and fish are dynamic materials, and they both change continuously *post mortem*.

5.g.2 Freezing

When muscle tissue is cooled below 0°C, ice does not usually begin to form until a temperature of -2 to -4°C is reached, because of the freezing-point-depressant effect of dissolved cellular solutes. As the temperature decreases, more and more water is frozen. In most commercial freezing operations the rate of freezing is generally not fast enough to effect intracellular freezing and therefore ice nucleates extracellularly and subsequent ice crystal growth occurs there.[42] As a consequence of this, the muscle cells shrink more and more as freezing progresses because the growth of extracellular ice promotes the transport of cellular water to the extracellular space via osmotic effects. If, during frozen storage, the muscle tissue experiences temperature cycling, then the size distribution pattern of ice crystals is altered significantly; this is known technically as ice crystal disproportionation.

In fish muscle, the effects of frozen storage can be extreme, and can depend on the physical state of the muscle at the time of freezing (i.e. whether the muscle is intact—e.g. whole fish/fillets—or in the form of a coarse comminute—e.g. commercially available fish mince) and on the temperature and time of frozen storage. In terms of the effect on the bulk material of fish muscle, the most obvious is the altered water distribution; on thawing of frozen stored fish there can be copious loss of cellular water in the form of 'drip'. In addition, the tissue itself becomes firmer and more elastic. We consider that the most likely explanation for this is that the freezing process (by inducing water redistribution) effectively concentrates the myofibrillar protein within the muscle fibre. If the tissue is frozen-stored for only a short time, most of this redistributed water is re-imbibed into the fibres. However, if there is a prolonged storage of the tissue, a physicochemical reaction that is not yet fully understood manifests itself as the often-quoted freeze denaturation mechanism. For a full description of the possible causes of this phenomenon, the reader is referred to Shenouda.[43] The best-supported theory is that in the fish genus known as *Gadidae* an enzyme-mediated reaction converts trimethylamine into dimethylamine

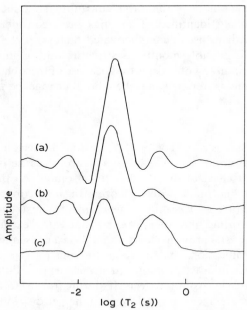

FIG. 15. The effect of $-7°C$ frozen storage on the distribution of T_2 relaxation times of water protons in cod muscle. Note the increase in water protons with longer relaxation times. (a) Fresh tissue, (b) 9 days storage and (c) 34 days storage. (Authors' unpublished data.)

plus formaldehyde. The latter compound then effects a cross-linking between adjacent myofibrillar proteins. This fixes the proteins in their freeze-concentrated state so that on thawing of the tissue the fibres are unable to swell and regain their lost water. Fortunately, this explanation can also accommodate the changes that occur in the mechanical properties of the tissue; the increased concentration and chemical fixing of the myofibrillar proteins would lead to the fibres becoming more rigid. There is strong evidence, from proton T_2 NMR studies, to support the assertion that water redistribution is one of the prime events occurring during frozen storage of fish. Such a study[22] has shown that prolonged frozen storage of coarsely comminuted fish induces irreversible structural alterations. These structural changes are considered to arise from aggregation of the myofibrillar components during the storage period. As a consequence of this, sufficiently large spaces in the system are created so that exchange between 'bound' and 'free' water cannot be averaged over the complete sample. Thus, 'freeze denaturation' of fish muscle can then be explained simply in terms of increasing aggregation, resulting in an increase in the number or size of interstitial spaces (see Fig. 15). An additional consequence of this aggregation, or 'freeze denaturation', of fish muscle is that the solubility of the constituent myofibrillar proteins is dramatically changed. From being readily soluble in 0·6M NaCl solution (see Section S.5.c.1) they become virtually insoluble after frozen storage conditions that are both time- and temperature-dependent. The consequences of this on the quality of fish muscle are described in Section 6.a.

6 APPLICATIONS

The description of the various intrinsic factors affecting the bulk properties of muscle tissue as a raw material could be considered as being largely irrelevant, since there are few muscle foods consumed in the uncooked state—i.e. the conversion of the raw material to food involves the superimposition of a cooking stage. Although we accept this, we would cite the consumption of the raw fish products sashimi and sushi, and the tendency of some meat eaters to consume very rare steaks, as examples in which the intrinsic and induced effects are the sole influences on product quality. Before considering these aspects we shall include here information

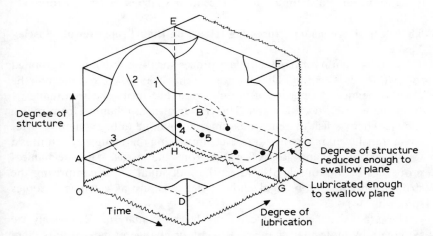

FIG. 16. The in-mouth texture model proposed by Hutchings and Lillford,[39] showing the breakdown routes for different food types. 1, Tender juicy steak; 2, tough dry meat; 3, dry sponge cake; 4, oyster; 5, liquids.

on food texture. The perception of food texture is a dynamic sensory monitor of changes made to a food by processes occurring in the mouth. Figure 16 shows a model, proposed by Hutchings and Lillford,[44] for in-mouth perceived texture of foods.

6.a Muscle Texture

In fresh fish muscle, the structure is as described in Section 2.a.2. If this system is eaten in the uncooked state, the sensory perception can be described as being similar to that elicited by a slightly chewy gel. There is very little perception of fibrosity at this stage because the tissue is mechanically, though not visually, isotropic. There is no differential mechanical response in its behaviour, whether forces are applied parallel to or orthogonal to the muscle fibre direction because the connective-tissue matrix dominates. If the muscle is fresh, then little free water will exude on the application of the forces described above and so the overall succulence perception is moist but not wet (a related example more recognizable to Western taste would be cold smoked salmon). However, if the fish has gone through *rigor mortis* or has been frozen and thawed, the changes in bulk properties effected by these processes (see Sections 5.f.1(iii) and 5.g.2) will manifest themselves; that is, the tissue will probably be perceived as firmer and chewier (as a result of protein aggregation within the fibre) and wetter during initial chewing, possibly becoming drier on subsequent chewing (because water redistributed to extracellular space is easier to express).

6.b Effect of Secondary Processing (Heat) on Bulk Properties of Muscle Tissue

The most common form of secondary processing (by our definition) applied to tissue foods is heating. The observation that heat has a gross effect on the physicochemical properties of muscle tissue has probably been made by everyone who has ever cooked. The colour of meat changes from red to brown, whereas fish in general assumes an opaque milky whiteness after being translucent in the raw state. Accompanying this change of colour is a visible shrinkage and loss of water from the tissue. The mechanical properties of the muscle also alter; the individual fibres comprising the structure become more rigid and yet the structure as a whole becomes somewhat weaker.

These physicochemical effects on the bulk material can easily be explained by molecular changes. The colour change in meat results from the thermal conversion of myoglobin into denatured metmyoglobin, whereas in fish thermal denaturation and aggregation of the tissue proteins creates molecular aggregates of a size sufficient to scatter light rather than allow its free passage. The influence of heat on the expulsion of water and tissue shrinkage occurs via the following sequence of events. As the tissue is heated, its temperature rises with little observable effect until the thermal denaturation temperature of myosin (about 40°C for meat, about 27°C for most fish) is reached. The myofibrils begin to shrink, causing (i) a spontaneous expulsion of myofibrillar water into extracellular space, (ii) an effective concentration of the myofibrillar protein, and (iii) an increase in the fibre rigidity. As the temperature continues to rise, the collagen in the connective tissue also begins to undergo thermal denaturation and shrinkage (at about 60°C for meat, about 45°C for fish), which compounds the expulsion of water from the cell, thus also increasing protein concentration. Finally, at about 70°C in meat, and about 65°C in fish, the actin component also denatures. This sequence of events, as followed by DSC, is depicted in Figs 7 and 13.

Why individual fibres become more rigid is adequately explained by the effects of heat on the myofibrillar proteins, but the decrease in the strength of the structure as a whole depends on the connective tissue. The effect is much more visible in fish tissue than in meat, since fish collagen possesses few heat-stable cross-links and as a result readily undergoes conversion to gelatine, which has low mechanical strength. The net effect is the ready breakdown of cooked fish muscle into flakes (the primary level of connective-tissue organization) and then into fibres (the secondary level of organization), on the application of even very gentle force.

In meat the same visual effect is not noticed because meat collagen possesses enough thermally stable cross-links to prevent the type of muscle breakdown described above from normally occurring. However, prolonged severe heat treatment can in some cases effect a response similar to that in fish and the meat tissue breaks down during processing into fibres or fibre bundles.

6.c Effect of Secondary Processing (Heat) on Sensory Properties

We have described the texture of raw muscle tissue as being that of a chewy gel, i.e. non-fibrous. However, it is probably stating the obvious to say that a key determinant in the sensory quality of cooked tissue foods is the degree of fibrosity perceived. How, then, is a non-fibrous chewy gel converted into a fibrous product by the action of heat? Our previous requirement that perception of fibrosity demands mechanical anisotropy, i.e. preferential planes of weakness in the tissue, is met as a result of cooking, since the fibres become stronger and the connective tissue weaker. Thus, when the forces exerted by the mouth during chewing are applied to cooked tissue, the structure preferentially fractures at the connective tissue level and the muscle fibres can then be sensed either individually or in the form of bundles, depending on the level of connective tissue that fails. The other sensory characteristics that depend on the mechanical properties of the tissue, i.e. toughness, chewiness, etc., are very much dependent on both the contribution of the residual strength of connective tissue and the fibre properties in meat (but only fibre properties in fish) resulting from the changes induced in these by heat.

The effects of induced or intrinsic factors on the muscle tissue (pH, freeze denaturation) are not removed by heat treatment, they are simply added to. In other words, the increased rigidity in fibres from *rigor mortis* or protein aggregation via chemical cross-linking is carried through into the cooked material. Muscle tissue of low pH is firmer/tougher and wetter after cooking than higher-pH material cooked under identical conditions. A point of interest arises in the effect of 'freeze denaturation' of fish muscle prior to heating. The interest arises in the use of the term 'denaturation'. If freeze-denatured and heat-denatured samples of muscle were assessed by some physicochemical tests (say, protein solubility or DSC) they would be almost indistinguishable. And yet they are totally different in sensory characteristics. This serves as a good example that denaturation mechanisms can be radically different and give rise to what must be different microstructures.

6.d Effect of Secondary Processing (Salting) on Bulk Properties of Muscle Tissue

The use of salts to modify the texture of muscle-based foods is common practice for both whole-muscle foods (e.g. ham) and comminutes (e.g. beefburger). Both sodium chloride and phosphate (often mixtures of orthophosphate and polyphosphates) are used. Although we will consider the salting of pork meat to produce ham, the principles and action of salt and phosphate apply to other salted muscle foods. It is usual practice, nowadays, for ham to be manufactured by what is known as the wet brining process. In this process solutions of sodium chloride (concentrations up to saturated) and polyphosphate mixtures (concentrations up to 0·5%) are injected into pork muscle, which is subsequently massaged and cooked.

The action of salt. Exposure to solutions of sodium chloride in excess of 0·5M causes myofibrils to swell and imbibe water,[11] the maximum water uptake occurring in the range 0·8–1·0M NaCl.[45] Whilst the phenomenon of salt-induced swelling of myofibrils, and thus of meat, is quite well documented, the detailed mechanism by which NaCl causes such effects is as yet not fully understood at the molecular level.

The action of phosphate. When considering the action of phosphates on meat we need to consider the type (i.e. ortho-, pyro-, tripoly-) employed, because pyrophosphate and tripolyphosphate have specific effects on the actin–myosin interaction. Pyrophosphate and tripolyphosphate, in the presence of Mg^{2+}, act by interacting with the ATP sites on the myosin molecule, as dissociation agents of actomyosin. This results in the formation of separated forms of actin and myosin. The overall effect is that the propensity of myosin to dissolve in salt is increased and restraints on the swelling of the myofibrils are decreased,[11] resulting in increased water holding in the muscle.

There are also other constraints on salt-induced swelling of meat. The collagenous endomysia, perimysia and epimysia that form the connective hierarchy throughout a muscle act as mechanical restraints on the swelling of muscle.[46] The role of the tumbling operation during ham processing may therefore be considered to be:

(i) to distribute the salts throughout the muscle;
(ii) to weaken/disrupt the connective-tissue sheaths and therefore remove constraints to swelling;
(iii) to expel dissolved myofibrillar proteins that, in the dissolved state, act as good binding agents, from the muscle fibres.

The overall result is an increase in the volume of the muscle, with a coincident decrease in the density of the fibrous components, and a weakening of the connective-tissue matrix.

6.e Effect of Secondary Processing (Salting) on Sensory Properties

Since most salted meats are eaten after cooking, we have to consider both the action of salts and the response of salted meats to heat. In Section 6.b we described the effects of heat on meat to be mainly twofold, i.e. it increases the fibre density and rigidity whilst weakening the connective-tissue matrix. If the myofibrils have undergone salt-induced swelling, then the increase in fibre density and rigidity on heating is reduced. The effects of heat on the connective tissue remain similar. As the structural contrast between the cooked fibres and the cooked connective-tissue matrix is reduced, the effect on product texture is an increase in the perceived fibrosity. If the salting and tumbling processes are carried out to extremes, then, after heating, the product has a texture of a rubbery gel (cf. Section G.5.c.3 on actomyosin gelation) with dispersed connective-tissue strands.

6.f Effect of Secondary Processing (Comminution) on Bulk Properties of Muscle Tissue

Comminution (i.e. size reduction of intact tissue) in the meat-processing industry historically arose almost certainly as a result of pressure to render previously unacceptable-quality meat palatable (i.e. meat containing intractable connective tissue). In fish processing in the Western world, the pressure was again economic, i.e. maximum utilization of available muscle tissue, whereas we suspect that in Japan the reason for fish comminution was the desire to utilize the gelling properties of the myofibrillar proteins.

In Section 6.a we described the bulk properties of intact muscle tissue in the raw state. These were relevant because they defined sensory properties (i.e. some of the intact-tissue products are eaten raw), whereas comminuted tissue is seldom eaten raw. Two exceptions to this exist: (i) steak tartare, where we consider that comminution is for presentation purposes only and not to render the meat acceptable, and (ii) salami-type products, which we classify as preserved products and are discussed in the following section. In this section, therefore, we consider only the properties of comminuted tissues that have undergone thermal processing either by the food processor or the consumer. The range of comminuted products we shall take as examples are:

- coarsely comminuted re-assembled meats (e.g. burgers, steaklets);
- finely comminuted meat products (e.g. paté, sausages);
- finely comminuted fish products (kamaboko, structured surimi products).

We shall only describe the technologies involved in relation to the molecular properties explained in earlier sections. In the product ranges described above, several scientific principles have to be considered in understanding what is done and why it is done.

6.f.1 Coarse Comminutes
Why comminute at all? The historical reason identified above is to reduce the contribution that connective tissues make to sensory properties by reducing its physical size. The reason for this is either that (i) the connective tissue need no longer be disrupted in the mouth to achieve a swallowable state or that (ii) the amount of connective tissue to be disrupted in any given piece is reduced. This results in the tissue being perceived as more tender. Coarse comminution is subdivided into different particle sizes and shapes.

6.f.2 Why Different Particle Sizes?

The two product types taken as examples, i.e. burgers and steaklets, can be characterized by their similarity to intact tissue texture. Burgers display little similarity, whereas steaklets are intended (as much as possible) to be identical to high-quality meat cuts (i.e. steaks). It is not unreasonable to assume that the larger the size of the meat particle the more meat-like will be the perceived texture. However, the size of particle used results from a compromise between the desired product texture and the inherent quality (i.e. toughness of the connective tissue) of the original meat.

6.f.3 Different Shapes

In this instance it is necessary to refer to the anisotropic nature of meat structure, i.e. an aligned fibrous composite, which as previously described results in tissue foods having their characteristic textures. As a result, it seems reasonable to assume that the recombination of particles of high aspect ratios will, especially if ordered, better simulate natural meat texture than will the reassembly of low-aspect-ratio particles. The former is a structural description of steaklets and the latter of burgers.

6.f.4 Assembly Method

To form a product, the particles (whatever their shape/size) must interact with each other to create a cohesive product. This is achieved most often by utilizing the molecular properties of the myofibrillar proteins described in Section 5, i.e. solubilization followed by their thermal gelation during cooking. To effect this phenomenon requires the judicious incorporation of salt and/or water into the particulate mix. The levels of salt required to solubilize the required proteins are of the order of 3·6% (see Section S.5.a.1), but this assumes that the salt 'sees' all the water in the product. However, this is not necessarily the case, since the diffusion of salt through meat[47] is slow relative to the process times usually involved. Because of this, salt will only act at the meat surfaces and thus the concentration required for interparticle adhesion will depend on available surface area, which is dictated by particle size and shape.

A further contributory factor to using a salt concentration that, on the face of it, is insufficient to dissolve actomyosin, is that the comminuted particles are often processed frozen or partially frozen, thus restricting the water available for salt dissolution. In practice, salt inclusion as low as 1% of the formulation weight can achieve the necessary degree of 'binding'. It is worth mentioning, however, that salt concentration is not the only factor controlling meat particle adhesion. By choosing a suitable salt and phosphate mixture, the proportion of extracted myosin can be enhanced, thus increasing the cohesive strength of the interparticle bonding.[48] Similarly, it has been shown that the geometry of the meat at the particle–particle interface (e.g. whether the particle fibres run parallel or perpendicular to the interface) also affects the bond strength.[49] How this is manifest in perceived textural properties still awaits clarification.

6.f.5 Fine Comminutes

In this product range the reasons for comminution are essentially those for coarse comminutes, but much greater use is made of the molecular properties of the constitutive proteins simply because the structure is disrupted to below the fibre level. Such products as frankfurters, English sausages, etc., are therefore based on the solubility and gel properties of muscle proteins, which were described in Section 5.

In this situation, to achieve solubility, the required salt addition should be based upon the total water content of the system, since the particles created by comminution are so small that diffusion of salt is effectively instantaneous and, as a result, all of the water is available.

Although the texture of these products is based on the gel properties of actomyosin, sensory differences are manipulated by changing the amount

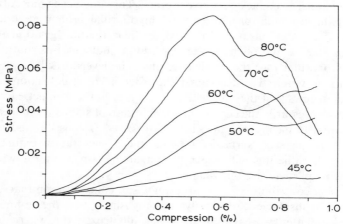

FIG. 17. The effect of heating on the compressive properties of a model product consisting of finely comminuted salted meat plus ungelatinized potato starch. Comparison with DSC and rheological data suggest that the meat protein gels between 40 and 70°C, and that the gelatinization of the starch occurs at about 60°C and higher temperatures. (Authors' unpublished data.)

and concentration of the gel phase and the amount, size distribution and type of the included phase. For example, in some sausage products fat as the included phase can range in size from micrometres (e.g. in frankfurters) to millimetres (e.g. in English sausages). Thus, these foods can also be considered as composite systems in which the origin of the sensory texture must arise from (i) the viscoelastic properties of the matrix (i.e. the actomyosin gel), (ii) the viscoelastic properties, shape, size and phase volume of the included phase, and (iii) interactions between components of the two phases. An example of the last is the use of starch granules dispersed in the matrix, which, upon heating, begin to gelatinize and take up water from the matrix. The end result is that the included phase changes its phase volume and viscoelastic properties, while the matrix coincidentally decreases in phase volume but undergoes an increase in concentration. Figure 17 shows the changes in mechanical properties of such a system.

The gelling properties of actomyosin are such, however, that in products where plastic, rather than elastic, properties are required (i.e. a spreading paté), then solubilized actomyosin should not form the continuous phase. In most products of this type the continuous phase is based on sodium caseinate or globular proteins, which do not form strong gels at the concentrations employed (i.e. liver proteins).

6.f.6 Comminuted Fish Products

The reasons for comminuting fish are twofold. In the West it was primarily economic, which has given rise to fish-mince products that comprise essentially entangled muscle fibres. The product range is normally seen as being of inferior textural quality, but the use of freshly generated fish mince can generate products that are barely distinguishable in texture (but not visually) from products made from intact fillets.[50] The poor quality image of fish mince arises not from the mincing *per se* but from the increased susceptibility of disrupted fish to deterioration during frozen storage (see Section 5.g.2).

In the Orient, the reason for comminuting is simply to attain the gel-forming properties of fish actomyosin to form kamaboko-type products. So much so, in fact, that the weak gel-forming proteins (the sarcoplasmic proteins) and much of the insoluble connective-tissue protein are removed by extensive water washing and straining prior to the use of salt to achieve actomyosin solubilization.

The gel-forming properties of fish actomyosin are also dependent on using fresh fish.[51] Traditionally, most kamaboko products were manufactured on land from 'inshore-caught' fish. As supplies began to decrease, methods for preserving the functionality of fish muscle proteins were developed. The methodology that has evolved involves the addition of low-molecular-weight compounds such as sorbitol, sucrose and phosphate, and in some instances NaCl, to protect the proteins from the freeze-denaturation phenomenon that occurs during frozen storage. Whilst we believe the mechanism of action is yet to be established conclusively, the reader is referred to Matsumoto[52] and to the more recent approach of Slade and Levine.[53] Freeze-denatured fish muscle is incapable of yielding soluble myofibrillar protein in the quantities demanded by products such as this.

The technology described here has given rise to the now-familiar fish commodity known as surimi. This raw material is basically a partially purified myofibrillar protein suspension that can be used to produce a range of products. These vary from tough rubbery gels (kamaboko) to products whose properties have been altered to achieve fibrosity by mechanical structuring (crab sticks) or by the inclusion of air and starch granules to alter the bulk mechanical properties.[54]

6.g Preservation Systems
In our introduction to this chapter we classified products into three groups. The third group was described as products that have evolved by virtue of the physicochemical changes they are subject to during preservation. There are a fairly large number of products in this category, but we will describe only two, since each is representative of a major preservation system.

6.g.1 High-Ionic-Strength/Low-pH (Example: Marinaded Herring)
Marinaded herring is a traditional food consumed in several North European countries where preservation is affected by curing the fish in a solution of sodium chloride and acetic acid. Inherent in the process, however, are the changes in the tissue properties caused by the increase in ionic strength and decrease in pH. These changes have been investigated and described previously,[55] but the origins of the effects observed when fish muscle is marinated are generally as follows. When we consider that the proteins that constitute muscle have functional groups that respond to changes in ionic strength and pH, it is not difficult to see why salt and acid have such an effect on tissue properties. The response of the myofibrillar proteins to acid can be explained by referring to their isoelectric point (see Section S.5.a.1). If the pH is shifted in any direction away from the isoelectric point, then the proteins attain either a net negative charge (alkaline side of the isoelectric point) or a net positive charge (acidic side of

the isoelectric point), resulting in increased electrostatic repulsion of the proteins. The result of this is that the structure swells and more water can be held in the structure (compare the effects of *rigor mortis* described in Section 5.f.1(iii)). Since the pH of fish flesh in rigor (6·2–7·0) is on the alkaline side of the isoelectric point, the progressive addition of acid to pH 4 should see the structure shrink as the pH approaches the isoelectric point and then begin to swell again. We mentioned in Section S.5.a.1 that the addition of salt to muscle proteins may cause them to dissolve and may cause myofibrils to swell at pH values above the isoelectric point. If salt is added to muscle tissue at pH values below the isoelectric point, then the opposite occurs (compare this effect with the solubility of myosin or actomyosin in Section 5). An explanation of this effect is that the presence of chloride ions decreases the pH at which the system has electrostatic balance, therefore decreasing the pH of minimum water holding.

An additional point worth mentioning is that if fish are cured in acid alone, they become exceedingly fragile because the connective tissue that holds the flakes or myotomes together is weakened and often partially soluble in dilute acid (see Section S.5.e.1). The effect of this loss of water from the muscle fibre is that the protein becomes more concentrated and probably aggregated, thus giving a firmer texture.

As anyone who has ever worked with a marinade knows, it is not a static product. There are progressive changes in its perceived texture, which becomes softer with storage time. These changes are most probably affected by enzymes present in the fish muscle itself. Although the specific enzymes have not been identified, only low activities would be necessary to bring about the observed changes over the time-scales of product storage/shelf-life.

6.g.2 Drying

The technology of drying to preserve food from spoilage is probably as old as mankind. The technique derives its efficacy by reducing the water content of the food to a level below which bacteria, moulds and fungi find it impossible to grow. In terms of convenience, it was undoubtedly a great discovery; not only was food able to be stored for long periods, it was very much lighter and easier to transport. However, drying, even under the mildest conditions, can lead to irreversible changes in the microstructure of muscle tissue, with concomitant loss in perceived textural quality on rehydration, so much so that dried muscle or its rehydrated form is, we would argue, an acquired taste. Although freeze drying is a gentle technique in relative terms, even freeze-dried muscle, on rehydration, does not recover its original texture. The mechanism behind these effects is, we consider, almost certainly concentration-induced protein aggregation effected by water removal. In air-dried products, the sample shrinks visibly as water is removed, and the myofibrillar proteins become more and more concentrated. At this point there is a close similarity with the effects of freezing, where removal of cellular water by osmosis causes structural changes in the muscle that we believe also to be concentration-induced aggregation. Effectively, these changes are irreversible within practical time-scales. Attempts to recover the original water-holding ability of air-dried tissue often requires extreme conditions, e.g. the use of dilute alkali as the rehydration medium for the Norwegian 'Lutefish'. However, this invariably results in severe modification of perceived texture. Whether rehydrated air-dried muscle would ever assume the texture of fresh if given enough time, i.e. if the rehydrated state were at equilibrium, has not to our knowledge been demonstrated.

Freeze drying does not cause visible product shrinkage, because the sublimation of ice fixes the tissue solids in the positions they occupied in the frozen state. In addition, freeze-dried materials may rehydrate to their original moisture content; however, the original sensory properties are seldom, if ever, recovered totally. We tentatively explain this by invoking the concept of water distribution within the tissue structure. Some of our NMR data show that the spectrum of water proton relaxation times, and hence the microstructure of the tissue, is different in fresh and rehydrated forms.

7 SUMMARY

Our intention in this chapter was not only to highlight those properties of muscle proteins that influence the sensory quality of muscle tissue as a food but also to describe the importance of the overall properties (mechanical/biophysical) of the structure that ultimately determine sensory properties. In addition, we hope to have covered the importance of various post-harvest processes as well as the intrinsic biological properties of muscle tissues on final sensory quality. In doing this we realize that the approach adopted has been a general overview rather than in-depth specific explanations, but feel that this was justified in this case. We also consider that the biggest gap in our knowledge of muscle proteins and their structures is our understanding of the effects induced by dehydration. Are the changes irreversible, and if so why? Twenty years ago the effects of freezing on muscle systems were considered in the same light, but as our understanding of freezing has grown so has the commercial success of frozen meat and fish. Will the same be true of dehydrated muscle tissue foods in the future?

ACKNOWLEDGEMENTS

The authors are grateful to Peter J. Lillford for consultation on content and to Simon Wilding for preparation of figures.

REFERENCES

1. Bailey, A. J. (ed.), *Recent Advances in the Chemistry of Meat*. Royal Society of Chemistry Special Publication No. 47, 1983.
2. Suzuki, T., *Fish and Krill Protein: Processing Technology*. Applied Science Publishers, London, 1981.
3. Schutt, J., Meat emulsions. In *Food Emulsions*, ed. S. Friberg. Marcel Dekker, New York, 1986, p. 157.
4. Asghar, A., Functionality of muscle proteins in gelation mechanisms of structured meat products. *Crit. Revs Food Sci. Nutrition*, **22** (1985) 27.
5. Stanley, D. W., Relation of structure to physical properties of animal material. In *Physical Properties of Foods*, ed. M. Peleg & E. B. Bagley. Avi Publishing Co., Connecticut, USA, 1983, p. 157.
6. Lawrie, R. A. (ed.), *Developments in Meat Science*, Vols 1–3. Applied Science Publishers, Barking, 1980–81.
7. Bailey, A. J., The chemistry of intramolecular collagen. In *Recent Advances in the Chemistry of Meat*, ed. A. J. Bailey. Royal Society of Chemistry Special Publication No. 47, 1983, p. 22.
8. Love, R. M., *Chemical Biology of Fishes*. Academic Press, London, 1980.
9. Bremner, H. A. & Hallett, I. C., Muscle fiber–connective tissue junctions in the fish Blue Grenadier (*Macruronus novaezelandiae*). A scanning electron microscope study. *J. Food Sci.*, **50** (1985) 975.
10. Tanaka, T., *Bull. Tokai Reg. Fish. Res. Lab.*, **20** (1958) 77.
11. Offer, G. & Trinick, J., On the mechanism of water holding in meat: the swelling and shrinking of myofibrils. *Meat Sci.*, **8** (1983) 245.
12. Huxley, H. E., *The Cell*, **4** (1960) 365.
13. Lawrie, R. A., Chemical and biochemical constitution of muscle. In *Meat Science*. Pergamon Press, New York, 1979, p. 75.
14. Sims, T. J. & Bailey, A. J., Connective tissue. In *Developments in Meat Science—2*, ed. R. Lawrie. Applied Science Publishers, Barking, 1981, p. 29.
15. Lowey, S., Myosin: molecule and filament. In *Subunits in Biological Systems*. Biological Macromolecules Series, Vol. 5, ed. S. N. Timasheff & G. D. Fasman, Marcel Dekker, New York, 1971, p. 201.
16. Elliott, A. & Offer, G. W., Shape and flexibility of myosin molecules. *J. Mol. Biol.*, **123** (1978) 505.

17. Oosawa, F. & Kasai, M., Actin. In *Subunits in Biological Systems*. Biological Macromolecules Series, Vol. 5, ed. S. N. Timasheff & G. D. Fasman, Marcel Dekker, New York, 1971, p. 261.
18. Bendall, J. R., The swelling effect of polyphosphates on lean meat. *J. Sci. Food Agric.*, **5** (1967) 468.
19. Brandts, J. F., In *Thermobiology*, ed. A. H. Rose. Academic Press, New York, 1976, p. 25.
20. Wright, D. J. & Wilding, P., Differential scanning calorimetric study of muscle and its proteins: myosin and its subfragments. *J. Sci. Food Agric.*, **35** (1984) 357.
21. Starbusvik, E. & Martens, H., Thermal denaturation of proteins in post rigor muscle tissue as studied by differential scanning calorimetry. *J. Sci. Food Agric.*, **31** (1980) 1034.
22. Lillford, P. J., Freeze-texturing and other aspects of the effects of freezing on food quality. In *Properties of Water in Foods in Relation to Quality and Stability*, ed. D. Simatos & J. L. Multon. Martinus Nijhoff, Dordrecht, 1985, p. 543.
23. Ishioroshi, M., Samejima, K. & Yasui, T., Further studies on the roles of the head and tail regions of the myosin molecule in heat-induced gelation. *J. Food Sci.*, **47** (1982) 114.
24. Samejima, K. & Yasui, T., Mechanism of heat-induced gelation of myosin. *Nippon Nogeikagaku Kkaishi J. Agric. Chem. Soc. Japan*, **62** (1988) 892.
25. Hermansson, A., Water and fatholding. In *Functional Properties of Food Macromolecules*, ed. J. R. Mitchell & D. A. Ledward. Elsevier Applied Science, Barking, 1986, p. 273.
26. Acton, J. C., Ziegler, G. R. & Burge, D. L., Functionality of muscle constituents in the processing of comminuted meat products. *Crit. Revs Food Sci. Nutrition*, **18** (1983) 99.
27. Yasui, T., Ishioroshi, M. & Samejima, K., Heat-induced gelation of myosin in the presence of actin. *J. Food Biochem.*, **4** (1980) 61.
28. Klement, J. T., Cassens, R. G., Fennema, O. R. & Greaser, M. L., Effect of direct acidification and heat on the solubility of protein extracts from a fermented sausage mix. *J. Animal Sci.*, **41** (1975) 554.
29. Samejima, K., Oka, Y., Yamamoto, K., Asghar, A. & Yasui, T., Studies on heat-induced gelation of cardiac myosin and actomyosin. 2. Effect of —SH groups, epsilon—NH_2 groups, ATP, and myosin subfragments on heat-induced gelling of cardiac myosin and comparison with skeletal myosin and actomyosin gelling capacity. *Agric. Biol. Chem.*, **52** (1988) 63.
30. Wright, D. J., Leach, I. B. & Wilding, P., Differential scanning calorimetric studies of muscle and its constituent proteins. *J. Sci. Food Agric.*, **28** (1977) 557.
31. Gustavson, K. H., *The Chemistry and Reactivity of Collagen*. Academic Press, New York, 1956.
32. Goll, D. E., Hoekstra, W. G. & Bray, R. W., Age-associated changes in bovine muscle connective tissue. II. Exposure to increasing temperature. *J. Food Sci.*, **29** (1964) 615.
33. Rodger, G. W. & Wilding, P., unpublished data.
34. Fjelkner, S. & Ruderus, H., The influence of exhaustion and electrical stimulation on meat quality of young bulls: Part 1—Postmortem pH and temperature. *Meat Sci.*, **8** (1983) 185.

35. Cornforth, D. P., Pearson, A. M. & Merkel, R. A., Relationship of mitochondria and sacroplasmic reticulum to cold shortening. *Meat Sci.*, **4** (1980) 103.
36. April, E. W., Brandt, P. W. & Elliott, G. F., The myofilament lattice: studies on isolated fibers. *J. Cell Biol.*, **51** (1971) 72.
37. Pearson, R. T., Duff, I. D., Derbyshire, W. & Blanshard, J. M. V., An investigation of rigor in porcine muscle. *Biochim. Biophys. Acta*, **362** (1974) 188.
38. Smulders, F. J., Eikelenboom, G. & Van Logtestijn, J. G., The effect of electrical stimulation and hot boning on beef quality. *Proc. 27th Meeting European Meat Research Workers, Vienna*, Vol. 1, 1981, p. 151.
39. Taylor, A. A., Shaw, B. G. & MacDougall, D. B., Hot deboning beef with and without electrical stimulation. *Meat Sci.*, **5** (1980) 109.
40. Penny, I. F., The enzymology of conditioning. In *Developments in Meat Science—1*, ed. R. Lawrie. Applied Science Publishers, Barking, 1980, p. 115.
41. Etherington, D. J., Collagen and meat quality: effects of conditioning and growth rate. In *Advances in Meat Research*, Vol. 4, ed. A. M. Pearson, T. R. Dutson & A. J. Bailey. Van Nostrand Reinhold, Wokingham, 1985, p. 351.
42. Calvelo, A., Recent studies on meat freezing. In *Developments in Meat Science—2*, ed. R. Lawrie. Applied Science Publishers, Barking, 1981, p. 125.
43. Shenouda, S. Y. K., Theories of protein denaturation during frozen storage of fish flesh. *Adv. Food Res.*, **26** (1980) 275.
44. Hutchings, J. B. & Lillford, P. J., The perception of food texture—the philosophy of the breakdown path. *J. Texture Studies*, **19** (1988) 103.
45. Hamm, R., Biochemistry of meat hydration. *Adv. Meat Res.*, **10** (1960) 355.
46. Wilding, P., Hedges, N. & Lillford, P., Salt-induced swelling of meat: the effect of storage time, pH, ion type and concentration. *Meat Sci.*, **18** (1986) 55.
47. Wood, F. W., Diffusion of salt in pork muscle and fat tissue. *J. Sci. Food Agric.*, **17** (1966) 138.
48. Turner, R. H., Jones, P. N. & Macfarlane, J. J., Binding of meat pieces: an investigation of the use of myosin-containing extracts from pre- and post-rigor bovine muscle as meat binding agents. *J. Food Sci.*, **44** (1979) 1445.
49. Purslow, P. P., Donnelly, S. M. & Savage, A. W. M., Variations in the tensile adhesive strength of meat myosin junctions due to test configurations. *Meat Sci.*, **19** (1987) 227.
50. Weddle, R. B., Texture profile panelling: a systematic subjective method for describing and comparing the textures of fish materials, particularly partial comminutes. In *Advances in Fish Science and Technology*, ed. J. J. Connell. Fishing News Books Ltd, Surrey, UK, 1980.
51. Lippencot, R. K., Effect of ice-storage time of whole red hake on surimi quality. Paper 21, 29th Atlantic Fish Technology Conference, 1984.
52. Matsumoto, J. J., Chemical deterioration of muscle proteins during frozen storage. In *Chemical Deterioration of Proteins*, ed. J. R. Whitaker & M. Fujimaki. American Chemical Society, Washington, DC, 1979, p. 95.
53. Slade, L. & Levine, H., Collapse phenomena—a unifying concept for interpreting the behaviour of low moisture foods. In *Food Structure—Its Creation and Evaluation*, ed. J. M. V. Blanshard & J. R. Mitchell. Butterworth, London, 1988.

54. Wu, M. C., Lavier, T. C. & Hamann, D. D., Thermal transitions of admixed starch/fish protein systems during heating. *J. Food Sci.*, **50** (1985) 20.
55. Rodger, G. W., Hastings, R., Cryne, C. & Bailey, J., Diffusion properties of salt and acetic acid into herring and their subsequent effect on the muscle tissue. *J. Food Sci.*, **49** (1984) 714.

Chapter 10

PECTIN

CLAUS ROLIN and JOOP DE VRIES
Copenhagen Pectin, DK-4623 Lille Skensved, Denmark

1 INTRODUCTION

Pectin is a general term for a group of natural polymers that occur as structural materials in all land-growing plants. It is thus part of the natural diet of man.

Commercial pectin is obtained either from citrus peel or from apple pomace. The pectin is released from these materials by an aqueous extraction under mildly acidic conditions, isolated by precipitation, and finally worked up to a powder of standardized properties.

Polymerized galacturonic acid partly esterified with methanol accounts for the major part of any commercial pectin. The percentage of the galacturonic acid subunits that are methyl esterified influences the functional properties of the pectin to a very large extent. Commercial pectins are divided into *low-ester pectins* and *high-ester pectins*. High-ester pectins may form gels at the low pH values of many fruit systems when the water activity is reduced by addition of sufficient amounts of sugar. Pectins may, however, form gels in the presence of calcium by another—fundamentally different—mechanism, which is in practice only used with low-ester pectins.

The ability of many fruit materials to form gels when boiled with sugar has been known for centuries. This traditional jam-making was based on the natural pectin content of the fruit. Today's commercial jam manufacturing consumes the major part of all pectin production, but other applications such as stabilization of acidified milk products are gaining importance.

This text will focus on the applications of *commercial* pectins and on functional or physical/chemical properties of relevance to the applications.

Other reviews exist;[1-4] the comprehensive texts of Kertesz[5] and Doesburg[6] are, in spite of their age, most valuable sources of information.

2 STRUCTURE

2.a Simplified Model of Pectin Structure

The polymerized, partly methanol-esterified $1 \rightarrow 4$ linked α-D-galacturonic acid, which accounts for the major part of the material of all commercial pectins, is depicted in Fig. 1. In some pectins part of the methyl ester groups

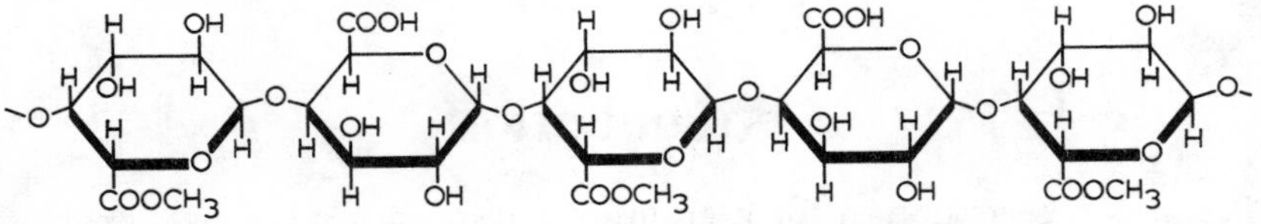

FIG. 1. Pectin, main component.

may be replaced by amide groups (Section 3b). The fraction of the subunits that are esterified may vary from approximately 80% maximum downwards. The sequence in which esterified and free acid groups are arranged along the molecule is not fixed.

The following section will review the present knowledge on the structure of pectins. Since it has not yet been established what effect constituents, other than the galacturonan part, have on the functional properties of commercial pectins, the reader who is primarily interested in applications may thus be satisfied with the simplified model of Fig. 1, and the following discussion may be skipped.

2.b Elaborate Model of Pectin Structure

Complete elucidation of the structure of a polysaccharide implies that knowledge is made available about sugar residue composition; presence and distribution of substituents; type of glycosidic linkages; anomeric and absolute configuration; sequence of the residues; possible linkages to other macromolecules, e.g. in a cell wall; molecular weight and polydispersity; conformation of the sugar residues (ring size, boat vs. chair); and the molecule (helix, random coil, etc.). As if this is not enough, an important complicating factor is the extraction from the natural environment. Only part of the pectin molecules can be extracted by non-degradative means. Usually, pectins are extracted by dilute acids from (processed) plant material in which enzymes might have been active. It is for these reasons not astonishing that proposals for the structure of pectic substances can be widely different.

Neutral sugars like galactose, glucose, rhamnose, arabinose and xylose are present in varying amounts, usually 5–10% of the amount of galacturonic acid. They can be bound to the galacturonate main chain as side-chains of araban, galactan, arabinogalactan or short xylose chains; they can be inserted into the main chain (rhamnose) or be part of contaminating polysaccharides (glucans and xyloglucans). Roughly the same neutral sugar composition is present in pectins from apple, citrus, cherry, strawberry, carrot, pumpkin, sugarbeet, potato, onion and cabbage,[7-15] but pectins from mountain pine pollen, Japanese kidney beans and duckweed contain large amounts of xylose or apiose.[16-18] Ripening is sometimes claimed to influence the neutral sugar composition,[13] sometimes not.[14,19] Differences found may reflect differences in extractability.[13]

Pectin from the primary cell wall may have more branches of neutral sugars than pectin from the middle lamella.[15] Neutral sugar side-chains are unevenly distributed along the main chain. A model based on mildly extracted apple pectin was proposed[20] in which the molecules are composed of 'smooth regions' (unbranched galacturonan) and 'hairy regions' (branched rhamnogalacturonans). The model also applies to citrus, sugarbeet, cherry and carrot pectic substances.[8-11]

Cell walls contain a pectic polysaccharide fraction with a backbone of alternating rhamnose and galacturonic acid residues.[21-24] Tobacco pectin can be separated into a fraction without neutral sugars, one with few, and one with a high proportion.[25] The evidence available points to a very uneven distribution of rhamnose units, although some authors claim that these residues are distributed regularly.[26]

The substituents are more important for the functional properties than are the neutral sugars. A proportion of the galacturonate residues are esterified with methanol, but other substituents, e.g. acetyl groups, can be important. Acetyl groups in potato and sugarbeet pectins[10,27] prevent gelation. The acetyl groups in sugarbeet pectins are located both in the hairy and the smooth regions.[10] Three positions could be distinguished by NMR,[28] probably the 2- and 3- positions of galacturonic acid and a position on one of the neutral sugars. In citrus and apple pectins, very small degrees of acetylation can be measured; in this case the acetyl groups are probably located in the hairy regions.[27] Sugarbeet pectin has another substituent that is relevant to functional properties: ferulic acid.[10,29] Sugarbeet pectin molecules can be cross-linked owing to this ferulic acid by addition of peroxide/peroxidase or ammonium persulphate.[10,30] Small amounts of protein[31,32] and silicate[33,34] are present in pectin samples, but it is not certain whether they are substituents of pectin.

The substituent that is most abundantly present is the methanol ester of the galacturonate residues. Degrees of esterification (DE) can be manipulated, but the pectin source is an important determinant. Apple and citrus pectins not subjected to de-esterifying processes have a high DE (about 70%). Pectins from sunflower heads have a low DE,[35] although this also depends on maturity.[36] Also, pectins from potato,[27] tobacco[25] and pear[27] are reported to be low in ester content. In tomato, DE of the pericarp pectin decreases with ripening, whereas no changes occur in the gel pectin.[37] In plum, DE decreases with ripening;[19] in apple no changes could be observed.[14] Again, differences found are influenced by differences in extractability.

The distribution of the ester groups depends on the source. In mildly extracted apple pectins, the intramolecular distribution is reported to be random,[38] although recent work points to some regularity.[39] In commercial pectins, the evidence available suggests a non-random distribution.[39-41] Fractionation on ion-exchange resins showed the presence of two populations of molecules (with DE values around 70% and 50%) in some pectins.[40] In cabbage, for example, a proportion of the molecules have a very low DE.[41]

Pectins subjected to the action of plant pectin esterase have blockwise-distributed galacturonate residues, whereas most fungal enzymes de-esterify at random.[42-44] The distribution pattern is an important factor governing the functional properties.[45]

Detailed information on other aspects of the structure of pectins can be found in recent reviews and articles on biosynthesis,[33] conformation in solution and gels,[45,46] molecular weight,[47,48] cell-wall structure[21,49,50] and cell-wall linkages.[50,51]

3 SOURCE AND PRODUCTION

3.a Raw Materials

The only significant sources of commercial pectins are citrus peel and apple pomace.

Amongst the *citrus fruits*, lemon is the major source, but lime, orange and grapefruit are also used. The peel is available to the pectin manufacturer subsequent to use of the fruit for juice manufacturing and extraction of flavour oil. At this stage the pectin in the peel is prone to degradation, and the peel is therefore immediately processed. It is washed in water in order to remove water-soluble materials other than pectin, and either used directly for pectin manufacturing or dried. Dried peel may be shipped to pectin production plants that are in many cases situated far from the citrus plantations.

Most European pectin plants developed in connection with apple or cider production. The original raw material, *apple pomace*, has to a large extent been replaced by citrus peel. The pectin content of apple pomace is typically 10–15% on a dry-matter basis, whereas citrus peel contains 20–30%. Citrus and apple pectins are essentially equivalent in application.

Waste *beet solids from sugar production*, i.e. the remains after extraction of the sugar, are rich in pectin. This material was used for pectin production to some extent during the Second World War,[5] and later in Russia and Sweden.[1,52] Sugarbeet pectin is no longer produced in the Western countries, because it is usually of inferior quality owing to the presence of acetyl ester (Section 2.b).

Other suggested sources include sunflower waste from edible-oil production[35,53–60] and mango.[61–62] Sunflower pectin also contains acetyl groups similar to those in sugarbeet pectin.[58]

3.b Process

Two commonly used processes for pectin production are described in the following. Other processes or variations of the described processes may exist, but most manufacturers regard this information as confidential. The process that is probably the most widespread is shown in Fig. 2.

The fresh or dried raw material is extracted in demineralized water that has been acidified by mineral acid. Typical conditions may be pH 2 temperature 70°C, duration 3 h. Some de-esterification takes place during the extraction, and the extraction conditions are thus chosen according to the intended product. More acidic conditions or longer processing times are used if a low-ester pectin type is desired.

The extracted peel is separated from the extract, e.g. by means of a drum-type vacuum filter. This used peel may be sold as cattle feed.

The extract is filtrated once more through diatomaceous earth or by a similar filtration aid in order to clarify the product.

In the *alcohol precipitation process* (Fig. 2) the clarified extract is mixed with an alcohol (e.g. isopropanol). The pectin is, in contrast to most of the other water-soluble materials in the extract, insoluble in alcohol, and it will thus precipitate. The used alcohol is recovered by distillation. Some pectin

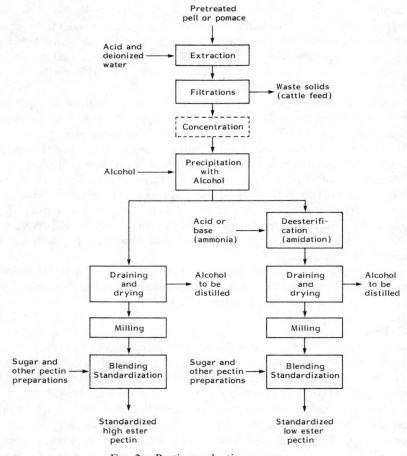

FIG. 2. Pectin production process.

manufacturers concentrate the extract by evaporation prior to the precipitation in order to save on distillation costs.

In the alternative *alumina precipitation process* (not shown in Fig. 2) pectin is separated as an insoluble salt by addition of aluminium(III) ions. Copper(II) may be used in place of alumina.[63-66] The metal ions are subsequently removed by washings with acidified alcohol.

The alcohol-wet pectin resulting from both of the above-mentioned processes may either be dried and milled directly or it may be de-esterified in alcohol suspension. De-esterification may be achieved with acid or with

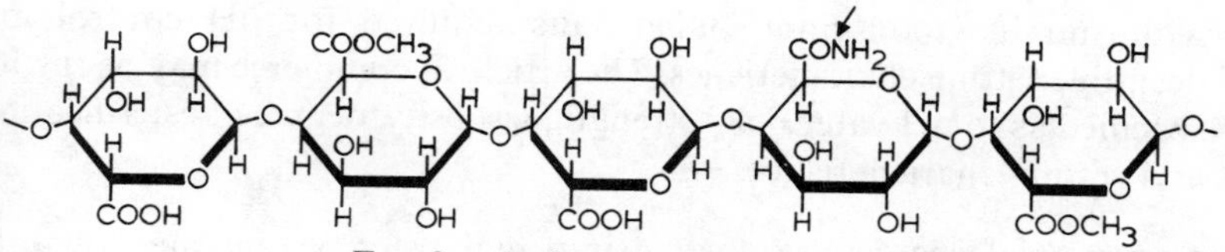

FIG. 3. Amidated pectin (arrow).

base. If ammonia is used for the de-esterification, some of the methyl ester groups are substituted by amide groups[67,68] (Fig. 3). The resulting product is referred to as 'amidated pectin'.

The last step of pectin production—blending and standardization—is crucial. The raw materials employed for pectin production may vary considerably in properties owing to, for example, weather conditions. From the consumer's point of view, reproducible performance from batch to batch is a must. In order to standardize, pectins from individual production batches are mixed, and the pectins are 'diluted' with sugar. Commercial pectins intended for gelation purposes are standardized to yield a defined firmness of a standard gel (Section 4.a). High-ester pectins are, in addition to this, standardized to solidify after a certain time when a standard gel is cooled under prescribed conditions (Sections 4.a and 6.d). Low-ester pectins may likewise be standardized to exhibit a certain calcium reactivity.

4 COMMERCIAL AVAILABILITY

4.a Definitions and Terminology for Commercial Pectins

Several definitions of pectin have been suggested. Complete and rigorous definitions have been given by the 1944 Committee for the Revision of the Nomenclature of Pectic Substances[69] and by Doesburg.[6]

Pectin for commercial use has been defined by the Joint FAO/WHO Expert Committee on Food Additives (JECFA)[70] as follows:

> Pectin consists mainly of the partial methyl esters of polygalacturonic acids and their sodium, potassium, calcium and ammonium salts. It is obtained by aqueous extraction of appropriate edible plant material, usually citrus fruits or apples. No organic precipitants shall be used other than methanol, ethanol and isopropanol. In some types, a portion of the methyl esters may have been converted to primary amides by treatment with ammonia under alkaline conditions. The commercial product is normally diluted with sugars for standardization purposes, and mixed

with suitable food-grade buffer salts required for pH control and desirable setting characteristics. The article of commerce may be further specified as to pH value, jelly strength, viscosity, degree of esterification, and setting characteristics.

Commercial pectins are divided into high-ester pectins and low-ester pectins according to the *degree of esterification*. The degree of esterification (abbreviated DE) is the percentage of galacturonic acid subunits that are methyl esterified. *High-ester pectins* are by definition pectins with DE = 50 or more. *Low-ester pectins* are pectins with more than 'negligible' but less than 50% esterification. *Pectate* is polymerized galacturonic acid with no or only negligible esterification.

The *degree of amidation* (DA) is the percentage of galacturonic acid subunits that are amidated.

High-ester pectins intended for gel-making are further subdivided into *rapid-set, medium rapid-set* and *slow-set* pectins according to the time taken by a standard gel prepared from the pectin to solidify. High-ester pectins are in general faster setting the higher the degree of esterification. Typical degrees of esterification are 60–64 for slow-set, 65–69 for medium rapid-set and 70–75 for rapid-set pectins.

Almost all high-ester pectins are standardized to 150 *grade USA-SAG*. This designation means that 1 part of the pectin is able to turn 150 parts of sucrose into a jelly prepared under standard conditions and with standard properties as follows:

(a) refractometer soluble solids: 65%;
(b) pH = 2·20–2·40;
(c) gel strength: 23·5% SAG (Section 6.c).

This method was implemented by the 1959 IFT Committee for Pectin Standardization, and a detailed procedure is available.[71]

4.b Purity

Specifications for the identity and purity of pectin have been set by acknowledged organizations.[70,72,73] The specifications that have been recommended by the Joint FAO/WHO Expert Committee on Food Additives (JECFA)[70] are summarized below. The relevant analytical methods may be found in the original.[70]

- *Galacturonic acid.* Not less than 65% on the ash-free and dried basis (after removal of sugars by washing with acid and ethanol)

- *Degree of amidation.* Not more than 25
- *Loss on drying.* Not more than 12% (105°C, 2 h)
- *Acid-insoluble ash.* Not more than 1%
- *Heavy metals* (in mg/kg). As max. 3; Pb max. 10; Cu max. 50; Zn max. 25
- *Alcohols.* Sum of methanol, ethanol and isopropanol not more than 1%
- *Nitrogen.* Not more than 2·5% after washing with acid and ethanol
- *Sulphur dioxide.* Not more than 50 mg/kg

A compilation of specifications including those of Food Chemicals Codex[72] and the Council of the European Communities[73] may be found in an article by Højgaard Christensen.[4]

4.c Toxicology

Pectin (including amidated pectin) was evaluated by the Joint FAO/WHO Expert Committee on Food Additives (JECFA)[74] in 1981. An ADI value (acceptable daily intake) 'not specified' was established, meaning that there are no limitations from a toxicological point of view to the use of pectin.

4.d Storage Stability

High methyl ester pectin (HM) loses up to 5% per year of its grade USA-SAG when stored at 20°C in dry atmosphere. The loss is much larger if the pectin is stored under warm or humid conditions.[75] Low-ester pectin is more stable: loss due to one year's storage at 20°C under dry conditions is not detectable.

5 SOLUTION PROPERTIES

5.a Solubility

Failure situations when a pectin is used for gel-making are often related to improper dissolution of the pectin or improper distribution of a pectin solution in the batch to be gelled. Pectin is soluble in pure water, but certain procedures must be adhered to when making a pectin solution. What is more, some of the materials that are typically present in application systems may suppress the solubility of pectin.

As a general rule—which is also true for other gelling polymers—pectin is not soluble in conditions under which it can form a gel.

A pectin solution can be made by gradual addition of the pectin to warm water in a *high-shear mixer* (blender). The pectin is added to the water, which should preferably be at not less than 60°C, while the mixer is running at reduced speed, and it is then stirred for 5 min or more at full speed. Solutions up to approximately 10% (limited by viscosity) can be made with an efficient mixer.

If pectin is added to water while the stirring is not sufficiently efficient, lumps may form. The lumps are sticky, and they are very difficult to dissolve. If a rapid mixer is not available, the problem may be overcome by *mixing the pectin powder with five parts of sugar*. This mixture can be dispersed by reasonably efficient stirring by ordinary means. Another way is to *disperse the pectin in a solution in which it is not soluble*, before adding this dispersion to the water. A concentrated (e.g. 65%) sugar solution or— for small-scale laboratory purposes—an alcohol may be used for the dispersion.

Pectin solutions that have not been made with a high-shear mixer should, as the last step of their preparation, be boiled for approximately 1 min in order to ensure full dissolution. It is not always possible to detect improper dissolution by observation.

5.b Rheology of Pectin Solutions

Pectin solutions are viscous, but pectin is not particularly efficient as a thickener compared to other water-soluble polymers. A relationship between the concentration of two commercial pectins in pure water solutions and the corresponding viscosities is shown as Fig. 4, but it must be emphasized that this serves only to illustrate typical orders of magnitude. The rheological properties of pectin solutions are very dependent on the presence of salts—in particular, salts of calcium or similar non-alkali metals—and on the pH. Other factors to be considered are chemical properties of the pectin: degree of ester and average molecular weight.

Dilute pectin solutions (for some high-ester pectins, up to approximately 0·5%) are almost Newtonian, and they are little affected by the presence of calcium. The viscosity of very dilute solutions increases with increasing pH.[76] This has been interpreted as an expansion of the dimensions of the molecule owing to repulsion between dissociated acid groups on the same molecule. Sodium chloride or other salts of monovalent cations reduce the viscosity of pectin solutions,[76,77] probably because the charge effects are reduced at high ionic strength (high dielectric constant of the solution).

Pectins with a high average molecular weight are more viscous in solution than are otherwise comparable pectins of lower molecular weight

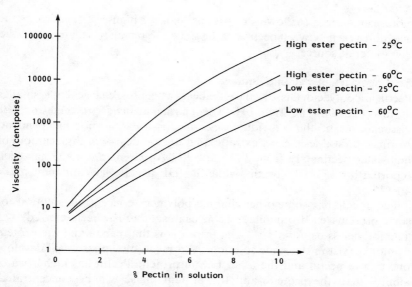

FIG. 4. Viscosity of pectin solutions; typical orders of magnitude; see text. (From Højgaard Christensen,[4] reproduced with permission.)

under the same circumstances. The molecular weight of pectin is often estimated using intrinsic viscosity methods.[78–80]

Solutions with more than approximately 1% pectin are pseudoplastic. Some pectins will, in the asbence of calcium, increase their viscosity if the pH is reduced within typical application ranges of pH 2·5–5·5. This is in contrast to the dilute solution behaviour mentioned above. Otherwise comparable high-ester pectin solutions will—still in the absence of metal interactions—show (slightly) higher viscosity the higher the degree of esterification.[81] This behaviour has been interpreted as a structuring of water around the hydrophobic methyl ester groups[46] (cf. Section 6.f Mechanisms of Gelation).

Most pectins form thixotropic solutions in the presence of calcium. The viscosity of such solutions will increase with increasing pH within the pH interval of typical applications of interest (c. 2·5–5·5). A continuum of textures ranging from that of water, through thixotropic solutions with a yield value, to stiff gels, may be obtained from different combinations of pectin concentration, calcium concentration, pH and pectin type. The extent to which metal ions like calcium can influence the viscosity is dependent on the specific pectin used. Calcium sensitivity is, in general, more pronounced for low-ester pectins than for high-ester pectins, but it may differ even between pectins of the same DE owing to differences in the raw material source.

5.c Chemistry of Pectin Solutions

Pectin is a polyacid. Dissolved pectin bears a negative charge that is smaller at low pH than at high pH. The charge attracts protons, and the dissociations of the individual acid groups are thus not independent. Intrinsic pK-values (i.e. pK extrapolated to zero degree of dissociation) of high-ester pectins are from $3 \cdot 1$[82] to $3 \cdot 3$.[83] Apparent p$K = 3 \cdot 55$ has been reported for 65% DE pectin, while 0% DE pectate had apparent p$K = 4 \cdot 10$.[84]

Pectin will, like some other charged polymers, concentrate solutions of some proteins (e.g. from milk) by a mechanism that has been referred to as 'membraneless osmosis'.[85,86] The principle is that pectin and the protein cannot coexist in the same solution. As a result, two phases will develop: one rich in pectin and the other rich in protein. Pectin has higher water affinity than the protein, and the protein phase will be concentrated, typically by a factor of 5–12.[86]

Addition of metal ions other than alkalis to a pectin solution will, in most cases, cause a viscosity increase or a gel formation or a precipitation of the pectin. Owing to its negative charge, pectin will form insoluble products with positively charged macromolecules.

Dissolved pectin has good stability at pH values that are typically encountered in acidulous food products. The stability optimum is about pH 4. Pectin is slowly degraded by depolymerization as well as by de-esterification when the pH is close to 4; degradation increases with the distance from the pH optimum. Depolymerization at low pH is a hydrolysis,[87] while the high-pH degradation is a beta-elimination.[88-91] The latter is significant at any pH above 5, where pectin solutions are only reasonably stable at room temperature. Glycosidic bonds to the C-4 of an esterified galacturonan subunit are more easily cleaved by beta-elimination than are bonds to the C-4 of unesterified subunits. High-ester pectins are consequently more vulnerable to this kind of degradation than are low-ester pectins.

Pectin-degrading enzymes are abundant in many natural materials, but they are only rarely a problem in pectin applications. However, fruit juices are often treated with technical preparations of pectin-degrading enzymes, in order, for example, to improve the clarity. Residual enzyme activity has in a few cases been seen to cause inferior storage stability of pectin gels made from enzyme-treated juice. Numerous reviews on pectic enzymes have been published.[92-98]

6 GEL PROPERTIES

6.a Preparation of High-ester Pectin Gels

High-ester pectin gels can be made in a variety of ways, but attention must be paid to certain rules if good results are to be achieved. In general, the pectin must be dissolved and evenly distributed in the gel batch prior to the gelation. This implies in most cases that conditions that are close to gelling conditions should be introduced as late as possible in the procedure and later than the distribution of the pectin.

As an example, a recommended procedure for making a jam could be as follows.

(a) Mix while heating fruit material and sugar in sufficient amounts to reach 65% soluble solids in the final batch. *Comment:* The sugar is added at an early stage in order to allow it to diffuse into the interior of fruit tissues. This is because osmotic equilibrium must be accomplished in the final product. Juice would otherwise exude from the interior of fruits to the surrounding gel, and thus reduce the strength of the gel.

(b) Add the pectin as an aqueous solution. Stir and boil under vacuum to mix. Adjust soluble solids to the desired value by evaporation or water addition. *Comments:* The pectin solution may be distributed in the batch because the pH is still too high for gelation. Concentration under vacuum is practised by most jam manufacturers in order to minimize heat destruction of fruit flavours and colours.

(c) Break the vacuum. Heat to pasteurize.

(d) Add citric acid in sufficient amounts to reduce pH to between 3·0 and 3·1. *Comment:* Gelling conditions have now been introduced except for the temperature, which is still too high.

(e) Cool to the filling temperature. Fill into the containers in which it is desired to solidify the gel. *Comment:* Filling must be completed before the onset of gelation.

This procedure reflects some characteristics of the high-ester pectin gelation: conditions necessary for the gelation are a fairly low pH and a high soluble solids concentration (details in Section 6.e). A critical temperature exists above which gelation will not take place, even though all other conditions for gelation are fulfilled. If the temperature is reduced below this limit, gelation commences after some time. The gelation temperature depends on the combination of pectin type and composition of the batch (Section 6.e); it can to a great extent be varied at will be selecting an appropriate pectin type.

It is not usually possible to melt a high-ester pectin gel once it has solidified.

If the gel batch is stirred or poured while gelation is in progress (it may still look liquid!), structures that have already formed may be broken. They will not reform, and the corresponding part of the pectin will not be utilized for the final gel structure. This phenomenon is known as *pre-gelation*. Pre-gelation is very similar to the situation of bad dissolution or distribution of the pectin. The results are similar: the gels will be weaker than expected, and gelation may, in severe cases, appear to be absent. The cause of failure is also basically the same in the two cases: only part of the pectin is utilized.

6.b Preparation of Low-ester Pectin Gels

Low-ester pectin gels may be prepared by the same sequence of events as for the high-ester pectin gel in the former section, but it is not necessary to achieve as high a solids content or as low a pH in order to introduce gelling conditions. The gelation is dependent on calcium, which is in most cases provided by the fruit material.

In analogy to high-ester pectin gels, the gel temperature may be varied by appropriate selection of combinations of pectin type and properties of the medium. The same considerations regarding dispersion of pectin prior to gelation and the possibility of pre-gelation apply to low-ester pectin gels as to high-ester pectin gels.

Two important differences between low-ester pectin gelation and high-ester pectin gelation must be mentioned. The difference between setting temperature and melting temperature of low-ester pectin gels is modest, and it is possible in many cases to re-melt a low-ester pectin gel. This is in contrast to the high-ester pectin gel. The low-ester pectin gel solidifies almost immediately when gelling conditions are introduced; a high-ester pectin gel system will show a time lag.

6.c Measurement of Gel Texture

Various devices for testing the 'strength' of gels are shown in Figs 5(a)–(c). The meaning of the word 'strength' is of course dependent on the method of measurement, and this is not just an academic subtlety. Results obtained from different methods are not easily correlated.[78,99,100] In particular, attention should be drawn to the fact that some of the methods measure a deformation of gels within the elastic limits, while other methods measure the force required to rupture gels. These two groups of measurements represent fundamentally different properties.

A review of methods for the measurement of gel strength has been made recently.[101]

In the *SAG method*[102] (Fig. 5(a)) the test gel is cast in a glass of standardized dimensions. The glass is extended at the top by adhesive tape. When the gel is to be measured, the tape is removed, and the part of the gel that was retained by the tape, and is now protruding above the plane defined by the edge of the glass, is cut in this plane by means of a stretched wire (as with a cheese cutter). The gel is then turned out on a glass plate to rest on the base that has been formed by cutting. The gel will sag under its own weight, and the loss in height is measured after a specified time.

A SAG measurement is part of the universally agreed method for grading of high-ester pectins (grade USA-SAG determination; see Section 4.a).

Various 'plunger methods' are available; the example shown in Fig. 5(b) is the *LFRA texture analyser*. Plungers of different shapes or sizes may be chosen; in operation of the instrument, they are brought into the gels at constant speed, and the stress-versus-deformation graph is automatically recorded. A maximum will appear on this graph, corresponding to the point at which the elastic limit of the gel was exceeded, i.e. at which the gel broke, and the breaking strength is thus obtained.

Another 'plunger instrument' is the Bloom gelometer, which is more often used for the classification of gelatines. It measures the forces required to press a piston of specified dimensions 4 mm into the gel. The original instrument measures and provides the force by the weight of shot pellets in a beaker; a modern device, the *Boucher Electronic Jelly Tester*, is based on the same principle, but the shot pellets have been replaced by an electronic measurement, and the result appears on a digital readout. The *Instron Universal Testing Machine* is an expensive instrument that can work like the LFRA, but it can also pull items apart, or it can be programmed for various load-versus-time functions. The *penetrometer* measures how deep a cone, which has been placed with the pointed end on a plane gel surface, will penetrate into the gel in a given time subsequent to its release. The cone is driven by gravity.

The *FIRA Tester* measures the torsional force required to deflect a spade-

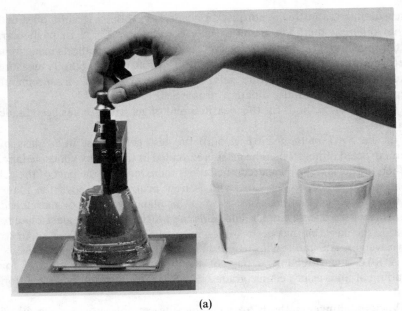
(a)

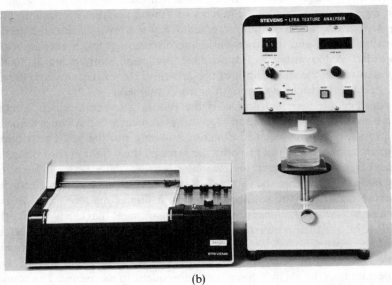
(b)

FIG. 5. Devices for texture measurements on pectin gels. (a) Determination of SAG; (b) LFRA Texture Analyser.

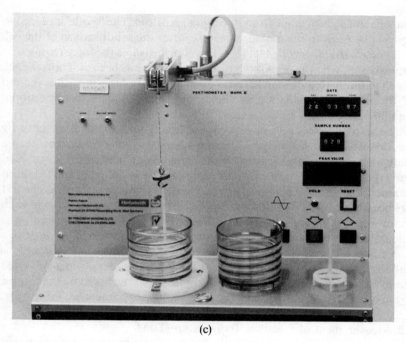

(c)

(d)

FIG. 5—*contd.* (c) Herbstreith Pectinometer; (d) Bostwick Consistometer.

shaped item a given angle from its original position. The 'spade' is placed in the gel just prior to the measurement, i.e. after the solidification of the gel.

The *Herbstreith pectinometer*[103] (Fig. 5(c)) measures the force required to rupture a gel, when a specially shaped item, which has been cast into the gel, is pulled out with a defined speed. This is a modern high-precision version of the older Luer/Lockmuller 'Pektinometer', which is the traditional standard instrument in Germany.[99,104]

The texture of soft gels may be characterized by the spreadameter or by the Bostwick Consistometer. The *spreadameter* is a transparent plastic plate with concentric circles drawn on the underside. The circles are divided into eight sectors by radii every 45°. The sample to be measured is placed symmetrically on the plate (e.g. turned out from the container in which it has been cast), and it is recorded how far it reaches in a given time in each sector when it spreads by gravity. The average distance of spreading is calculated from the individual values of each sector. The *Bostwick Consistometer* (Fig. 5(d)) is a sloping surface with a scale. The sample is initially stemmed up behind a wall that can be lifted suddenly in an inverted guillotine manner, thus releasing the sample. The time taken by the sample to move a certain distance down the scale is used to characterize the texture.

6.d Measurement of Gelation Temperature/Time

The most widespread method for the measurement of setting rates is that of Joseph and Baier,[105] which is used for the classification of commercial high-ester pectins into the categories 'rapid-set', 'medium rapid-set' and 'slow-set'. The test gel in this procedure is the same as is used for the USA-SAG procedure (Section 4.a), and the two determinations may even be done on the same gel batch. A standard glass for the SAG procedure is placed in a 30°C water bath and allowed to reach thermal equilibrium. The 95°C gel batch is poured into the glass, which is subsequently carefully twisted by jerks—while still keeping it in the water bath—in order to make the contents move and stop repeatedly. The setting time is defined as the interval from the time of pouring to the first time it is observed that the liquid stops its motion by an elastic reversion. Typical values range from 50 s (rapid-set) to 250 s (slow-set).

This procedure describes the properties of pectins in test gels, not the setting characteristics of a particular application system. Such systems may, however, be characterized by analogous procedures. But what property is measured? The method yields a *setting time*, which is strictly dependent on the conditions (heat conduction, etc.). The *gelation temperature* is a more fundamental property, but owing to the time lag of high-ester pectin gelation the only direct way to determine it would involve an infinitely slow cooling, which would in turn require an infinite amount of patience! A procedure for the determination of true gelling temperatures by extrapolation of results from several cooling rates has been devised by Hinton.[106]

6.e Factors Affecting Gelation

When pectin is used for the preparation of a gel, one would usually wish to control the texture of the gel and the gelation temperature. The system parameters that accordingly must be considered are summarized in Table 1. They can be subdivided into two main groups: (a) molecular properties of the pectin, i.e. the degree of methyl esterification and degree of amidation, and (b) composition of and conditions in the gel system, i.e. pH, water activity (sugars), presence of salts and pectin concentration.

It must be emphasized that these effects are interdependent. The way in which one of them contributes to either the texture or the gelling temperature will depend on the values of all system parameters. Thus, the statements given in the following discussion and in Table 1 imply that all parameters but the one referred to are kept constant.

The *degree of methyl esterification* influences the functional properties of high-ester pectins as well as low-ester pectins, but in an inverse manner. High-ester pectins gel at higher temperature the higher the degree of esterification. Low-ester pectins gel at higher temperature and need less calcium the lower the degree of esterification. Typical DE values for high-ester pectins range from 60–75, while low-ester pectins usually have DE values from 20 to 40. The DE = 50 borderline, which by definition distinguishes between low-ester pectins and high-ester pectins, is theoretically speaking not a sharp limit. Pectins of DE less than 50 may gel by the high-ester pectin mechanism, while many high-ester pectins may be gelled as low-ester pectins with calcium; however, the latter gels have presently little or no practical use.

Amidation has in a similar way different consequences for high-ester pectins and low-ester pectins. Amidated high-ester pectins[107] set more slowly than non-amidated pectins of the same degree of esterification.

The major part of commercial low-ester pectin types are amidated; typical degrees of amidation for these pectins range from 15 to 20. Low-ester pectins of the same degree of esterification gel at higher temperature and need less calcium the higher their degree of amidation. The consequence of the lower requirement for calcium is that the amount of calcium provided by most fruit materials is sufficient to produce the

TABLE 1
COMPARISON OF HIGH- AND LOW-ESTER PECTIN GELATIONS

	High-ester pectin gelation	Low-ester pectin gelation
General conditions	Low water activity provided by soluble solids (or water-miscible solvents); pH < 3·5	Presence of non-alkali metal ions, e.g. calcium
Other phenomenological characteristics	Gel is not thermoreversible. Time lag from introduction of gelling conditions to gelation	Gel is thermoreversible. Immediate gelation when gelling conditions are introduced
Mechanism	Hydrophobic interaction and hydrogen bonding (Fig. 6; Refs 46, 108, 109)	Calcium bridging between chains (Fig. 7; Refs 45, 111, 112, 115, 116)
Influence of molecular properties of the pectin		
DE (%)	The highest-esterified pectins give the fastest gelation and the highest gelling temperature	Lowest-esterified pectins have least calcium requirement and give highest gelling temperature
DA (%)	Amidated high-ester pectins set slower and at lower temperature than non-amidated pectins	Amidated pectins need less calcium than non-amidated pectins. They have a more reproducible performance and a broader calcium working range
Influence of system conditions/co-solutes		
Soluble solids concentration	Lower limit = 55% soluble solids. Typical use level = 65% soluble solids. Positive correlation between per cent soluble solids and gelling temperature or gel strength	No lower limit. Positive correlation between per cent soluble solids and gelling temperature or gel strength
Calcium ion concentration	No requirement. Gel strength may be positively affected or (owing to pre-gelation) negatively affected if pectin is calcium-sensitive	Required. Positive correlation between calcium concentration and gelling temperature or gel strength
pH	Upper limit = 3·5, typical level = 3:1. Inverse correlation between pH and gel strength or gelation temperature	Typical level = 3–4·5. Inverse correlation between pH and gel strength or gelation temperature

optimal gel strength. The gel strength will thus be easier to reproduce irrespective of variations in the calcium content of the fruit material. The gelation temperature will likewise be only modestly dependent on the calcium content of the fruit material, because the pectin is already 'saturated' with calcium. The amidated pectin can be said to have a broad working range in this respect. Non-amidated pectins with low degree of esterification may match the calcium sensitivity of amidated pectins, but the consequence of the low DE is that such pectins give rise to higher—and more calcium concentration-dependent—gelation temperatures than their amidated counterparts. Gel systems made with these pectins are consequently more prone to pre-gelation.

The *pectin concentration* has not been mentioned in Table 1. There is a positive correlation between pectin dosage and gel strength, and between pectin dosage and gelling temperature. Use levels range typically from 0·2% (of 150-grade USA-SAG) to 0·7%. The low dosages are used with high-ester pectin gels at high soluble solids levels, while the high dosages are used with low-ester pectins gelling at relatively low soluble solids levels. It is to a limited extent possible to compensate, for example, the effect of a slight change in soluble solids or pH by an appropriate change in the pectin dosage in order to arrive at the same gel strength.

Gelation by the high-ester pectin mechanism requires a low *water activity*. This may be accomplished either by addition of sugars or (for laboratory purposes) by addition of a water-miscible solvent. Almost all applications depend on sugars as water activity-reducing substances. The absolute lower limit is around 55% soluble solids, and 65% is usual.

Low-ester pectins may form gels even when no other soluble solids than the pectin itself and calcium are present. The gel strength is positively correlated with the soluble solids concentration. Pectin is rarely used as the sole gelling agent in systems with less than 20% soluble solids, because such gels have a large tendency for syneresis. Artificially sweetened fruit gels or jams are often gelled by carrageenan, either together with pectin or in the place of pectin.

The presence of *calcium ions* (or other metal ions with valence greater than 1) is required for the formation of a low-ester pectin gel. The gel strength is positively correlated to the calcium concentration at low concentrations; it levels off at higher concentrations. The setting temperature is also positively correlated to the concentration of calcium.

Some high-ester pectin solutions may become viscous or gelled in the presence of calcium. If the pectin is used for making a gel, the presence of calcium may reinforce this gel, but it is more often seen that gels made from calcium-sensitive high-ester pectins in the presence of calcium are weaker than corresponding gels made in the absence of calcium. This effect is due to pre-gelation.

A fairly low *pH* is critical to high-ester pectin gelation. Gels will not form if the pH is higher than 3·5; the typical pH is about 3·1. The gel strength as well as the gelation temperature is inversely correlated to the pH. Slow-setting pectins need slightly lower pH than rapid-setting pectins in order to produce gels of the same strength.

Low-ester pectins may form gels at any pH. There is an inverse correlation between pH and gel strength as well as gelation temperature.

6.f Mechanisms of Gelation

Some evidence points to a combination of hydrogen bonding and hydrophobic interaction as the mechanism responsible for high-ester pectin gel formation.[46,108,109] The ester groups are the hydrophobic parts of the high-ester pectin molecules. An energy contribution is associated

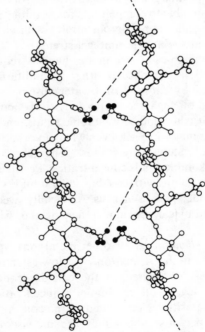

FIG. 6. Junction zone in high-ester pectin gel. The dotted lines represent hydrogen bonds. (From Oakenfull and Scott,[108] reproduced with permission.)

with the contact between these hydrophobic areas and water; the hydrophobic areas will tend to aggregate in order to minimize the contacting surface, in analogy to the coalescence of oil drops in water. Hydrogen bonds that form between adjacent galacturonan chains contribute even more to the energy decrease of junction-zone formation. However, it is theorized that the energy contribution of the hydrophobic interaction is necessary in order to make the sum of energy contributions that favour the gelation, large enough to exceed energy contributions that resist gelation. A junction zone according to the above hypothesis has been proposed[108] as shown in Fig. 6.

It has been suggested that the rigidity of the pectin molecule is positively correlated to the DE and the concentration of sugars in the solution, and that this is also an important factor that plays a role in high-ester pectin gel formation.[82,110]

The 'egg-box' model (Fig. 7) has been widely accepted as an explanation of the low-ester pectin/calcium ion gelation, although another model has been suggested.[46] The egg-box model was originally developed to describe the gelation of calcium alginate[111,112] (see Section 6.a of Chapter 2). Pectins and pectates are configured in helices with three subunits per turn, as determined by X-ray studies on the dried substances.[46,113,114] However, the egg-box junction zones are organized with pectin chains in helices with only two subunits per turn. These two-fold helix structures are joined by calcium ions bridging two opposing carboxyl groups. The model forces the assumption that the helix structure changes from twofold to threefold when the gel is dried to a powder. Changes in circular dichroism spectra that occur when a gel is dried have been taken as evidence that this is indeed the case.[45,115,116] The egg-box hypothesis is further supported by equilibrium dialysis studies that concluded that about 50% of the calcium ions could not be removed when calcium pectate was exposed to very large

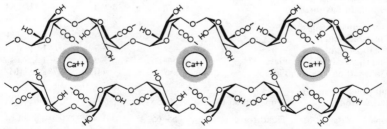

FIG. 7. Junction zone in low-ester pectin gel. (From Højgaard Christensen,[4] reproduced with permission.)

concentrations of univalent cations. It was interpreted that these ions were specifically bound (inside the egg box). It is expected from the stoichiometry of the complex of Fig. 7 that this proportion should amount to 50%.

7 APPLICATIONS

7.a Food Gels

The traditional, and still prevailing, use for pectin is the making of jams, jellies and similar acidulous, sweet gels. The way of making such gels and the conditions governing the gelation were described in Section 6. This section will review some of the possible variations within this class of food products. More detailed descriptions may be found in the commercial literature that is issued by the major pectin producers (see, e.g., Ref. 117).

Some main characteristics of different pectin gels are summarized in Table 2.

Ordinary jam is made almost exclusively with high-ester pectin, although low-ester pectins may be used if a soft, spreadable texture is wanted. The

TABLE 2
FOOD GELS

Product	Texture or appearance	Typical % SS	Typical pH	Pectin type[a] and typical dosage
Jam	Gel with suspended fruit material			
Ordinary		60–70	3.0–3.3	0.2–0.5% HM
Low-sugar		30–55	3.1–3.5	0.5–0.8% LM
Fruit jelly	Brilliant gel			
Ordinary		60–70	3.0–3.3	0.4–0.8% s.s.HM
Low-sugar		30–55	3.1–3.5	0.6–1.0% LM
Confectionery jelly	Firm gel			
Fruit flavoured		78	3.5	1.5% s.s.HM
Non-fruit flavoured		78	4.2–4.8	2.0–2.5% LM
Bakery jams or jellies				
Heat-resistant	Gel	65–75	3.3	0.6–1.0% r.s.HM
Heat-resistant	Gel	45–75	3.5	0.6–1.0% n.a.LM
Heat-reversible	Gel	64–65	3.3–3.6	0.8–1.5% a.LM
Cold-setting	Gel	61	4.0	0.7% r.s.HM
Fruit preparation for dairy products	Semigel/thixotropic	30–65	3.6–4.0	0.3–0.6% LM
Ripple	Thixotropic	55–65	3.0–4.0	0.3–0.6% LM

[a] Abbreviations: HM = high-methyl ester pectin, LM = low-methyl ester pectin, r.s.HM = rapid-set pectin, s.s.HM = slow-set pectin, a.LM = amidated low-ester pectin, n.a.LM = non-amidated low-ester pectin.

type of high-ester pectin is chosen so that a desired gelation temperature will result from the combination of pectin type and relevant system parameters (solids content, pH, etc.). This gelation temperature will in turn often be selected according to the size of the containers in which the jam is to be sold. Solidification immediately subsequent to the filling is desired in order to prevent uneven distribution of fruit particles (berries) owing to flotation. This is best obtained with a gel system possessing high gelling temperature, because it solidifies relatively soon after cooling to below this temperature. On the other hand, a lower filling temperature—and thus lower gelling temperature—is preferred if the jam is filled in large containers. This is in order to minimize heat destruction of fruit flavours and colours. The centre part of gels in large containers cannot be cooled rapidly.

Low-sugar jams have in most people's opinion a better and more intense fruit flavour than ordinary jams, because they are less sweet; they may also be preferred from a nutritional point of view. Low-sugar jams cannot be solidified with high-ester pectins, but a range of low-ester pectin types may be used. The gelation temperature may be varied by appropriate selection of more or less calcium-reactive types.

Jellies should be clear and homogeneous. The absence of particles allows for a long time lag from filling to solidification; this is even an advantage, because air bubbles are allowed to escape and the risk of pre-gelation is minimized. Slow-setting pectin types are accordingly preferred for these purposes.

Confectionery jellies are usually gelled with slow-setting high-ester pectin. Special pectin types with specific dissolution characteristics have been designed for continuous manufacture of confectionery jellies in so-called jet cooking equipment. The pectin is typically dissolved at 140°C in the presence of all of the solids of the final product, i.e. about 75–80%.

Some flavours, like liquorice, vanilla or toffee, which are sometimes used in confectionery jellies, are not compatible with the low pH that is necessary for high-ester pectin gelation; low-ester pectin is used in these cases. It is often necessary to add calcium-sequestering salts like sodium hexametaphosphate in order to reduce the gelation temperature. Minute amounts of calcium are sufficient for the gelation because of the high solids content of confectionery gels.

Bakery goods often contain jam or jellies either inside or on the top, or they may be covered by a glazing. Gel systems can be formulated in different ways in order to fulfil various requirements.

A high-ester pectin gel system with a high gelation temperature may form a *heat-resistant gel*, i.e. it will not melt at ordinary baking temperature.

This is because the difference between setting temperature and melting temperature of high-ester pectin gels is so large than in practice they are considered thermoirreversible. On the other hand, heat-resistánt gels may also be made from low-ester pectins if, for example, calcium citrate is used in the formulation. Low-ester pectin is said to be thermoreversible, because the gap between solidification temperature and melting temperature is small. However, the calcium citrate slowly releases calcium to the pectin, and this has the effect that the system gradually increases its gelation temperature subsequent to the setting. Non-amidated low-ester pectin is superior, because the gelation temperature of gels made from this kind of pectin is very dependent on the calcium activity in the gel system.

Amidated low-ester pectin may be used for making *heat-reversible bakery jellies*. These products are available to the baker as paste-like, pregelled products, which are melted and diluted with hot water before they are poured over the bakery goods.

High-ester pectin may be used for *cold gelation* owing to the time lag between initiation of gelling conditions and solidification. A gel system may be formulated where gelling conditions exist, except for pH, which is too high. This jelly base is used, for example, for glazing tarts by acidifying it with citric acid and then pouring it over the items to be glazed before it solidifies.

Fruit preparations for dairy products, such as fruit yoghurts, are sold to the dairy industry in relatively large containers. These preparations must tolerate transportation and storage, while the berries must continue to be uniformly distributed. On the other hand, they must also be pumpable, and for some applications it must be possible to mix them into the dairy product. These requirements are best satisfied with a thixotropic or semi-gelled texture. This kind of texture is obtained with low-ester pectin in lower dosages than those that would form an ordinary, firm gel. *Ripples* and *fruit sauces* of various kinds are often thickened with low-ester pectin in a similar way.

The viscosity of *stirred yoghurt* or the firmness of *set yoghurt* gel may be increased by addition of about 0·10–0·25% calcium-reactive low-ester pectin. The mechanism behind this effect is not known; it is not merely a gelation of the whey phase. The pectin seems to reduce the charge of the casein particles, thus increasing the tendency for gelation.

7.b Beverages

High-ester pectins of DE more than approximately 70 are used for the stabilization of pasteurized or sterilized, acidified milk products.[118-121]

The products may have been acidified either by fermentation or by the addition of fruit juice. The precipitated casein of such systems will form aggregates during heat treatment and during the subsequent storage if it is not stabilized. An unstable product has an unpleasant 'grainy' mouth-feel and an unappealing appearance owing to excessive whey separation.

Stabilization is accomplished by adding about 0·5% pectin just before homogenization. Acidification may take place either before the homogenization (for fermented products) or immediately subsequent to the homogenization (for directly acidified products). A mild heat treatment such as pasteurization or UHT sterilization may then be done as the next step without serious adverse effects on the product.

The stabilization is only effective within the pH interval 3·5–4·2, and it works better the higher the titre of the milk system (millilitres 0·1N NaOH required to titrate 100 ml milk product to the phenolphthalein point of change). Typical titres of pectin-stabilized milk systems are 100–120; organoleptic demands set an upper limit.

If properly acidified milk systems are prepared in the above-mentioned way, but with a range of small pectin additions, it will be seen that the viscosity of the systems increases with the pectin dosage at small dosages, passes through a maximum, and eventually declines to a level far below that of the unstabilized system when the pectin dosage becomes sufficiently high.[120,121] It may be theorized that high viscosity is connected with instability while low viscosity means stability, i.e. too-small pectin dosages decrease the stability. It may further be interpreted that the pectin is adsorbed to the casein particles, which have a weak positive charge in the unstabilized milk.[120] Small pectin dosages neutralize this charge, and they increase the tendency for collapse of the system owing to removal of repulsive forces between particles. When even more (negatively charged) pectin is added, a new repulsive force is introduced that is stronger than the original one and is strong enough to stabilize the system.

Other products, such as soy-bean milk, may be stabilized by the same mechanism.

Pectin finds various other uses in beverages such as fruit juices or soft drinks. It may be used for stabilization of fruit pulp[122] or flavour oils, and as a mouth-feel improver. The cloud stability of orange juices is dependent on the amount of pectin present and the nature of the pectin;[123-126] it may be improved by enzymatic removal of some of the pectin content followed by substitution of this with high-ester pectin of high molecular weight. Sedimentation of fruit tissue may be prevented in orange juice concentrates with more than 45% soluble solids by making a weak pectin gel of the concentrate, which is subsequently stirred; the product then appears as a viscous liquid.

REFERENCES

1. Towle, G. A. & Christensen, O., Pectin. In *Industrial Gums*, 2nd edn, ed. R. Whistler. Academic Press, New York, 1973, p. 429.
2. Nelson, D. B., Smit, C. J. B. & Wiles, R. R., Commercially important pectic substances. In *Food Colloids*, ed. H. D. Graham. Avi, Westport, Connecticut, 1977, p. 418.
3. Pedersen, J. K., Pectins. In *Handbook of Water-Soluble Gums and Resins*, ed. R. L. Davidson. McGraw-Hill, New York, 1980, p. 15-1.
4. Højgaard Christensen, S., Pectins. In *Food Hydrocolloids*, Vol. III, ed. M. Glicksman. CRC Press, Boca Raton, Florida, 1986, p. 205.
5. Kertesz, Z. I., *The Pectic Substances*. Interscience Publishers, New York, 1951.
6. Doesburg, J. J., Pectic substances in fresh and preserved fruits and vegetables. *IBVT Commun. No. 25*, Institute for Research on Storage and Processing of Horticultural Produce, Wageningen, The Netherlands, 1965.
7. Amado, R. & Neukom, H., Isolation and partial degradation of pectic substances of potato cell walls in phosphate buffer. *Abstracts: 9th Triennial Conf. Eur. Ass. Potato Res.*, 1984, p. 103.
8. Konno, H., Yamasaki, Y. & Katoh, K., Enzymatic degradation of pectic substances and cell walls purified from carrot cell cultures. *Phytochemistry*, **25** (1986) 623.
9. Thibault, J. F., Enzymatic degradation and beta-elimination of the pectic substances in cherry fruits. *Phytochemistry*, **22** (1983) 1567.
10. Rombouts, F. M. & Thibault, J. F., Feruloylated pectic substances from sugar beet pulp. *Carbohydrate Res.*, **154** (1986) 177.
11. Guillon, F., Thibault, J. F., Rombouts, F. M., Voragen, A. G. J. & Pilnik, W., Structural features of the neutral sugar side chains of beet pulp pectins. *Proc. Cell Walls 86*, Paris, 1986, p. 112.
12. El Tinay, A. H., El Sharif, A. S. & Nour, A. A., A chemical study of pumpkin pectic substances. *Tropical Sci.*, **24** (1982) 173.
13. Huber, D. J., Strawberry fruit softening: the potential roles of polyuronides and hemicelluloses. *J. Food Sci.*, **49** (1984) 1310.
14. De Vries, J. A., Voragen, A. G. J., Rombouts, F. M. & Pilnik, W., Effect of ripening and storage on pectic substances. *Carbohydrate Polymers*, **4** (1984) 3.
15. Redgwell, R. J. & Selvendran, R. R., Structural features of cell wall polysaccharides of onion. *Carbohydrate Res.*, **157** (1986) 183.
16. Matsuura, Y., Chemical structure of polysaccharide of cotyledons of kidney beans. *J. Agric. Chem. Soc. Japan*, **58** (1984) 253.
17. Bouveng, H. D., Polysaccharides in pollen. II: The xylogalacturonan from mountain pine pollen. *Acta Chem. Scand.*, **19** (1965) 953.
18. Mascaro, L. J. & Kindell, P. K., Apiogalacturonan from *Lemna minor*. *Arch. Biochem. Biophys.*, **183** (1977) 139.

19. Boothby, D., Pectic substances in developing and ripening plum fruit. *J. Sci. Food Agric.*, **34** (1983) 1117.
20. De Vries, J. A., Den Uyl, C. H., Voragen, A. G. J., Rombouts, F. M. & Pilnik, W., Structural features of the neutral sugar side chains of apple pectic substances. *Carbohydrate Polymers*, **3** (1983) 193.
21. McNeil, M., Darwill, A. G., Fry, S. C. & Albersheim, P., Structure and function of the primary cell wall of plants. *A. Rev. Biochem.*, **53** (1984) 625.
22. Lau, J. M., McNeil, M., Darvill, A. G. & Albersheim, P., Structure of the backbone of RG I, a pectic polysaccharide in the primary cell walls of plants. *Carbohydrate Res.*, **137** (1985) 111.
23. York, W. S., Darvill, A. G., McNeil, M. & Albersheim, P., 3-Deoxy-D-manno-2-octulosonic acid is a component of RG II, a pectic polysaccharide in the primary cell walls of plants. *Carbohydrate Res.*, **138** (1985) 109.
24. Melton, L. D., McNeil, M., Darvill, A. G. & Albersheim, P., Structural characterization of oligosaccharides isolated from the pectic polysaccharide RG II. *Carbohydrate Res.*, **146** (1986) 279.
25. Siddiqui, I. R., Rosa, N. & Woolard, G. R., Structural investigations of water-soluble tobacco polysaccharides: pectic polysaccharide. *Tobacco Int.*, **186** (1984) 29.
26. Powell, D. A., Morris, E. R., Gidley, M. J. & Rees, D. A., Conformations and interactions of pectins. II. Influence on residue sequence and chain association in calcium pectate gels. *J. Mol. Biol.*, **155** (1982) 517.
27. Voragen, A. G. J., Schols, H. A. & Pilnik, W., Determination of the degree of methylation and acetylation of pectins by HPLC. *Food Hydrocolloids*, **1** (1986) 65.
28. Dea, I. C. M. & Madden, J. K., Acetylated pectic polysaccharides of sugar beets. *Food Hydrocolloids*, **1** (1986) 71.
29. Fry, S. C., Feruloylated pectins from the primary cell wall. *Plants*, **157** (1983) 111.
30. Rombouts, F. M. & Thibault, J. F., Enzymic and chemical degradation and the fine structure of pectins from sugar beet pulp. *Carbohydrate Res.*, **154** (1986) 189.
31. Hodges, L. C., Deutsch, H. M., Green, K. & Zalkow, L. H., Polysaccharides of *Cannibis sativa* active in lowering intraocular pressure. *Carbohydrate Polymers*, **5** (1985) 141.
32. Stevens, B. J. H. & Schendran, R. R., Pectic polysaccharides of cabbage (*Brassica oleracea*). *Phytochemistry*, **23** (1984) 107.
33. Stoddart, R. W., *The Biosynthesis of Polysaccharides*. Croom Helm, London, 1984.
34. Khalil, N. F. & Duncan, H. J., The silica content of plant polysaccharides. *J. Sci. Food Agric.*, **32** (1981) 415.
35. Turmucin, F., Ungan, S. & Yilder, F., Pectin production from sunflower. *METU J. Pure Appl. Sci.*, **16** (1983) 263.
36. Pathak, D. K. & Shukla, S. D., Quantity and quality of pectin in sunflower at various stages of maturity. *J. Food Sci. Technol.*, **18** (1981) 116.
37. Huber, D. J. & Lee, J. H., Pectin changes in tomato pericarp and gel tissue. In *Chemistry and Function of Pectins*, ed. M. L. Fishman & J. J. Jen. ACS Symposium Series **310**, 1986, p. 141.

38. De Vries, J. A., Rombouts, F. M., Voragen, A. G. J. & Pilnik, W., Distribution of methoxyl groups in apple pectic substances. *Carbohydrate Polymers*, **3** (1983) 245.
39. De Vries, J. A., Hansen, M. E., Glahn, P. E., Søderberg, J. & Pedersen, J. K., Distribution of methoxyl groups in pectins. *Carbohydrate Polymers*, **6** (1986) 165.
40. Anger, H. & Dongowsky, G., Über die bestimmung der Estergruppeverteilung in Pektin durch Fraktionierung an DAEA-cellulose. *Nahrung*, **28** (1984) 199.
41. Anger, H. & Dongowsky, G., Distribution of free carboxyl groups in native pectins from fruit and vegetables. *Nahrung*, **29** (1985) 397.
42. Kohn, R., Marbovich, D. & Machova, E., Deesterification mode of pectin by pectin esterase of *Aspergillus foetidus*, tomatoes and alfalfa. *Collect. Czech. Chem. Commun.*, **48** (1983) 790.
43. Kohn, R., Dongowsky, G. & Bock, W., Die Verteilung der freien und veresterten Carboxylgruppen im Pektinmolekul nach Einwirkung von Pektinesterasen aus *Aspergillus niger* und Orangen. *Nahrung*, **29** (1985) 75.
44. Markovic, O. & Kohn, R. G., Mode of pectin deesterification by *Trichoderma reesei* pectin esterase. *Experientia*, **40** (1984) 842.
45. Rees, D. A., Polysaccharide conformation in solutions and gels—recent results on pectins. *Carbohydrate Polymers*, **2** (1902) 254.
46. Walkinshaw, M. D. & Arnott, S., Conformations and interactions of pectins. 2: Models for junction zones in pectinic acid and calcium pectate gels. *J. Mol. Biol.*, **153** (1981) 1075.
47. Fishman, M. L. & Pepper, L., Difference in number average MW by end group titration and osmometry for LM pectins. In *Proc. Conf. New Devel. Ind. Polysacch.*, ed. V. Cressarzi, I. C. M. Dea & S. S. Stivala. Gordon & Breach, New York, 1984, p. 159.
48. Fishman, M. L., Pepper, L., Damert, W. C., Phillips, G. G. & Barford, R. A., A critical reexamination of molecular weight and dimensions for citrus pectins. In *Chemistry and Function of Pectins*, ed. M. L. Fishman & J. J. Jen. ACS Symposium Series **310**, 1986, p. 22.
49. Stephen, A. M., Other plant polysaccharides. In *The Polysaccharides*, Vol. II, ed. G. O. Aspinall. Academic Press, New York, 1983, p. 97.
50. Selvendran, R. R., Developments in the chemistry and biochemistry of pectin and hemicellulose polymers. *J. Cell. Sci., Suppl. 2* (1985) 51.
51. Fry, S. C., Cross-linking of matrix polymers in the growing cell walls of angiosperms. *A. Rev. Plant Physiol.*, **37** (1986) 165.
52. Karpovich, N. S., Telichuk, L. K., Donchenko, L. V. & Totkailo, M. A., Pectin and raw materials resources. *Pishch. Promst.* (Kiev), **3** (1949) 85.
53. Sabir, M. A., Sosulski, F. W. & Campbell, S. J., Polymetaphosphate and oxalate extraction of sunflower pectins. *J. Agric. Food Chem.*, **24** (1976) 348.
54. Lin, M. J. Y. & Humbert, E. S., Extraction of pectins from sunflower heads. *Can. Inst. Food Sci. Technol. J.*, **9** (1976) 70.
55. Lin, M. J. Y., Sosulski, F. & Humbert, E. S., Acidic isolation of sunflower pectin. *Can. Inst. Food Sci. Technol. J.*, **11** (1978) 75.
56. Sosulski, F., Lin, M. J. Y. & Humbert, E. S., Gelation characteristics of acid-precipitated pectin from sunflower heads. *Can. Inst. Food Sci. Technol. J.*, **11** (1978) 113.

57. Campbell, S. J., Sosulski, F. W. & Sabir, M. A., Development of pectins in sunflower stalks and heads. *Can. J. Plant Sci.*, **58** (1978) 863.
58. Kim, W. J., Sosulski, F. & Campbell, S. J., Formulation and characteristics of low-ester gels from sunflower pectin. *J. Food Sci.*, **43** (1978) 746.
59. Kim, W. J., Sosulski, F. & Lee, S. C. K., Chemical and gelation characteristics of ammonia-demethylated sunflower pectins. *J. Food Sci.*, **43** (1978) 1436.
60. Pathak, D. K. & Shukla, S. D., A review on sunflower pectin. *Indian Food Packer* (1978 May–June), p. 49.
61. Srirangarajan, A. M. & Shrikhande, A. J., Mango peel waste as a source of pectin. *Current Sci.*, **45** (1976) 620.
62. Beerh, O. P., Raghuramaiah, B. & Krishnamurthy, G. V., Utilization of mango waste: peel as a source of pectin. *J. Food Sci. Technol.*, **13** (1976) 96.
63. Kausar, P. & Nomura, D., A new approach to pectin manufacture by copper method, Part 1. *J. Fac. Agric., Kyushu Univ.*, **25** (1980) 61.
64. Kausar, P. & Nomura, D., A new approach to pectin manufacture by copper method, Part 2. *J. Fac. Agric., Kyushu Univ.*, **26** (1981) 1.
65. Kausar, P. & Nomura, D., A new approach to pectin manufacture by copper method, Part 3. *J. Fac. Agric., Kyushu Univ.*, **26** (1982) 111.
66. Michel, Florence, Thibault, J.-F. & Doublier, J.-L., Characterization of commercial pectins purified by cupric ions. *Sciences des Aliments*, **1** (1981) 569.
67. Joseph, G. H., Kieser, A. H. & Bryant, E. F., High-polymer ammonia-demethylated pectinates and their gelation. *Food Technol.*, **3** (1949) 85.
68. Reitsma, J. C. E., Thibault, J. F. & Pilnik, W., Properties of amidated pectins. I. Preparation and characterization of amidated pectins and amidated pectic acids. *Food Hydrocolloids*, **1** (1986) 121.
69. Kertesz, Z. I., Baker, G. L., Glenn, H. J., Mottern, H. H. & Olsen, A. G., Report of the committee for the revision of the nomenclature of pectic substances. *Chem. Eng. News*, **22** (1944) 105.
70. FAO Food and Nutrition Paper 31/2, Rome (1984).
71. Final Report of the IFT Committee, Pectin standardization. *Food Technol.*, **13** (1959) 496.
72. Food Chemicals Codex, 3rd edn. National Academy Press, Washington, DC, 1981, p. 215.
73. EEC Council, Council directive of 25 July 1978 laying down specific criteria of purity for emulsifiers, stabilizers, thickeners and gelling agents for use in foodstuffs. Official Journal of the European Communities, No. L223/7, 14 August 1978.
74. Joint FAO/WHO Expert Committee on Food Additives, Evaluation of certain food additives. World Health Organization Technical Report Series, No. 669, Geneva, 1981.
75. Padival, R. A., Ranganna, S. & Manjrekar, S. P., Stability of pectins during storage. *J. Food Technol.*, **16** (1981) 367.
76. Michel, F., Doublier, J. L. & Thibault, J. F., Investigations on high-methoxyl pectins by potentiometry and viscometry. *Prog. Food Nutr. Sci.*, **6** (1982) 367.
77. Kawabata, Akiko, Studies on chemical and physical properties of pectic substances from fruits. *Mem. Tokyo Univ. Agric.*, **19** (1977) 115.
78. Christensen, P. E., Methods of grading pectin in relation to the molecular weight (intrinsic viscosity) of pectin. *Food Res.*, **19** (1954) 163.

79. Smit, C. J. B. & Bryant, E. F., Properties of pectin fractions separated on diethylaminoethyl-cellulose columns. *J. Food Sci.*, **32** (1967) 197.
80. Anger, H. & Berth, G., Gel permeation chromatography and the Mark–Houwink relation for pectins with different degrees of esterification. *Carbohydrate Polymers*, **6** (1986) 193.
81. Smit, C. J. B. & Bryant, E. F., Ester content and jelly pH influences on the grade of pectins. *J. Food Sci.*, **33** (1968) 262.
82. Michel, E., Thibault, J.-F. & Doublier, J.-L., Viscometric and potentiometric study of high-methoxyl pectins in the presence of sucrose. *Carbohydrate Polymers*, **4** (1984) 283.
83. Rinaudo, M., Comparison between experimental results obtained with hydroxylated polyacids and some theoretical models. In *Polyelectrolytes*, ed. E. Sélegny. Reidel, Dordrecht, The Netherlands, 1974, p. 157.
84. Plaschina, I. G., Braudo, E. E. & Tolstoguzov, V. B., Circular-dichroism studies of pectin solutions. *Carbohydrate Res.*, **60** (1978) 1.
85. Antonov, Yu. A., Grinberg, V. Ya., Zhuravskaya, N. A. & Tolstoguzov, V. B., Concentration of the proteins of skimmed milk by membraneless, isobaric osmosis. *Carbohydrate Polymers*, **2** (1982) 81.
86. Zhuravskaya, N. A., Kiknadze, E. V., Antonov, Yu. A. & Tolstoguzov, V. B., Concentration of proteins as a result of the phase separation of water–protein–polysaccharide systems. Part 2. Concentration of milk proteins. *Nahrung*, **30** (1986) 601.
87. BeMiller, J. N., Acid-catalyzed hydrolysis of glycosides. *Adv. Carbohydrate Chem.*, **22** (1967) 25.
88. Kenner, J., The alkaline degradation of carbonyl oxycelluloses and the significance of saccharinic acids for the chemistry of carbohydrates. *Chemistry and Industry* (1955 June 25) 727.
89. Whistler, R. L. & BeMiller, J. N., Pectin and alginic acid. *Adv. Carbohydrate Chem.*, **13** (1958) 289.
90. Neukom, H. & Deuel, H., Alkaline degradation of pectin. *Chemistry and Industry*, **77** (1958) 683.
91. Albersheim, P., Neukom, H. & Deuel, H., Splitting of pectin chain molecules in neutral solutions. *Arch. Biochem. Biophys.*, **90** (1960) 46.
92. Pilnik, W. & Rombouts, F. M., Pectic enzymes. *Easter Sch. Agric. Sci. Univ. Nottingham (Proc.)*, **27** (1979) 109.
93. Pilnik, W. & Rombouts, F. M., Utilization of pectic enzymes in food production. *Develop. Food Sci.*, **2** (1979) 269.
94. Pilnik, W. & Rombouts, F. M., Pectic enzymes. In *Enzymes Food Process*, ed. G. G. Birch, N. Blakebough & K. J. Parker. Applied Science Publishers, Barking, 1981, p. 105.
95. Gierschner, K., Pectin and pectic enzymes in fruit and vegetable technology. *Gordian*, **81** (1981) 171.
96. Gierschner, K., Pectin and pectic enzymes in fruit and vegetable technology. *Gordian*, **81** (1981) 205.
97. Voragen, A. G. J. & Pilnik, W., Pectin-degrading enzymes in fruit and vegetable processing. ACS Symposium Series (Biocatal. Agric. Biotechnol.) **389**, 1989, p. 93.

98. Rombouts, F., Pectic enzymes, their biosynthesis and roles in fermentation and spoilage. *Current Developments in Yeast Research (Proc. Int. Yeast Symp.)*, **5** (1981) 585.
99. Steinhauser, von J., Otterbach, G. & Gierschner, K., Vergleich von Methoden zur Bestimmung der Gelierkraft von Pektin. *Industrielle Obst- und Gemüseverwertung*, **64** (1979) 179.
100. Beach, P., Davis, E., Ikkala, P. & Lundbye, M., Characterization of pectins. In *Chemistry and Function of Pectins*, ed. M. L. Fishman & J. J. Jen. ACS Symposium Series **310**, 1986, p. 103.
101. Crandall, P. G. & Wicker, L., Pectin internal gel strength: theory, measurement, and methodology. In *Chemistry and Function of Pectins*, ed. M. L. Fishman & J. J. Jen. ACS Symposium Series **310**, 1986, p. 88.
102. Cox, R. E. & Higby, R. H., A better way to determine the jelling power of pectins. *Food Ind.*, **16** (1944) 441.
103. Fox, G., Zur Wirtschaftlichkeit der Trocknung von Apfeltrester. *Confructa-Studien*, **2** (1984) 174.
104. Uhlenbrock, W., Die Gelierkraftbestimmung von Pektinen. *Gordian*, **83** (1983) 148.
105. Joseph, G. H. & Baier, W. E., Methods of determining the firmness and setting time of pectin test jellies. *Food Technol.*, **3** (1949) 18.
106. Hinton, C. L., The setting temperature of pectin jellies. *J. Sci. Food Agric.*, **1** (1950) 300.
107. Ehrlich, R. M. & Cox, R. E., Slow-set pectin and process for preparing same. US Patent 3 835 111 (1974).
108. Oakenfull, D. & Scott, A., Hydrophobic interaction in the gelation of high methoxyl pectins. *J. Food Sci.*, **49** (1984) 1093.
109. Oakenfull, D. G. & Scott, A. G., Gelation of high methoxyl pectins. *Food Techn. Australia*, **37** (1985) 156.
110. Plashchina, I. G., Semenova, M. G., Braudo, E. E. & Tolstoguzov, V. B., Structural studies of the solutions of anionic polysaccharides. IV. Study of pectin solutions by light-scattering. *Carbohydrate Polymers*, **5** (1985) 159.
111. Morris, E. R., Rees, D. A., Thom, D. & Boyd, J., Chiroptical and stoichiometric evidence of a specific primary dimerisation process in alginate gelation. *Carbohydrate Res.*, **66** (1978) 145.
112. Grant, G. T., Morris, E. R., Rees, D. A., Smith, P. J. C. & Thom, D., Biological interactions between polysaccharides and divalent cations: the egg-box model. *FEBS Lett.*, **32** (1973) 195.
113. Palmer, K. J. & Hartzog, M. B., An X-ray diffraction investigation of sodium pectate. *J. Am. Chem. Soc.*, **67** (1945) 2122.
114. Walkinshaw, M. D. & Arnott, S., Conformations and interactions of pectins. *J. Mol. Biol.*, **153** (1981) 1055.
115. Gidley, M. J., Morris, E. R., Murray, E. J., Powell, D. A. & Rees, D. A., Spectroscopic and stoichiometric characterisation of the calcium-mediated association of pectate chains in gels and in the solid state. *JCS Chem. Commun.*, **22** (1979) 990.
116. Morris, E. R., Powell, D. A., Gidley, M. J. & Rees, D. A., Conformations and interactions of pectins. *J. Mol. Biol.*, **155** (1982) 507.

117. Copenhagen Pectin, *Handbook for the Fruit Processing Industry*, 1984.
118. Doesburg, J. J. & Vos, L. de, Pasteurized mixtures of fruit juices and milk, with a long shelf life. *V. Intern. Fruchtsaftkongress, Wien 1959*, p. 32.
119. Exler, H., Verfahren zur Herstellung saurer Milchmischgetränke. German Patent, DE 1,270,938, 1968.
120. Glahn, P.-E., Hydrocolloid stabilization of protein suspensions at low pH. *Prog. Food Nutr. Sci.*, **6** (1982) 171.
121. Lohmann, R., Einsatzmöglichkeiten von Pektin in Milchergeugnissen. *Gordian*, **82** (1982) 148.
122. Röcken, W. von, Die Bedeutung von Pektin und Pektinasen für die Trubstabilität von Orangenlimonaden. *Brauwelt*, **8** (1979) 224.
123. Kanner, J., Ben-Shalom, N. & Shomer, I., Pectin–hesperidin interaction in a citrus cloud model system. *Lebensm.-Wiss. u. -Technol.*, **15** (1982) 348.
124. Whitaker, J. R., Pectic substances, pectic enzymes and haze formation in fruit juices. *Enzyme Microbiol. Technol.*, **6** (1984) 341.
125. Shomer, I. & Merin, U., Recovery of citrus cloud from aqueous peel extract by microfiltration. *J. Food Sci.*, **49** (1984) 991.
126. Ben-Shalom, N., Pinto, R., Kanner, J. & Berman, M., A model system of natural orange juice cloud: effect of calcium on hesperidin–pectin particles. *J. Food Sci.*, **50** (1985) 1130.

Chapter 11

WHEY PROTEINS

R. C. BOTTOMLEY, M. T. A. EVANS and C. J. PARKINSON

*Express Foods Group Limited, R&D Department,
430 Victoria Road, South Ruislip, Middlesex HA4 0HF, UK*

1 INTRODUCTION

Whey protein is now a significant source of functional protein for the food industry world-wide. Produced mainly as a by-product of hard-cheese and casein manufacture in Europe, North America and Australasia, thousands of tonnes of whey protein concentrate (WPC) and isolate (WPI) powders are now available from major manufacturers, who offer ranges of products that vary both in functionality profiles and protein content (25–95% protein) to meet the particular applications needs of food manufacturers. This aspect of whey protein utilization has developed rapidly over the last 10 years, out of a realization of the valuable properties of whey proteins, the availability of new fractionation technologies such as ultrafiltration and ion-exchange, and a continuing search for a better exploitation of whey than the base disposal methods of dumping, animal feeding or the manufacture of low-protein whey powders.

In this chapter we shall first outline the present knowledge of whey protein composition and structure. We shall follow this with practical details of production and purification methods, commercial availability and product types. Finally, we shall provide information on functional properties such as solubility, viscosity and especially gelation, together with a selection of food applications for commercial WPCs. We hope these will give some idea of the versatility of whey protein products and their potential for use in new applications.

Since there have been a number of good and detailed reviews of the whole field of whey protein composition, structure, functionality and manufacture in recent years,[1-7] we shall confine ourselves in this account to providing essential basic information, with updates where appropriate, while also giving more recent results of a practical and commercial type.

2 STRUCTURE

2.a Composition

Whey proteins can be defined in general terms as those proteins remaining soluble at pH 4·6 and 20°C after casein removal from skim milk or whole milk. However, as commercially available whey protein concentrates and isolates may be manufactured from different whey types, their overall chemical composition and relative protein content can vary somewhat, reflecting their origin. The main whey sources are sweet whey, e.g. from rennet casein or Cheddar-cheese manufacture, where the casein fraction of milk is removed by the action of rennet, and acid whey, as in the production of acid casein or cottage cheese, where the casein is precipitated by the direct addition or the generation of acid *in situ*.

The principal whey proteins in all whey types are β-lactoglobulin and α-lactalbumin (Table 1). The other proteins normally present in whey are bovine serum albumin and the immunoglobulins that are also found in blood serum. Other significant components are the so-called proteose-peptones, soluble intact caseins and a number of other enzymes and proteins in small amounts, including lactoperoxidase, lysozyme, lactoferrin, lactollin and many others.

It has become clear in recent years that the proteose-peptone components 5, 8-slow and 8-fast are in fact fragments of β-casein, peptides generated by the proteolytic action of plasmin in milk.[8-10] This raises the

TABLE 1
BOVINE WHEY PROTEIN COMPOSITION AND SOME COMPONENT PROPERTIES

Protein	Approximate concentration in whey (g/l)	Molecular weight (daltons)	Isoelectric point	Approximate percentage of total whey protein
β-Lactoglobulin	3·0	18 363	5·35–5·49	50
α-Lactalbumin	0·7	14 175	4·2–4·5	12
Immunoglobulins	0·6	$(1·61-10) \times 10^5$	5·5–8·3	10
Bovine serum albumin	0·3	66 267	5·13	5
Proteose-peptones, minor proteins and caseins	1·4	4 100–40 800	3·3–3·7	23

question of the use of the term proteose-peptones, and their classification as whey proteins.[11] Obviously, depending on the type of whey used to make commercial WPCs and isolates, and especially with rennet whey, further constituents of this type must be present, e.g. intact caseins, random peptides formed by the stray proteolytic activities in starters and rennet, and also the glycomacropeptide formed by the specific cleavage of κ-casein during curd formation. On the other hand, where direct acidification is used for curd formation, the whey produced will contain little or none of the latter species. The effect of the presence or absence of these components on the total functionality of commercial WPC and isolates is not yet clear. Some of the recent work relating to this will be discussed later. Typical compositional data for commercial WPCs will also be given, indicating variation in mineral as well as protein content related to whey source and production method.

2.b Whey Protein Structure
2.b.1 β-Lactoglobulin

The detailed three-dimensional structure of bovine β-lactoglobulin has recently been determined by X-ray crystallography at 0·28 nm resolution.[12] While the report notes that one external loop and the N- and C-terminal residues are poorly defined and require further resolution, the molecule consists of antiparallel β-sheet, formed by nine strands wrapped round to form a flattened cone. The core of the molecule is an eight-stranded, antiparallel β-barrel. This is an unusual structure, which has been observed in ten binding proteins, including plasma retinal-binding protein.[13] Quantitatively, the secondary structure contains about 20–30% β-sheet, with 50–60% unordered plus turn, and the balance α-helix. Although β-lactoglobulin is the major protein constituent of whey, and occurs only in milk, there has been no clear evidence for any function for it other than a nutritional role. The present data suggest a possible transport role for the protein.

Structurally, β-lactoglobulin is a globular protein of 162 amino-acid residues in a single chain, with a monomer molecular weight of 18 363 for β-lactoglobulin A, the most abundant genetic variant.[11] The protein has two disulphide bridges, between residues 106–119 and 66–160, and a free thiol group at 121. These provide a potential for inter- and intramolecular disulphide link interchange during conformational changes associated with pH alterations or heat treatment. β-Lactoglobulin is sensitive to heat denaturation and pH and ionic strength changes. While the protein is very acid-stable, resisting denaturation at pH 2,[14] the structurally stable monomer associates to form a complete dimer in the region pH 3·5–5·2 at room temperature and moderate concentrations. In this pH range, β-lactoglobulin dimers associate to an octomer at temperatures close to 0°C, with a maximum between pH 4·4 and pH 4·7. The octomer is a compact cyclic structure and conformational changes accompany these association reactions.[15,16] In the alkaline region the protein undergoes a conformational change at pH 7·5, accompanied by a molecular expansion. Above pH 7·0 there is a rapid increase in the reactivity of the thiol group; dissociation of the dimer also occurs in this region. At pH 8·0 and above, the protein can be regarded as unstable, since it will form aggregates of denatured protein.[17] On heating, the protein undergoes denaturation and aggregation and this step is very sensitive to pH and calcium ions. Under heating, the β-lactoglobulin dimer will dissociate to a monomer above 30°C, while above 55°C there is a progressive unfolding of the globular structure, exposing cystine and cysteine groups and hydrophobic surfaces, allowing primary and secondary aggregation.

2.b.2 α-Lactalbumin

Three genetic variants (A, B and C) exist of α-lactalbumin, though only the B variant is found in the milk of Western cattle, and this has a molecular weight of 14 175. The complete sequence of 123 amino-acid residues is known[18,19] and the structure is stabilized by four disulphide bonds. α-Lactalbumin is a globular protein with a very stable conformation between pH 5·4 and pH 9·0. While the complete three-dimensional structure of bovine α-lactalbumin has not yet been reported, a recent low-resolution X-ray study of baboon α-lactalbumin[20] has added more weight to the proposal that α-lactalbumins and Class C lysozymes are homologous, with α-lactalbumin evolving from a lysozyme precursor. The role of α-lactalbumin as modifier protein with galactosyl transferase in the synthesis of lactose is well known. The hydrodynamic properties of α-lactalbumin suggest that it is an almost spherical, very compact globular protein. Circular dichroism studies indicate that its secondary structure at physiological pH consists of 26% α-helix, 14% β-structure and 60% unordered structure.[21,22] It has a low solubility in water between pH 4·0 and pH 5·0, and above pH 9·0 and below pH 4·0 conformational changes occur at ambient temperature, without causing irreversible aggregation.[17,23] Similarly, on heating, conformational changes can be detected, but in buffer solutions at neutral pH no precipitation occurs, in contrast to the behaviour of other globular proteins.[24] α-Lactalbumin is therefore the most stable of the principal whey proteins. It has been found to possess a strong cation-binding site, and much of the recent research on the protein has been concerned with its calcium-binding properties and the large conformational changes associated with metal binding.[25]

2.b.3 Bovine Serum Albumin (BSA)

The serum albumin protein of whey is identical to blood serum albumin. As the major protein of plasma, easily obtained in pure crystalline form, its structure, physicochemical and functional properties have been extensively studied over many years. Early work has been summarized by Joly,[26] and typical later work was carried out by Jaenicke[27,28] and his co-workers. More recent studies[29-32] have provided further information on the solubility and gelation properties of BSA. These are examples only; the literature on this protein is very large, and no attempt is made to provide a comprehensive review. The function of bovine serum albumin as a transport protein for fatty acids in the circulatory system is well known.[33]

A complete amino-acid sequence for bovine serum albumin has been reported,[34,35] and it consists of 582 amino-acid residues in a single chain, with a molecular weight of 66 267; of the whey proteins, it has the longest polypeptide chain primary structure. To stabilize its tertiary structure it has 17 disulphide bonds, but there is only one free sulphydryl group at position 34. Structurally, bovine serum albumin is a classic globular protein. It is essentially monomeric, although dimers and higher polymers can occur in small quantities. The protein also displays some microheterogeneity; this has been ascribed variously to fatty-acid binding[33] and to disulphide interchange.[36] At pH 4·0 the molecule undergoes acid denaturation owing to charge repulsion,[37] while at high pH an increase in Stokes radius can be detected owing to a similar effect.[38] The current view of the structure of bovine serum albumin[35] is that the molecule exists in three major domains, each consisting of two large double loops and a small double loop, with the overall shape as a 3:1 ellipsoid.

2.b.4 Immunoglobulins

The properties of the bovine immunoglobulins have been reviewed frequently and in detail.[11,39,40] In cow's milk and whey the principal immunoglobulin class is IgG, being about 80% of the total; IgG_1 is the main protein, with IgG_2 as the second subclass, and IgM and IgA are also present. As with serum albumin, the immunoglobulins are not unique to whey, but are part of the immune defence system of bovine serum, being produced in response to the exposure of the animal to foreign antigens. Structurally they are all large globular protein molecules, easily denatured

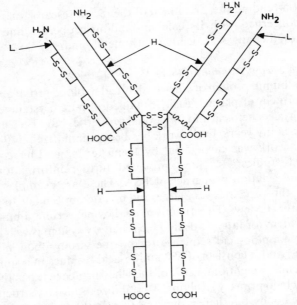

Fig. 1. Diagrammatic representation of an immunoglobulin monomer. H = heavy chains ($c.$ 50 000 daltons); L = light chains ($c.$ 23 000 daltons).

by heat, e.g. a treatment of 70°C for 30 min is reported to denature them completely.[17] The immunoglobulins exist either as monomers or as polymers of a basic structural unit. In this brief discussion we shall ignore the enormous amount of heterogeneity that occurs in these molecules as a result of their function, and which is the basis of their immunological characterization. As our interest is mainly in physicochemical properties, we shall confine our description to common structural features and functions.

The basic unit is a four-polypeptide chain molecule; this is made up of two identical light (L) chains and two identical heavy (H) chains (Fig. 1). The light chains can be of two types (κ and λ) and have molecular weights of about 20 000. The heavy chains can be of several types ($\gamma, \alpha, \mu, \varepsilon$) with molecular weights in the range 50 000–70 000. The chains are linked by disulphide bonds to form the monomer unit; these can be linked intermolecularly to form polymers. Antigen binding is carried out in the regions of variable amino-acid sequence at the N-terminal ends of the light and heavy chains. Classes of immunoglobulins are distinguished by differences in the amino-acid sequences in the constant regions of the heavy chains, at the C-terminal ends.

Functionally, IgG and IgM are the major serum antibodies against infection; IgM is the serum antibody to Gram-negative bacteria, e.g. *Escherichia coli*. IgA and IgM combine to provide the local immune response of the alimentary tract, while IgA confers passive resistance by its presence in maternal colostrum.

In more detail, IgG_1 and IgG_2 are monomeric units as described earlier, with two light and two heavy chains, the latter of the γ type. The molecular weights are about 150 000. IgA is present as a dimer of the basic unit, linked covalently by a polypeptide termed the J-component. IgM is a pentamer of the basic unit, also joined by the J-component.

Most studies of immunoglobulins have naturally concentrated on structural elucidation and antigenic properties. The functional properties of these large, globular, heat-denaturable proteins have received little attention. Therefore, their contribution to the functionality of commercial whey protein products is unclear. It is only as further systematic work proceeds on fractionated whey proteins that their role in the gelation or foam formation by whole-whey protein will emerge.

3 SOURCE AND PRODUCTION

3.a Whey Types
As already indicated, there are a number of types of whey that are potential sources for the commercial manufacture of whey protein products. These wheys have different compositions and processing characteristics, which give them various advantages and disadvantages when considered for the manufacture of specific whey protein products having definite compositional and functional specifications.

In simple terms, there are two main types of whey, sweet whey and acid whey (Table 2), within which types some further variants also occur.

3.a.1 Sweet Whey
Sweet whey is mostly derived from the large-scale manufacture of hard-cheese varieties, principally Cheddar but also Gouda, Edam, Mozzarella, Colby, Monterey Jack and some soft-cheese varieties such as Brie and Camembert. These cheeses are formed by the rennet coagulation of milk, where starter addition is to develop flavour rather than significant acidity (see Chapter 6 for more detail). The resultant wheys have pH values of

TABLE 2
COMPOSITION OF ACID AND SWEET WHEYS[4]

	Average composition (g/l)			
	Rennet casein whey	Lactic casein whey	Mineral acid casein whey (H_2SO_4)	Cheddar cheese whey
Total solids	66	64	53	67
pH	6.4	4.6	4.7	5.9–6.2
Whey protein as (TN − NPN) × 6.38	6.2	5.8	5.8	6.2
NPN	0.37	0.40	0.30	0.27
Lactose	52.3	44.3	46.9	52.4
Minerals (as ash)	5.0	7.5	7.9	5.2
Fat	0.2	0.3	0.3	0.7
Ca	0.5	1.6	1.4	0.4
Mg	0.07	0.10	0.11	0.08
Na	0.53	0.51	0.50	0.50
K	1.45	1.40	1.40	1.50
Cl	1.02	0.9	0.9	1.0
SO_4	0.7	0.5	2.8	0.6
Lactate	—	6.4	—	2.0
Phosphate	1.0	2.0	2.0	0.5

about 5.8–6.3; much of the calcium present is in colloidal form, and where the cheese is made from whole milk the fat content of the whey can be in the range 0.06–0.1%. Whey from the manufacture of rennet casein is also a sweet whey, and as it is made from skim milk it has a low fat content (0.02%). The use of rennet (chymosin) in curd formation means that the κ-casein in milk is split, and the glycomacropeptide formed as a result is found in the whey. Some proteolysis may also occur because of starter activity. It is also possible that whey will contain some casein fines, remaining after the various separation processes (30–100 ppm).

3.a.2 Acid Whey

Acid whey is obtained either as a by-product of acid casein manufacture or from the production of cottage cheese or similar soft curd cheese types. Acid casein whey can be a mineral-acid product, made by the addition of hydrochloric acid or sulphuric acid to skim milk, or be formed by starter culture addition to generate a lactic casein whey (see Chapter 6 for more details). Cottage cheese whey is normally formed by the latter method. In all cases the pH of the whey is about 4·6 (4·2–4·8), and the calcium present is essentially ionic because of the pH. Ash contents can be high because of mineral-acid addition, and solids can be lower than in sweet whey, especially if significant curd washing is carried out. While some proteolysis may occur if starters are used, there is no glycomacropeptide from κ-casein in the whey, as rennet is not normally used. As the acid wheys are derived from skim milk, their fat contents are normally low (0·02%).

3.b Production Methods
3.b.1 Simple Concentration and Drying
For the manufacture of standard low-protein whey powders, the conventional but highly developed large-scale processes of evaporation, crystallization and spray drying are used (see K. Masters in Ref. 41, p. 331, for details). To produce a non-caking, non-hygroscopic and stable whey powder for human food use, good-quality sweet whey is pre-concentrated, preferably to 55–60% total solids (TS), using large multistage evaporators (6 or 7 effect for efficiency). The whey concentrate is then pre-crystallized either by seeding, cooling and agitation or by a combination of these methods. The lactose present is then in a crystalline form that prevents hygroscopicity in the powder. The pre-crystallized concentrate is then spray-dried, either using a two-stage process with a vibrating fluid bed or by a belt process followed by fluid beds. Use of one-stage drying with a pneumatic conveying system gives a dusty, hygroscopic powder suitable only for animal feeding.

3.b.2 Whey Protein Concentrates
For the commercial production of whey protein concentrates, the most widely used process is now that of ultrafiltration (UF). This is selective membrane filtration, in which whey protein is concentrated and low-molecular-weight salts, other molecular species, and some of the lactose are removed by passage through a porous membrane under mild conditions of pH, temperature and pressure (see the schematic Fig. 2). Very large plants are now in operation in the dairy industry world-wide, using various modular membrane configurations such as the spiral modules of Abcor and the plate-and-frame flat-sheet system of Pasilac/DDS. These are continuous, automatic, high-volume throughput, stage-in-series plants (see Fig. 3) that can process 0·5–1·0 million litres of whey per day. The basis of the large-scale industrial processes now in place has been the development of robust, asymmetric composite membranes, resistant to extremes of pH and chemical attack, easily cleanable in hygienic systems and with long

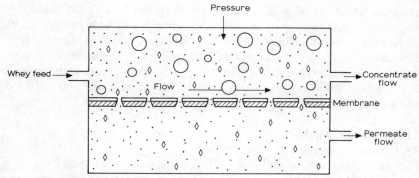

FIG. 2. Schematic of ultrafiltration mechanism: ○, proteins; ◇, lactose and mineral salts; •, water.

operational lives. Developments in design configuration and membrane systems are continuing to improve selectivity and efficiency.

To summarize, the total UF production process consists of whey pretreatment, separation and clarification to remove fat and fines, usually a pasteurization, and sometimes a conditioning step (e.g. heat-and-hold) prior to the ultrafiltration step itself.

For the manufacture of lower-protein WPC (35–55% protein), the ultrafiltration method consists of a number of concentration stages giving the final liquid concentrate. When higher-protein WPC is required (65–85% protein), stages of diafiltration are required, in which deionized

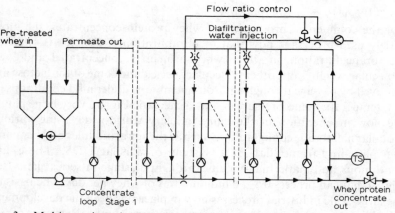

FIG. 3. Multistage-in-series UF plant, flow ratio control and diafiltration water ratio control. Total solids control at product outlet.

water is injected to remove further salts and lactose, to give the required final protein-to-solids ratio in the concentrate (see the schematic Fig. 3). The final liquid concentrate can be subjected to evaporation and spray drying in a conventional system under controlled conditions, or may be directly dried in a conventional spray drier or a box drier of the Rogers, Marriott Walker or the Filtermat type.

For those interested, there are now many practical and theoretical reviews of this technology. For instance, Marshall[4] gives a recent general review, while Madsen[42] provides a comprehensive and detailed practical and theoretical review of UF in the dairy industry. Matthews[43] has given a very useful description of the practical side of designing and operating a large-scale UF plant for WPC manufacture.

3.b.3 Industrial-scale Whey Protein Fractionation

On the laboratory scale, whole whey protein has been fractionated into its individual pure protein components for many years by a variety of chromatographic and classical techniques.

To produce whey protein fractions, or high-protein fat-free whey protein isolates, on a significant commercial scale, a number of processes have been tried over the years with limited success (see Matthews, Ref. 7, p. 2682, for example, or Ref. 44 for a more recent review).

However, very few processes have a track record for producing fractionated whey proteins on an industrial scale for any length of time. Gel filtration was used in the 1970s[4] and abandoned, but two other methods have shown some recent and more enduring success. Both are based on adsorption chromatography. One is the Vistec process operated by Bioisolates,[4,7,45-47] which uses a modified cellulose ion-exchanger in a stirred-tank system. It has been operated for a number of years in the UK and Ireland, and produces several hundred tonnes per year of a high-protein whey isolate ($c.$ 95% protein) with a very low fat and lactose content, and very good whipping, heat-setting and nutritional properties.

The other process is the Spherosil method, developed by Rhône-Poulenc and operated commercially by Fromageries Bel.[48,49] Here silicate-based ion-exchangers such as Spherosil QMA are used in fixed-bed columns. These types of adsorption/desorption processes, and particularly the latter, have the capability to fractionate the whey proteins into enriched or substantially pure separate components[49] by manipulation of process conditions. Problems with this type of process relate to the large volumes of deproteinated whey generated, the volumes of rinse solutions required, low resin capacity for protein, losses of protein and resin in processing, cleaning and sanitizing difficulties, and the need to reconcentrate diluted solutions by UF before drying.

Two further closely related processes of interest are currently actively being developed to industrial scale. They are based on classical and physical treatments, and may therefore be more economic at large scale than ion-exchange chromatography. The one developed at CSIRO by Pearce[50] uses the precipitation of α-lactalbumin from 12% total solids whey at pH 4·2 and 64°C. The sediment is conditioned to assist separation, concentrated and dried, and contains fat and α-lactalbumin. Pearce has advocated its use in processed cheese or for humanizing infant feeds. The supernatant fraction is of more functional interest, since it contains the β-lactoglobulin at high protein level and very little fat, and gives a clear solution and has good foaming and heat-setting properties. Pearce has suggested its use in soft-drink and juice fortification.

Maubois[51] has developed a process in which the first stage is whey defatting by calcium addition, pH adjustment and heating followed by microfiltration. Fractionation of α-lactalbumin and β-lactoglobulin-enriched products is effected from the clarified whey after concentration by ultrafiltration. Similar physical methods to those of Pearce are employed.

Developments of methods such as those described, especially if they can be combined with selective membrane fractionation and microfiltration for proteins, will potentially provide an array of protein products with selective functionalities. Modified whey proteins with enhanced gelling characteristics are conceivable from β-lactoglobulin-enriched products.

Quite pure whey protein fractions highly enriched in β-lactoglobulin, α-lactalbumin and immunoglobulins can therefore already be made at the large pilot plant or small industrial scale, but at a considerable cost. Their eventual commercial availability will require both high added functionality and added-value applications.

4 COMMERCIAL AVAILABILITY

A range of traditional whey products has been manufactured for many years. We will give a brief description of them before we cover the newer whey protein products in more detail.

4.a. Traditional Whey Products

The traditional products comprise relatively low-protein powders (10–25% protein) such as whey powders, delactosed whey powders and their demineralized derivatives produced by ion-exchange or electrodialysis.

Another product of this type is 'lactalbumin', a high-protein powder made by the heat precipitation of whey protein. It is essentially insoluble and can contain anything from 50% to 90% protein. These products tend to be used as fillers, binders or nutritionals, and do not have heat-setting or aeration properties. Most are obtainable in large volumes, world-wide from many suppliers. The powders are obtainable as animal-feed grade, human grade, infant-nutritional grade, and in a range of specialized specifications. They can be supplied in 25 kg, 50 kg or even 1000 kg bags, and in container loads of 20 tonnes or more. Detailed information on these products is readily available from dairy companies and elsewhere (see review, Refs 2 and 41).

4.b Whey Protein Concentrates

The production of higher-protein-content, more functional whey protein products, commonly referred to as whey protein concentrates (WPC), has increased greatly in the last decade, with significant further expansion in the last 3–4 years. Most major dairy countries (except the UK) have substantial installed capacity for the manufacture of WPC with a range of protein contents and functional properties. Some companies have now been making these products for over a decade, and volumes are now counted in thousands of tonnes. This area can now be regarded as a thoroughly established though still-developing business. A list of some of the major producing companies world-wide is given in Table 3, together with some typical WPC products available.

The range of whey protein concentrates now marketed includes 25% and 30% products, through the skim powder replacer types at 35% WPC, to 55–60% WPC for nutritional use, and 75–80% WPC both for nutritional and functional use, the latter including aeration, gelation and viscosity enhancement in specific applications. Now on the market, but currently at low volume, are some whey protein 'isolates' (WPI), at 90% whey protein content or more. These are usually low-fat products, and can be partially fractionated whey protein. They can be used as nutritionals in high-protein dietary supplements, or as whipping or gelling agents as in egg-white substitution.

The advance of processing technology has made other high-protein products possible from whey. These are being actively researched and a number are at pilot-plant production level. Such products include WPC enriched in immunoglobulin, high in α-lactalbumin or high in β-lactoglobulin. Products of this type have potential in veterinary protection, specific nutrition and an altered range of protein functionality from the normal range of WPCs.

TABLE 3
SOME EXAMPLES OF COMMERCIALLY AVAILABLE WHEY PROTEIN CONCENTRATES AND THEIR SUPPLIERS[a]

Standard WPC products (% protein)	25	30	35	55	65	75	80	75/80 High gel	85	95	Special products supplied
Suppliers											
CMP[b]	WPC25	WPC30	WPC35	WPC55				75% HG			42% WPC; 70% high-gel; low-gel temp. 70%; lactose reduced 60% and 65%; low pH, high viscosity 55% and 75%; heat-treated WPC 35% and 55%
Melei[c]	—		WPC35		WPC65	WPC75	WPC80	75% HG			Standard 50–65% WPC; instant 80%; sour 70%; lactose reduced 75%; 70% heat stable; 60% egg substitute, etc.
NZDB[d]			WPC35			WPC75		75% HG			75% range includes gel enhanced and reduced products. Also lactose-reduced, enhanced whipping, heat- and chill-stable products. 35% WPC: heat stable, low-Na and instantized
Danmark[e]			WPC35			WPC75	WPC80	80% HG			WPC60
Meggle[f]											
Bioisolates[g]										95–97%	Globulates of 70% WPC
GCCC[h]				WPC55 50–54%		WPC75		75% HG			
Valio[i]											20–80% WPC, heat treatments as requested

[a] Limited selection of some major producers.
[b] Carbery Milk Products Ltd, Ballineen, County Cork, Republic of Ireland. *Carbelac* range of WPC products.
[c] Milei (Sales), Rosensteinstrasse 20, 7000 Stuttgart, FRG. Lactalbumen range of WPC products.
[d] New Zealand Dairy Board, PO Box 417, Wellington, New Zealand (UK: New Zealand House, Haymarket, London SW1Y 4TD). *Alacen* range of WPC.
[e] Danmark Proteins a/s Sumatravej, DK-8000 Aarhus C, Denmark. *Lacprodan* range of WPC products.
[f] Meggle Milchindustrie GmbH & Co. KG, 8094 Reitmehring, Sf. 40 Bst Wasserburg/Inn-Bahnhof, FRG. Globulal range of WPC products.
[g] Bioisolates Ltd, 10 Adelaide Street, Swansea, West Glamorgan SA1 1SE, UK.
[h] Golden Cheese Company of California, 1138 West Rincon Street, Corona, California 91720, USA. There are a large number of USA suppliers of 35% and higher-protein products. GCCC is a major producer of 55–85% WPC.
[i] Valio Finnish Cooperative Dairies Association, PO Box 390, SF-00101, Helsinki 10, Finland.

The broad classification of standard WPCs given above is not restrictive. The present method of WPC manufacture using ultrafiltration allows for any level of protein between 25% and 80% to be achieved, allowing for production costs. Accordingly, for example, 32%, 47% or 69% WPCs can be made should a customer require them. Furthermore, process manipulations can provide variations at or between protein levels in a product range. For example, low heat treatment and careful drying conditions can give a very soluble, almost undenatured WPC. On the other hand, controlled heat treatment, pH variations and sequestrant addition can give reproducible modifications in solubility, dispersibility and heat stability. Changes in mineral balances, levels and physical state, e.g. colloidal to ionic calcium, Ca:P ratio, sodium and potassium levels, etc., can be obtained by altering process conditions.

More complex process operations can materially change the basic functionality of the whole-whey protein present, inducing changes in viscosity characteristics and improving the gelling characteristics of the final product.

Other product variations that can be combined with the above include reduction in lactose content by treatment with β-galactosidase, instantization to improve rehydration properties, or partial hydrolysis to improve heat stability.

It can be seen from this that the nature of whey protein concentrate production technology allows a wide choice of product compositional and functional specifications. Indeed, while generic types such as 35%, 55% or 75% WPC are habitually quoted in the literature or manufacturer's published information, many commercial WPCs actually supplied are delivered against precise 'non-standard' specifications, and are tailored to meet the customer's particular application. This is regularly done by a close collaboration between customer and supplier, in order to define the modification in properties necessary to provide a WPC that will give the required cost-effective performance in application.

In common with other milk and whey powder products, WPCs can be supplied in 25 kg bags or in bags up to 1 tonne. Container loads of 20 tonnes are also commonly supplied. Except where non-standard specifications are required, the microbiological and compositional specifications met by other milk-derived powders are normally achieved. Different bulk densities can be obtained. Since ultrafiltration is a concentration/filtration process, the accumulation of contaminants is technically possible. However, provided whey is from cheese that is treated with no significant additives during manufacture (e.g. nitrates), and particularly if the milk is from grass-

5 SOLUTION PROPERTIES

5.a Dissolution

Standard spray-dried whey protein concentrates and isolates do not dissolve or disperse readily in either warm or cold water. Vigorous agitation is required: this not only aids dispersion by eliminating coagulation, but also forces the powder into close contact with the water and thereby accelerates the rate of dissolution. It is important that the whey protein powders are added—gradually—to water, and not vice versa. In practice, a stream of the powder is best mixed into a vortexing solution; the solution temperature should be relatively low (55°C or below) to avoid denaturing the protein.

Dispersion can be improved to some extent by agglomeration of the powders, but significant improvement in dissolution can only be achieved by adding a water-active compound such as lecithin to agglomerated powders. Such 'instantized' powders are available commercially.

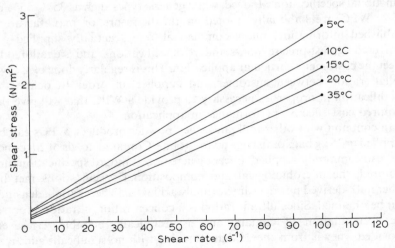

FIG. 4. Flow curves for 20% solutions of WPC 75.

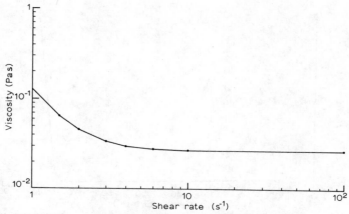

FIG. 5. Viscosity of 20% WPC solutions at 5–30°C.

5.b Solution Viscosity

At temperatures below 60°C, and protein concentrations below 10%, whey protein solutions have low viscosity[52] and are relatively shear-stable. Current measuring techniques, however, indicate that even at low solids they are not Newtonian fluids, but exhibit Bingham flow with a measurable, albeit small, yield stress (Figs 4 and 5). At higher concentrations, solutions become increasingly viscous,[53] especially at lower temperatures (2–10°C), and more shear-sensitive, exhibiting pseudoplastic flow behaviour (Fig. 6). As the solution temperature is increased, the viscosity decreases slightly

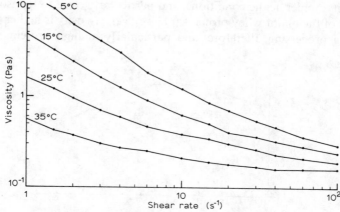

FIG. 6. Viscosity of 30% WPC 75 solutions at 5–35°C.

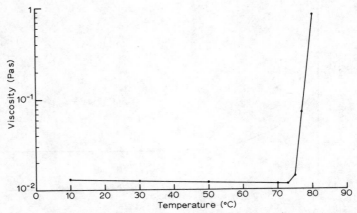

FIG. 7. Viscosity of 15% WPC 75 solution with increasing temperature.

until it reaches 70–80°C, when—as long as the protein content is sufficiently high—gelling occurs (Fig. 7). At lower protein concentrations, high viscosity may be achieved if the temperature is kept below 80°C. However, excessive heat treatment when the protein concentration is too low for true gelation to occur will result in complete denaturation of the proteins, followed by aggregation and precipitation.[54] This in turn will result in much lower solution viscosities.

5.c Factors Affecting Viscosity

At processing temperatures below 60°C, pH exerts an influence on viscosity: under acidic conditions, and particularly close to the isoelectric points of the major whey proteins (pH 4·5–5·2), viscosity is higher (Fig. 8). During processing, therefore, and particularly at high protein concen-

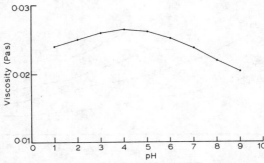

FIG. 8. Viscosity of 15% WPC 75 solutions at 25°C over a range of pH values.

trations, high viscosity is more likely to cause problems in this pH range. Similarly, the presence and concentration of certain ionic species can affect viscosity. Anions such as phosphate and citrate, and divalent cations such as calcium and magnesium, exert substantial effects.[55] Specific heat modifications of whey proteins, including heating at both alkaline pH[56] and acidic pH,[57,58] have been shown to increase viscosity substantially, to the extent in one instance of generating a semi-solid with thixotropic properties at temperatures below 50°C.[56] In these modifications, the precise pH, the degree of heat treatment and the protein concentration at the time of heat treatment are critical: if heat treatment is excessive, complete denaturation of the whey proteins is the likely result, with concomitant reduced viscosity.[52]

5.d Solubility

In general, normal whey protein concentrates show good solubility in water across the full pH spectrum. Minimum solubility is in the pH 4–5 region, close to the isoelectric points of the main whey proteins: the protein insoluble at these pH values is usually that proportion that has been denatured during processing.[59] Concentrates produced from acid wheys generally contain a higher proportion of denatured protein than those from sweet wheys.[60] Similarly, isolates prepared using the acidic 'Vistec' process contain relatively high levels of protein insoluble at pH 4·6,[61] much more than those prepared using anionic resins at neutral pH—as long as bound protein has been eluted from the latter resin using mild solvents such as NaCl rather than the often used 0·1M HCl.[49,60]

A number of whey protein products, however, have much lower solubilities at most pH values. These include traditional lactalbumin, which is produced by heat denaturation of the whey proteins (as discussed earlier) and some specifically heat-modified WPCs, such as Carbelac WB 700 (Carbery Milk Products), which has enhanced water-binding properties.

5.e Emulsifying Properties

Whey proteins are moderately good emulsifiers—better than milk protein in the form of skim milk powder but not as good as soya proteins, milk protein concentrates or caseinates.[53,61] WPCs maintain their emulsifying capacity over a wide pH range,[62] unlike caseinate, skim milk and soya-protein products, which are very pH-dependent. Highly denatured whey protein products are inferior to native, soluble WPCs, though a small degree of heat damage may be beneficial. Data in the literature on the emulsification properties of WPCs is at best inconsistent. This may be due

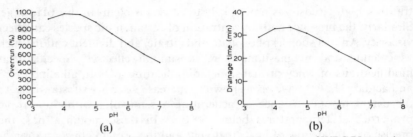

FIG. 9. (a) Overrun and (b) drainage time for a 15% solution of WPC 75 at various pH values.

to the varied methods of evaluation of emulsification properties, which can be misleading.[61,63]

5.f Whippability

The surface activity of whey protein solutions is dramatically reduced in the presence of relatively low concentrations of lipid. Hence, only fat-free whey protein isolates (such as those obtained by ion-exchange techniques) and 'defatted' WPCs with fat contents below 1% form truly stable foams. These can be usefully employed in low-fat food formulations: low-fat, high-protein WPCs have been used very successfully as replacers of egg white in recipes such as meringues and marshmallows.[63,64] Interestingly, best overrun and foam stability tends to occur at low pH (4–4·5), as can be seen in Fig. 9.

6 GEL PROPERTIES

Most of the information in this section refers to model systems comprising purely whey protein in aqueous solution. It provides a guide to what can be expected in applied food formulations, especially those in which the whey protein content is high. In these model systems, two different types of whey protein gel are encountered: a soft, white 'curdy' gel prone to syneresis (formed at 'neutral' whey pH by unmodified WPCs) and a firm, translucent gelatine-like gel with very good water-retention properties (formed by non-standard WPCs).

6.a Preparation

Whey protein gels are prepared by heating solutions to or above their gelling temperature—generally around 80°C for UF WPCs[63]—for between 10 and 60 min. Gel formation does not occur at protein

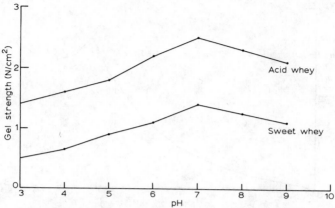

FIG. 10. Variation of gel strength with pH for two different WPC 75 solutions at 15% TS.

concentrations below 5%, and physically firm gels are not normally formed if the protein concentration is below 8%.[65,66] The precise gelling temperature and subsequent gel strength are affected by most solution parameters: ionic strength, protein content, pH, the type of dissolved ions and total solids content all exercise effects (for detailed review, see Ref. 66). To maximize gel strength in model systems and certain applications such as 'surimi' (Japanese fabricated fish products), any air entrapped during solubilization or transfer to the gelling vessel should be removed (before heating) by ultrasonic vibration or evacuation. Gelation will occur over a wide pH range, but pH values around 7 for higher gel strengths appear to be optimal (Fig. 10). At or near the predominant isoionic pH (4·5–5·0), heating whey protein solutions tends to result more in protein denaturation than in gel formation. At more acidic pH values, gelling time/temperature is increased[67] and the type of gel formed is rather different, being translucent—possibly owing to the absence of divalent metal ion in the structure—not the curdy-white type formed at pH values near 6. Similarly, translucent gels are formed, at high gelling temperatures, at pH values above 7.[63] At a given gelling temperature, gelling time increases with pH at these alkaline pH values.[65]

6.b Measurement of Gel Strength

The strength of whey protein gels is normally measured by some form of compression or penetration testing, although the precise methodology used to quantify their properties is confusingly varied.

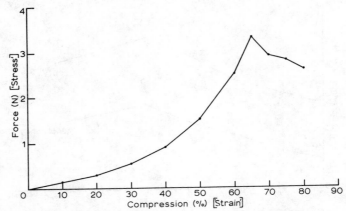

FIG. 11. Stress–strain relationship for a typical high-gel WPC at 15% TS.

Whey protein gels are complex viscoelastic materials and as such do not behave as Hookean solids. A typical stress–strain curve is shown in Fig. 11, the stress increasing non-linearly with increasing strain. The break point can often be characteristic of the type of gel.

The relationship of the measuring geometry to the size of the sample causes considerable variance in any 'absolute' measurements. If measuring probes are small (1–5 mm) then measurements are of a 'puncture' type and are really concerned with the ability to fracture the surface of the gel. A more rigorous test is to have a measuring system with a considerably larger geometry than that of the sample under test. Although this type of test will result in a more quantifiable 'modulus', it may not necessarily give a potential user the information he requires. Various intermediate geometries exist between these two measuring extremes. Figure 12 illustrates three types of measurement.

The two extremes of test can both contribute towards a full appreciation of the functional properties of the gel in a particular application and should not be treated in isolation.

For compression testing, Instron 'Universal' and similar types of equipment have been used;[63] a specifically designed mould was used by Langley et al.[68] in conjunction with an Instron model 1140. This type of equipment generates results in terms of rigidity (modulus of rigidity, newtons per square metre) and fracture force (newtons), and can also be used to measure gelling temperature.[63] The puncture or penetrometer-style apparatus (e.g. the LFRA Stevens Texturometer) is also widely used. This equipment, based on the 'Bloom' gelometer (originally designed for testing

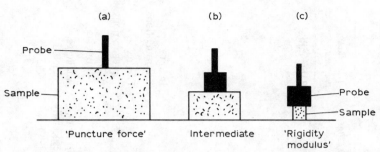

FIG. 12. Various 'gel strength' geometries in current use. (a) Probe diameter 3 mm. Sample depth 40 mm, diameter 70 mm. (b) Probe diameter 13 mm. Sample depth 22 mm, diameter 37 mm. (c) Probe diameter 15 mm. Sample depth 10 mm, diameter 10 mm.

the strength of gelatine gels) most frequently generates results expressed as the weight (in grams) necessary to cause a probe of defined surface area to fracture the gel matrix.

6.c Factors Affecting Gel Strength

The whey protein gel, produced at 'natural' pH (6–6·5) from a source containing largely undenatured protein, is of the curdy type, which will sustain a considerable amount of compression (80%) without fracture and recover to something resembling its original shape. This type of gel, however, does not exhibit a significant gel 'strength', being easily compressed. A more rigid gel can be produced by the addition of mineral salts, notably phosphates. When salts are added to give an optimum ionic strength and the pH is adjusted to about 7, depending upon the type of salt being used, a very rigid gelatine-type gel may be produced. This type of gel is, however, very susceptible to fracture and needs careful control of the ionic balance to produce the required properties.

6.c.1 Effect of Heat

Carefully and specifically applied heat treatments can also be used to modify the gel strength of whey protein products. Heat treatments at both alkaline pH[56] and acidic pH[57] have been shown to enhance gelling characteristics. Acidic pH heat treatments can lead to improved gel quality, although the apparent gel strength may be reduced.[57]

6.c.2 Effect of Additives

The addition of other high-molecular-weight components does not present too much of a problem. Lipids can be incorporated at a ratio of 1:4 with the

TABLE 4
SUMMARY OF APPLICATIONS OF WHEY PROTEINS

Protein type	Flour confectionery	Sugar confectionery	Meat and fish	Soups, sauces, dressings	Desserts	Beverages	Dietary preparations
15–30%		Chocolate					Baby 'milks'
30–45%			Burgers Patés Processed meats				
High-heat 35%	Biscuits Bread Cakes	Caramels	Freeze–thaw stability Retort stability	Freeze–thaw stability			
45–60%	Doughnuts Waffles	Fudge					
60–80%	Pasta			Salad dressings Salad creams Mayonnaise			Infant formulations Dietary preparations Athletic formulas
Low-fat 90–95%		Meringues				Soft drinks, especially acidic	Anallergenic diets
High-gel 75–80%			Analogues		Jellies Mousses		

protein, with little loss of structure, although an inherent loss of strength is experienced. Carbohydrates can be effectively used within the protein gel network providing that the protein concentration is maintained.

6.c.3 Effect of Temperature

Once a stable whey protein gel has been produced, it should be stable to temperature variations. High temperatures in excess of 100°C will not affect the condition of the gel. Freezing at domestic deep-freeze temperatures ($-18°C$) does not significantly affect the gel properties, i.e. the gel appears to be freeze–thaw stable (RCB and CJP, unpublished observations). The resistance to shear is dependent initially upon the method of manufacture. A soft, curdy gel will be more resistant to high strain forces than will a hard, brittle gel of the gelatine type.

7 APPLICATIONS

The application of whey proteins in the food industry is characterized by their versatility across a wide spectrum of food products. Their water-binding, emulsifying, aerating and gel-forming properties can be utilized at various levels to provide useful functional applications in a wide variety of foods. This versatility means that a list of potential applications would be tediously long. In order to simplify our assessment of whey protein applications, we have classified them in terms of the protein content of the powdered product. Table 4 is a fairly representative spectrum of all types of whey protein application. In the following descriptions, however, we concentrate more on those applications that utilize the functional properties of whey proteins.

7.a Whey Powders

Whether natural or demineralized, whey powders produced simply by concentration and drying are traditionally used for their nutritional properties, and in roles as fillers and extenders rather than for their functional properties. Uses in infant formulations, dietetic foods and animal feeds are common.

7.b 30–40% Whey Protein Concentrates

These are normally produced by ultrafiltration, and are frequently utilized as an effective and economic replacement for skim milk solids.

7.b.1 Standard Soluble Form

Utilization in meat products, in particular burgers and patés, has been established for several years. In this instance, it is their ability to emulsify fat and bind water that make whey proteins a useful functional ingredient. They are also starting to be used, as a cost-effective replacement for caseinates, in processed meats, where their water-binding properties can be used effectively in injection solutions. These concentrates can be successfully used in yoghurt formulations.[69,70]

7.b.2 Heat-treated, Reduced Solubility Type

Heat treatment of these concentrates denatures the whey proteins, but this also increases their water-binding capacity. This characteristic can be utilized in several applications. Flour confectionery, especially in American cake and biscuit recipes, which traditionally contain milk solids, has utilized heat-treated whey protein concentrates to increase dough strength and improve product quality. However, improvement in freeze–thaw stability is arguably the most advantageous functional property that heat-treated whey proteins possess. This is utilized to good effect in such systems as cook-in-the-bag sauces and freezable mousses. Further applications include cream soups and use in condensed-milk and caramel manufacture, an as yet under-utilized application, where the improved water-binding properties contribute towards higher viscosities and firmer products.

7.c 50–60% Whey Protein Concentrates

It is this protein content at which whey protein products begin to be excellent cost-effective alternatives to egg solids. At levels of about 2% protein they can be used effectively in chemically raised formulations of such products as doughnuts and waffles.

The utilization of higher levels of denaturation in 50–60% WPCs has not yet been fully exploited. The potential, however, must exist, just as it does for 30–40% whey protein products.

7.d 70–80% Whey Protein Concentrates

It is when whey proteins are concentrated to this level, at the expense of lactose and minerals, that they really achieve the unique advantage of their gelation properties. However, their emulsifying properties also achieve a greater potential at this protein content.

Complete or partial replacement of whole egg by high-protein WPCs has been exploited successfully in both egg products, such as omelettes and scrambled egg, and in applications where the egg acts as a binding agent.

For example, WPCs can be used in flour-based products, such as 'babas' and oriental-style pasta products (noodles), and can contribute positively to the properties of gelled desserts, even to the extent of providing the primary gel network.

An application of growing importance is in extending the Japanese fish-based product, surimi.[71,72] Surimi has traditionally been produced from relatively low-value fish catches, utilizing the properties of fish muscle proteins to generate high-value analogues of crustacean (crab, prawn) flesh. Egg solids, and more particularly egg-white solids, are used to extend and enhance the gel properties of surimi. Whey protein concentrates can also fulfil such a role; within the last few years, specially modified WPCs with enhanced gelling properties have proved particularly effective in replacing egg white in surimi products (see Section 7.e).

The use of high-protein WPCs in nutritional fortification of pasta products such as spaghetti is well established. However, more functional uses of whey proteins in this context does not appear to be restricted to high-protein WPCs, and traditional lactalbumin can perform a useful role.[73]

The emulsifying properties of WPC 75 have been exploited in the production of a wide range of dressings from relatively low-viscosity salad dressings for products such as coleslaw, through to thick, spoonable mayonnaise types. These applications again utilize the ability of high-protein WPC to replace egg solids, often at a lower usage level.

7.e High-gel Whey Protein Concentrates (70–80% Protein)

Recent developments have led to the manufacture of WPCs with their natural gel properties enhanced, such that they produce a gel with strength superior to that of any other available food-protein gel. Their use has been demonstrated in the production of meat and fish analogues, including the surimi application referred to in Section 7.d, cheese analogues and a range of chilled desserts with a variety of consistencies. High-gel WPC is a comparatively new product whose full potential has probably not yet been fully realized, although import levels into Japan indicate that a relatively large market has already been established there—presumably in fish analogue products.

7.f Low-fat 90% Whey Protein Concentrates

The removal of fat to achieve concentrations below 1% on a dry basis results in a specialist product that has extremely good aeration properties. The use in meringues as a total substitute for egg white has been well established.[63] An additional use for defatted product is in the fortification of beverages, particularly acidic ones, where the clarity, pH stability and nutritional value are all utilized.

7.g Fractionated Whey Protein Products

Such products are just becoming available. The most abundant of these, and the most significant in functional terms, are β-lactoglobulin-rich and β-lactoglobulin-depleted products. The latter are likely to be utilized in infant formulations, whereas the former—which should retain most of the gelling characteristics of whey protein concentrates—are more likely to find functional applications. Both products will, of course, be substantially more expensive than standard whey protein concentrates.

Whey protein concentrates are now firmly established as products in the spectrum of functional food ingredients. Their supply is backed by a large, world-wide manufacturing capacity, and considerable scientific and technical support for further development for food applications. We have already shown that, specifically for heat-gelling applications, higher-protein WPC products have gained a significant role. From recent research and development, it is clear that modification and fractionation procedures can enhance the gelling properties of whey protein products to a significant extent, providing opportunities for their use in a wide range of food products, where specific functionality is required, or where more expensive proteins need to be replaced.

REFERENCES

1. Smith, G., Whey proteins. *World Review of Nutrition and Dietetics*, **24** (1976) 88–116.
2. Evans, M. T. A. & Gordon, J. F., Whey proteins. In *Applied Protein Chemistry*, ed. R. A. Grant. Applied Science Publishers, London, 1980, p. 31.
3. Swaisgood, H., Chemistry of milk proteins. In *Developments in Dairy Chemistry—1*, ed. P. F. Fox. Applied Science Publishers, London, 1982, p. 1.
4. Marshall, K. R., Industrial isolation of milk proteins: whey proteins. In *Developments in Dairy Chemistry—1*, ed. P. F. Fox. Applied Science Publishers, London, 1982, p. 339.
5. Morr, C. V., Functional properties of milk proteins and their use as food ingredients. In *Developments in Dairy Chemistry—1*, ed. P. F. Fox. Applied Science Publishers, London, 1982, p. 375.
6. de Wit, J. N., New approach to the functional characterisation of whey protein for use in food products. In *Milk Proteins '84*, Proceedings of the International Congress on Milk Proteins, Luxemburg, 1984, ed. T. E. Galesloot & B. J. Tinbergen. Padoc, Wageningen, 1985, p. 183.

7. Symposium: Production and utilisation of whey and whey components. *J. Dairy Sci.*, **67**(11) (1984) 2621–774.
8. Andrews, A. T., The composition, structure and origin of proteose-peptone component 5 of bovine milk. *Eur. J. Biochem.*, **90** (1978) 59.
9. Andrews, A. T., The composition, structure and origin of proteose peptone component 8F of bovine milk. *Eur. J. Biochem.*, **90** (1978) 67.
10. Eigel, W. N. & Keenan, T. W., Identification of protease peptone component 8-slow as a plasmin-derived fragment of bovine β-casein. *Int. J. Biochem.*, **10** (1979) 529.
11. Eigel, W. N., Butler, J. E., Ernstrom, C. A., Farrell, H. M. Jr, Harwalkar, V. R., Jenness, R. & Whitney, R. McL., Nomenclature of proteins of cow's milk: fifth revision. *J. Dairy Sci.*, **67** (1984) 1599–631.
12. Papiz, M. Z., Sawyer, L., Eliopoulos, E. E., North, A. C. T., Findlay, J. B. C., Sivaprasadarao, R., Jones, T. A., Newcomer, M. E. & Kraulis, P. J., The structure of β-lactoglobulin and its similarity to plasma retinol-binding protein. *Nature (London)*, **324** (1986) (27 Nov.) 343.
13. Sawyer, L., One fold among many. *Nature (London)*, **327** (1987) (26 June) 659.
14. Aschaffenburg, R. & Drewry, J., An improved method for the preparation of crystalline β-lactoglobulin and α-lactalbumin from cow's milk. *Biochem. J.*, **65** (1957) 273–7.
15. Townend, R., Herskovits, T. T., Timasheff, S. N. & Gorbunoff, M. J., The state of amino acid residues in β-lactoglobulin. *Arch. Biochem. Biophys.*, **129** (1969) 567.
16. Timasheff, S. N., Mescanti, L., Basch, J. J. & Townend, R. J., Conformational transitions of bovine β-lactoglobulins A, B and C. *J. Biol. Chem.*, **241** (1966) 2496.
17. Lyster, R. L. J., Review of the progress of dairy science, Section C. Chemistry of milk proteins. *J. Dairy Res.*, **39** (1972) 279.
18. Brew, K., Castellino, F. J., Vanamen, T. C. & Hill, L. R., The complete amino acid sequence of α-lactalbumin. *J. Biol. Chem.*, **245** (1970) 4570.
19. Vanaman, T. C., Brew, K. & Hill, L. R., The disulphide bonds of bovine α-lactalbumin. *J. Biol. Chem.*, **245** (1970) 4583.
20. Smith, S. G., Lewis, M., Aschaffenburg, R., Fenna, R. E., Wilson, I. A., Sundaralingam, M., Stuart, D. I. & Phillips, D. C., Crystallographic analysis of the 3-dimensional structure of baboon α-lactalbumin at low resolution. *Biochem. J.*, **242**(2) (1987) 353–60.
21. Barel, A. O., Prieels, J. P., Maes, E., Loose, Y. & Leonis, J., Comparative physiochemical studies of human α-lactalbumin and human lysozyme. *Biochim. Biophys. Acta*, **257** (1972) 288.
22. Robbins, R. M. & Holmes, L. G., Circular dichroism spectra of α-lactalbumin. *Biochim. Biophys. Acta*, **221** (1970) 234.
23. Shukla, T. P., Chemistry and biological function of α-lactalbumin. *CRC Critical Reviews in Food Technology*, **3**(3) (1973) 241.
24. Baer, A., Droz, M. & Blanc, B., Serological studies on heat induced interactions of α-lactalbumin and milk proteins. *J. Dairy Res.*, **43**(3) (1976) 419.
25. Desmet, J., Hanssens, I. & Cauveleart, F. van, Comparison of the binding of Na^+ and Ca^{2+} to bovine α-lactalbumin. *Biochim. Biophys. Acta* (Protein Structure and Molecular Enzymology), **912**(2) (1987) 211–19.

26. Joly, M., *A Physico-chemical Approach to the Denaturation of Proteins*. Academic Press, New York, 1965.
27. Jaenicke, R., Intermolecular forces in the process of heat aggregation of globular proteins and the problem of correlation between aggregation and denaturation phenomena. *J. Polymer Sci.*, Part C, **16** (1967) 2143–60.
28. Jaenicke, R., Volume changes in the isoelectric heat aggregation of serum albumin. *Eur. J. Biochem.*, **21** (1971) 110–15.
29. Gumper, S., Hegg, P. O. & Martens, M., Thermal stability of fatty acid serum albumin complexes studied by differential scanning calorimetry. *Biochim. Biophys. Acta*, **574** (1979) 189.
30. Hegg, P. O., Conditions for the formation of heat induced gels of some globular food proteins. *J. Food Sci.*, **47** (1982) 1241–4.
31. Lin, V. J. C. & Koonig, J. L., Raman studies of bovine serum albumin. *Biopolymers*, **15** (1976) 203.
32. MacRitchie, F., Effects of temperature on dissolution and precipitation of proteins and polyamino acids. *J. Colloid Interface Sci.*, **45** (1973) 235.
33. Spector, A. A., Fatty acid binding to plasma albumin. *J. Lipid Res.*, **16** (1975) 165.
34. Brown, J. R., Structure of bovine serum albumin. *Fed. Proc.*, **34** (1975) 591.
35. Brown, J. R., Structure and evolution of serum albumin. In *Albumin Structure; Biosynthesis; Function. Proc. FEBS Meeting*, **50** (1977) 1.
36. Sogami, M., Petersen, H. & Foster, J., The microheterogeneity of plasma albumins. V. Permutations in disulphide pairings as a probable source of microheterogeneity in bovine albumin. *Biochemistry*, **8**(1) (1969) 49.
37. Haurowitz, B., Albumins, globulins and other soluble proteins. In *The Chemistry and Function of Proteins*, Chap. 8, ed. F. Haurowitz. Academic Press, New York, 1963.
38. Jones, A. & Weber, G., Partial modification of bovine serum albumin with dicarboxylic anhydrides. Physical properties of modified species. *Biochemistry*, **9**(24) (1970) 4729–35.
39. Butler, J. E., Bovine immunoglobulins, an augmented review. *Veterinary Immunology and Immunopathology*, **4**(1/2) (1983) 43–152.
40. Whitney, R. McL., Brunner, R. J., Ebner, K. E., Farrell, H. M., Josephson, R. V., Morr, C. V. & Swaisgood, H. E., Nomenclature of the proteins of cow's milk: fourth revision. *J. Dairy Sci.*, **59**(5) (1976) 795.
41. Evans, M. T. A., Electrodialysis and ion exchange as demineralisation methods in dairy processing. In *Evaporation, Membrane Filtration, Spray Drying in Milk Powder and Cheese Production*, ed. R. Hansen. North European Dairy Journal, Vanlose, Denmark, 1985, pp. 55–78.
42. Madsen, R., Theory of membrane filtration and membrane filtration in the dairy industry. In *Evaporation, Membrane Filtration, Spray Drying in Milk Powder and Cheese Production*, ed. R. Hansen. North European Dairy Journal, Vanlose, Denmark, 1985, pp. 179–286.
43. Matthews, M. E., Practical considerations in the design and operation of a commercial UF plant. *N.Z. J. Dairy Sci. Technol.*, **15** (1980) A73.
44. Morr, C. V., Fractionation and modification of whey protein in the US. In *Trends in Whey Utilisation*, Bulletin of the IDF No. 212 (1987), p. 145.
45. Palmer, D. E., High purity protein recovery. *Process Biochem.*, **12**(5) (1977) 24.

46. Ayers, J. S. & Petersen, M. J., Whey protein recovery using a range of novel ion exchangers. *N.Z. J. Dairy Sci. Technol.*, **20** (1985) 129.
47. Ayers, J. S., Elgar, D. F. & Petersen, M. J., Whey protein recovery using Indion S, an industrial ion exchanger for proteins. *N.Z. J. Dairy Sci. Technol.*, **21** (1986) 21.
48. Mirabel, B., UK Patent 1 563 990 (1980).
49. Skudder, P., Evaluation of a porous silica based ion exchange medium for the production of protein fractions from rennet and acid whey. *J. Dairy Res.*, **52** (1985) 167–81.
50. Pearce, R. J., Fractionation of whey proteins. In *Trends in Whey Utilisation*, Bulletin of the IDF No. 212 (1987), p. 150.
51. Maubois, J. L., Pierre, A., Fauquant, J. & Piot, M., Industrial fractionation of main whey proteins. In *Trends in Whey Utilisation*, Bulletin of the IDF No. 212 (1987), p. 154.
52. Schmidt, R. H., Smith, D. E., Packard, V. S. & Morris, H. A., Compositional and selected functional properties of whey protein concentrates and lactose-hydrolysed whey protein concentrates. *J. Food Protection*, **49**(3) (1986) 192–3.
53. Delaney, R. A. M., Compositional properties and uses of whey protein concentrates. *J. Soc. Dairy Technol.*, **29**(2) (1976) 91–101.
54. Mulvihill, D. M. & Donovan, M., Whey proteins and their thermal denaturation—a review. *Irish J. Food Sci. Technol.*, **11** (1987) 43–75.
55. Johns, J. E. M. & Ennis, B. M., The effect of the replacement of calcium with sodium ions in acid whey on the functional properties of whey protein concentrates. *N.Z. J. Dairy Sci. Technol.*, **15** (1981) 79–86.
56. Phillips, D. J. & Evans, M. T. A., Process for lowering gelling temperature of whey proteins obtained from milk. UK Patent 2 055 846B (1981).
57. Modler, H. W. & Emmons, D. B., Properties of whey protein concentrate prepared by heating under acidic conditions. *J. Dairy Sci.*, **60**(2) (1977) 177–89.
58. Modler, H. W. & Harwalkar, V. R., Whey protein concentrate prepared by heating under acidic conditions. Recovery by ultrafiltration and functional properties. *Milchwissenschaft*, **36**(9) (1981) 537–42.
59. Donovan, M. & Mulvihill, D. M., Thermal denaturation and aggregation of whey proteins. *Irish J. Food Sci. Technol.*, **11** (1987) 87–100.
60. De Wit, J. N., Klarenbeek, G. & Adamse, M., Evaluation of functional properties of whey protein concentrates and isolates. 2. Effects of processing history and composition. *Neth. Milk Dairy J.*, **40** (1986) 41–56.
61. Fox, P. F. & Mulvihill, D. M., Milk proteins: molecular, colloidal and functional properties. *J. Dairy Res.*, **49**(4) (1982) 679–93.
62. Delaney, R. A. M. & Donnelly, J. V., Functional properties of membrane filtered whey and skim milk. *Maelkeritende*, **88** (1975) 55.
63. De Wit, J. N., Functional properties of whey proteins in food systems. *Neth. Milk Dairy J.*, **38** (1984) 71–89.
64. De Boer, R., De Wit, J. N. & Hiddink, J., Processing of whey by means of membranes and some applications of whey protein concentrate. *J. Soc. Dairy Technol.*, **30**(2) (1977) 112–20.
65. Hillier, R. M. & Cheeseman, G. C., Effect of proteose peptone on the heat gelation of whey protein isolates. *J. Dairy Res.*, **46** (1979) 113–23.

66. Mulvihill, D. M. & Kinsella, J. E., Gelation characteristics of whey proteins and β-lactoglobulin. *Food Technol.* (Sept. 1987) 102–11.
67. Harwalkar, V. R. & Kalab, M., Thermal denaturation and aggregation of β-lactoglobulin at pH 2·5. Effect of ionic strength and protein concentration. *Milchwissenschaft*, **40**(1) (1985) 31–4.
68. Langley, K. R., Millard, D. & Evans, E. W., Determination of tensile strength of gels prepared from fractionated whey proteins. *J. Dairy Res.*, **53** (1986) 285–92.
69. Broome, M. C., Willman, N., Roginski, H. & Hickey, M. W., The use of cheese whey protein concentrate in the manufacture of skim milk yoghurt. *Aust. J. Dairy Technol.*, **37** (1982) 139–43.
70. Greig, R. W. & Van Kan, J., Effect of whey protein concentrates on fermentation of yoghurt. *Dairy Ind. Internat.*, **49**(10) (1984) 28–9.
71. Chang, P. K., Fabricated shellfish products containing whey protein concentrate: composition and method of preparation. US Patent 4 411 917 (1983).
72. Burgarella, J. C., Lanier, T. C. & Hamann, D. D., Effects of added egg white or whey protein concentrate in rigidity of Croaker surimi. *J. Food Sci.*, **50** (1981) 1588–94.
73. Towler, C., Utilisation of whey protein products in pasta. *N.Z. J. Dairy Sci. Technol.*, **17** (1982) 229–36.

INDEX

Acanthopeltis japonica, 5, 7
Acetate, 322
Acetic acid, 256, 394
N-Acetylgalactosamine, 124
N-Acetylneuramine acid, 124
Actin, 366, 368, 387, 389
Actinin, 366
Actomyosin, 368, 374, 389, 391, 392, 393
Adenosine triphosphate, 368
Adipic acid, 256, 257
Agar
  bar-style, 13, 14–15
  BSA and, 310, 311–12
  calcium and, 5
  commercial availability, 14–18
  contamination possibilities, 17
  extraction methods, 9
  gelation, mechanism of, 28–9
  gelation and, 310, 311
  gel properties, 22–40
    additions to, effects of, 34–7
    applications, 40–3
    breaking load, 25
    irreversibility by freezing, 34
    junction zones, 29–31
    melting point, 27–8, 31
    rigidity coefficient, 25
    strength, 23–7, 31
    syneresis, 32, 33–4
    turbidity, 34
  grade, 16–17

Agar—*contd.*
  jellies, condensed, 43
  jellies, home-made, 42–3
  laboratory uses, 2
  LBG and, 336
  plain gel, 42
  prices, 17
  sol setting point, 21–2
  solution properties, 18–22
  stringy, 13, 15–16, 40
  structure, 2–6
  sugar skeleton, 3
  sulphate substituents, 339
  viscosity, 18, 19–21
Agarobiose, 2
Agaropectin, 2, 3, 11
Agarophytes, 9, 11, 31
Agarose, 2, 3–4, 5, 13, 17, 22, 23, 37–8, 325, 329, 335, 336
Agarose–agaropectin gels, 37
Agarose–gelatine gels, 36
Ahnfeltia plicata, 8
Algae. *See* Seaweed
Alginates:
  applications, 70–6
  bacterial, 58
  calcium and, 53, 59, 61, 62, 64, 66–8, 69
  commercial availability, 58–9
  enzyme modification, 58
  gelation, mechanism of, 63–4
  gel preparation, 64–8

Alginates—contd.
  gel properties, 63–70
    factors affecting, 69–70
    measurement, 68
    strength, 68
    syneresis, 69
  molecular weight, 61, 62
  precipitate, as, 299
  production, 56–8
  solution properties, 59–63
  source, 55
  structure, 54–5
Alginate–pectin gels, 344–52
Alginate poly-L-guluronate, 347, 348, 349
Alginic acid, 53, 54, 57, 59, 70, 270
Aluminium, 186, 406
Aluminium chloride, 186
Amidation, 408, 409, 419–21
Amino acids, 234, 235, 236, 238
Aminopeptidases, 125
Ammonia, 248, 407
Ammonium alginate, 59
Amorphophallus konjac. See Konjac mannan
Amylopectin, 303, 304, 305
Amylose, 303, 304, 305
Anhydrides, 189
Anhydrous dicalcium phosphate, 71
Apiose, 403
Apple pectin, 403, 404, 405
Apple purée, 351
Arginine, 125, 235, 238, 293
Ascophyllum nodosum, 53, 55
Aspargine, 238, 248
Aspartic acid, 238, 296
Aspics, 298
Azotobacter vinelandii, 58

Bananas, 75
Beef, 371
Beef steaks, 75, 76
Beverages, 426–8
Biozan welan gum, 203
Blackcurrants, 73, 74
Blancmange, 97, 106

Blood plasmin, 125
Blueberries, 73
Borax, 63
Boron, 5–6
Bovine serum albumin (BSA), 439
Brick cheese, 148
Buffalo, 131
Burgers, 391, 460
Butterfat, 108

Cabbage, 404
Caesalpinia spinosa, 315
Cakes, 175, 193
Calcium acetate, 67
Calcium alginate, 59, 63, 65–6, 69
Calcium carbonate, 67
Calcium chloride, 36, 57, 67, 136, 145, 146, 163, 243, 331
Calcium citrate, 426
Calcium hydrogen orthophosphate dihydrate, 73
Calcium lactate, 67
Calcium phosphate, 107
Calcium sulphate, 67
Calf rennet, 133, 134, 143, 148
Calgon, 351
Camembert cheese, 149
Campylaephora hypnaeoides, 5, 8
Capsules, 226
Carboxymethylcellulose, 36, 298, 299, 304, 350
Carob gum. See Locust-bean gum
Carrabiose, 80, 82
Carrageenan
  agar and, 110
  commercial availability, 85–7
  definition, 79
  gelation, mechanism, 92–4
  gel preparation, 91–2
  gel properties, 91–104
    applications, 104–13
    hysteresis, 94–5
    proteins and, 96–104
    strength, 92, 96
    syneresis, 96, 106, 108
    transition temperatures, 95
  ham and, 112

Carrageenan—*contd.*
  LBG and, 110, 338, 339, 340, 343
  milk and, 104–9
  molecular weight, 86
  production, 83–5
  regulatory aspects, 113
  solution properties, 87–91
  source, 83
  standardization, 86–7
  structure, 79–83
  sulphated, 269
  types available, 86
Casein
  associations, 128
  carrageenans and, 168
  commercial availability, 131–2
  concentration, 158, 160
  flocculation, 139, 141
  gelation, 139, 140, 141
  gel properties, 132–64
    assembly, 137–41, 149
    renetted milk, 132–52
    strength, 137, 140, 142–4, 155, 157–65
    syneresis, 144–7, 157, 162
  heat treatment, 166–7
    see also Milk
  hydrolysis, 135
  ion binding, 128–9
  micelles, 127, 128, 129–31, 135, 137, 138, 139, 140, 141, 147, 150, 153, 154, 155, 156, 167
  molecular characteristics, 125–7
  phosphorylation, 123
  production, 131
  proteins in, 122
  solution properties, 132
  source, 131
  structure, 121–31
Casein–alginate, 306, 307
Caseinate, 158
Cassia gum, 221
Cassia tora, 315–16
Cattle, 124, 131, 438
Caustic soda, 243
Cellulase, 11
Cellulose, 330, 334
Ceramium spp., 4, 8, 26, 28

Cheddar cheese, 144, 146, 147, 148, 436, 441
Cheese
  cheddaring, 147, 148
  fresh, 122
  imitation, 108
  lypolysis, 152
  microstructure, 147–9
  pressing, 149
  ripening, 140, 149
  salting, 144, 146–7, 149
  scalding, 148, 152
  stretching, 147
  see also under specific names
Cheshire cheese, 148
Chocolate dessert mix, 282
Chocolate milk, 97, 99, 103, 106
Chocolate syrups, 106
Cholesterol, 182
Chondrus crispus, 79, 80, 84, 102
Chymosin, 126, 127, 132, 133, 134, 143, 150
Citric acid, 156, 157, 163, 214, 256, 276, 351, 413
Citrus pectin, 403, 404, 405
Cocoa, 97, 99, 101, 102, 106
Cod, 371
Collagen
  bone, 235
  fish muscle, 377–8
  heat and, 387
  hydrolysis, 261
  muscle, 363, 369, 377–8
  skin, 235
  structure, 234–7
Colloidal calcium phosphate (CCP), 129, 130, 135, 136, 137, 143, 145, 148, 152, 153, 154, 155
Conalbumin, 186, 188, 192
Confectionery, 274, 425
Confectionery jellies, 425
Copper, 406
Copper sulphate, 186
Cottage cheese, 108, 153, 155, 160, 436, 442
Coulommier, 145
Crab sticks, 394
Cream, 108

Cream, whipping, 73, 108
Creamed cheese, 108
Custards, 73, 106, 175, 193
Cysteine, 124, 237, 238

Desserts, 72, 109, 267, 282
Dextran, 203, 300, 301, 302, 303–4
Dextrose, 267, 303
Difco agar, 31
Digenea simplex, 7
Disodium hydrogen orthophosphate, 68
Dutch cheese, 144
see also Edam; Gouda

Edam cheese, 148
EEC, 113, 233, 271, 409
Egg albumin
  composition, 175
  demand for, 192
  gels, 182, 183–8, 190
  modification, 188–9
  solution properties, 181–2
  thermal properties, 187
Egg jerky, 193
Eggs
  aging of, 175
  commercial availability, 179–81
  freezing, 181
  gelation, 186–7
  gel properties, 182–92
    brittleness, 185
    factors affecting, 183–92
    salts and, 185, 190
    strength, 183, 186, 187
  gels, application, 192–3
  heat treatment, 182, 186
  protein compositions, 175, 176
  solution properties, 181–2
  source, 179
  storage, 185
  structure, 175–9
Egg yolk
  composition, 175, 177, 178
  gels, 182, 190
  solution properties, 181

Emmenthal cheese, 147
Endothia parasitica, 133
Ethanol, 61, 166, 167, 409
Euchemia spp., 8, 26, 27, 81, 84

FAO, 407, 408
FDA, 107, 113, 271
Fenugreek, 335
Ferrous sulphide, 191
Ferulic acid, 403
Feta cheese, 149
Fish
  drying, 395–6
  frozen, 112, 388, 394
  minced, 75
  preserving, 394–6
  uncooked, 386
Fish muscle, 364–5
  age and, 370
  comminuted, 393
  denaturation, 384, 385, 387, 394
  freezing, 383–5
  heat and, 387
Flans, 106, 350
Flavoprotein, 187
Formaldehyde, 167, 384
Frankfurters, 112, 392
Fructose, 267
Fruit, structured, 71
Fruit drink mixes, 110–11
Fruit gels, 350
Fruit juices, 427
Fumaric acid, 256, 276
Furcellaran, 82, 83, 85, 94, 101, 112, 339
Furcellaria spp., 82, 84

Gadidae, 383
Galactomannans, 96, 266, 292, 313–44
α-L-Galactopyranose, 2
β-D-Galactopyranose, 2
Galactose, 124, 315, 316, 331, 333
β-D Galactose, 335
D-Galactose, 1
L-Galactose, 3
α-Galactosidase, 325, 331, 332

β-Galactosidase, 449
Galactosyl, 331
Galacturonic acid, 401, 403, 408
GDL, 351
Gelatine
  agar and, 270
  alginates and, 270
  analysis of, 249–50
  carrageen and, 269, 270
  celluloses and, 270
  commercial availability, 245
  compatibility with other food gels, 265–71
  derivation of, 233, 258
  derivatives, 281–4
  dissolving, 253
  filled gels, 271–3
  gel formation, 258–61
  gellan and, 271, 297–8
  gel properties, 258–74
    factors affecting, 263–4
    strength, 244, 246, 251, 260, 261–3, 268, 269
  gels
    applications, 274–84
  glucose and, 277
  gum arabic and, 266, 270, 277, 297
  hydrocolloids and, 266–71
  hydrolized, 282–4
  isoelectric point, 247–9, 280
  melting and setting points, 252, 259, 265
  molecular weight, 238–41, 265
  pectins and, 269–70
  protein, as, 233
  sodium alginate and, 296
  solution properties, 252–8
  sources, 242
  stabilizer, as, 278–80
  structure, 238–42
  sulphur dioxide and, 251–2
  surface-active properties, 257–8
  viscosity, 253–7
GDL, 156, 157, 159, 161, 162, 163, 346
Gelidium spp., 1, 4, 5, 6, 7, 10, 13, 16, 26, 33, 41, 42

Gellan gum
  agar and, 220
  commercial availability, 206
  gelatine and, 225
  gelation, 219
  gel properties, 209–25
    applications, 226
    melting and setting points, 217–18
    strength, 209, 214
    syneresis, 220
    texture, 209–17, 220, 221, 222–3
  guar gum and, 225
  other hydrocolloids and, 218–25
  safety studies, 207
  solution properties, 206–9
  source, 205
  structure, 204–5
  xanthan gum and, 221
Gelrite, 205, 206, 211, 212, 213, 214, 215, 216, 217
  agar and, 226, 227
Gigartina spp., 80, 81, 84
Globulin, 187
Glucano-δ-lactone, 345
Glucose, 204, 205, 267, 268, 282, 334
Glucuronic acid, 204
D-Glucuronic acid, 3
Glutamic acid, 235, 238, 295, 296
Glutamine, 238, 248
Glutaraldehyde, 297
Gluten, 295
Glycerin, 35
Glycerol, 61
Glycogen, 382
Glycoprotein, 175
Goats, 131
Goiopeltis spp., 339
Gouda cheese, 148
Gracilaria spp., 4, 5, 7, 8, 10, 11, 16, 33, 337
Grapefruit, 404
Ground almonds, 272
Guar gum, 219, 266, 304, 314, 315, 317, 321, 331
  agarose and, 331
  enzymic de-branching, 331–4
  LBG and, 331

Guar gum—*contd.*
  xanthan and, 325, 329, 331
Guluronate, 345, 350
Guluronic acid, 54, 55, 59, 64
Gum tara, 339, 343

Ham, 269, 283, 389
Herring, marinaded, 394–5
Hide gelatines, 235, 237, 240, 241, 242, 255, 279, 280
Histidine, 237
Hydrochloric acid, 156, 159, 162, 163, 243
Hydrogen sulphide, 186
Hydroxylysine, 234, 235, 236, 237, 293, 369
Hydroxyproline, 234, 235, 258, 369
Hypnea spp., 82–3

Ice-cream, 73, 107
Imino acids, 234, 258, 259, 261
Immunoglobulins, 436, 439–41, 446, 447
Irish Moss, 79
Isoleucine, 236
Isopropanol, 297, 409

Jam, 424–5
Jam-making, 401, 413
Jellies, 233, 267, 274–6, 276–8, 343, 424, 425
Joint Export Committee on Food Additives (JECFA), 113, 407, 408

'Kainic acid', 7
Kamaboko, 393, 394
Kelcogel, 205, 229
Kelco Inc., 206, 271, 322
Konjac glucomannan, 335
Konjac mannan, 221, 331, 334, 335

α-Lactalbumin, 124, 137, 157, 166, 436, 438–9, 446, 447
Lactic acid, 152, 256, 380, 381
β-Lactoglobulin, 103, 124, 137, 156, 166, 167, 436, 446, 447, 462
Lactose, 144, 152, 156, 299
Laminaria hyperborea, 53, 55
Lemon, 404
Leucania leucophala, 325
Leucine, 133, 235, 236
L-Leucine dodecyl ester, 257
Lime, 243, 244, 248, 404
Lipoproteins, 176, 178, 189, 190–1
Livetins, 189
Locust-bean gum, 35–6, 102, 103, 108, 208, 219, 221, 222, 230, 266, 314, 315, 317, 330, 331, 333
  xanthan and, 341, 342, 343, 344
'Lutefish', 396
Lysine, 125, 188, 236, 270, 293, 295
Lysozymes, 187, 188, 438

Macrocystis pyrifera, 53, 55
Malic acid, 256, 276
Maltodextrins, 268, 303
Mango, 405
Mannanase, 331
β-Mannanase, 325, 326
Mannans, 96, 332, 333
Mannose, 315
Mannosidase, 331
Mannosyl, 331
Mannuronate, 293, 295, 346
Mannuronic acid, 54, 55, 59, 64
Marine Colloids, 23
Marshmallow, 278–80
Mayonnaise, imitation, 111
'Membraneless osmosis', 412
Meringues, 461
Methanobacterium, 227
Methanobrevibacter, 227
Methanol, 403, 404, 409
Methionine, 236
Methylgalacturonate, 350
Microcapsules, 226, 292, 296
Milk
  calcium, 136

Milk—*contd.*
  fat content, 145
  gelation of, 121
  heat treatment, 132, 136, 137, 143, 146, 156, 160–1, 166, 167
  powders, 132
  proteins, 96, 97, 122
    see also Casein
  proteolysis, 124–5
  sodium and, 136
  sterilization, 121
  sterilized, 167
  technological properties, 129
  ultrafiltration, 136, 149, 158, 159
Milk puddings, 108
Milkshake, 114
Mitsuname, 42
Mixed polymer gels
  application, 342–4
  biphasic gels, 309–12
  covalent interactions, 293–6
  gelation mechanism, 327–31
  gel formation, 291
  gel strength, 291, 329, 333, 341, 343
  melting point, 345
  phase separation, 305–8
  polyanion–polycation interactions, 296–300
  polymer exclusion, 300–8, 310
  product implications, 312–13
  thermodynamic incompatibility, 300
  thermoreversibility, 344
Mousses, 282, 460
Mozzarella, 148
Mucor, 133
Muscle
  animal age and, 380
  animal gender and, 380
  connective tissues, 367
  factors affecting, 378–82
  freezing, 383–5
  hanging, 382–3
  structure, 363–4, 365–7
  texture, 386
  treatment of, 382–5
Muscle protein
  applications, 385–96
  classes of, 367

Muscle protein—*contd.*
  comminution, 390–4
  denaturation, 370, 371, 387
  freezing, 372–3
  heat and, 371, 373–4, 375, 377–8, 386–7, 387–8
  properties, 369–85
  salt and, 371, 373, 374, 388–9, 391
  structural organization, 363–5
Myoglobin, 367
Myosin, 366, 367–8, 371–4, 387, 389

Nitrogen, 122, 251, 409
Nitrogen, non-protein, 150

Oblate, 43
Oleic acid, 188, 189
Oligomers, 258
Omelettes, 193, 175
Orange, 404
Ornithine, 238
Ossein gelatines, 235, 240, 241, 242, 243, 247, 248, 279, 280
Ovalbumin, 185, 187
Ovalbumin–BSA, 308
Ovomucin, 175, 176
Oxalic acid, 156, 157, 162

Papain, 257
Paper, edible, 43
Pasta, 461
Patés, 460
Peach, internally set, 71
Pear pectin, 404
Pectate, 408, 423
Pectic acid, 348
Pectin poly-D-galacturonate, 347, 348, 349
Pectins
  applications, 424–8
  availability, 407–9
  calcium, 401, 410, 411, 414, 421
  definition, 407–8
  gelation, factors affecting, 419–22
  gelation, measurement, 418–19

Pectins—*contd.*
  gel formation, 401
  gel mechanisms, 422–4
  gel properties, 413–24
    strength, 421
    texture, 414–18
  molecular weight, 410, 411
  production, 405–7
  protein and, 404
  purity, 408–9
  solution properties, 409–13
  source, 401, 404–5
  stabilizing properties, 401, 427
  storage, 409
  structure, 402–4
  toxicology, 409
Pepsins, 132, 133, 134, 136, 141, 150
Peptides, 135, 258, 260, 261
Peroxide/peroxidase, 404
Petfoods, 111, 226, 227–9, 298, 299
Phaeophycea, 53
Phe–Ile bond, 134
Phe–Met bond, 133, 134, 135
Phosphoric acid, 243
Phosvitia, 189
Photographic films, 226, 274
Pigskin gelatine, 235, 237, 238, 240, 241, 242, 243, 245, 248, 268, 276, 279, 281, 302
Plum pectin, 404
Pizza, 111
Plasmin, 122
Polyethylene glycols, 268, 300, 301, 303
Polygalacturonate, 351
Polyguluronate, 349, 350, 351
Polymannurate, 350
Polyvinyl alcohol, 295
Pork pies, 269
Potassium, 83, 93, 94, 96
Potassium alginate, 59
Potassium chloride, 80, 81, 85, 91, 100, 112, 143, 146, 373
Potassium phosphate, 373
Potassium sorbate, 351
Potassium sulphate, 34
Potatoes, 75, 403, 404
Precollagen, 237

Processed cheese, 129
Prolines, 126, 127, 234, 235, 258
Propylene glycol, 226
Propylene glycol alginate (PGA), 57, 62, 64, 270, 293, 294, 295, 299, 308, 352
Propylene glycol pectate, 293
Propylene oxide, 57
Proteinases, 124, 125, 127
Proteins, extrusions, 75
Proteose peptones, 123, 125, 436, 437
Provolone cheese, 148
Pseudomonas spp., 58, 205
Pterocladia spp., 6, 7
Pumpkin pie, 111
Pyrolidine, 258
Pyrophosphate, 389
Pyruvate, 80, 322
Pyruvic acid, 3

Quarg cheese, 149, 153, 155, 157
Quiche, 193

Radishes, 111
Renett, 126, 132–3
  action of, 133–6
  coagulation time, 135, 136, 137, 138, 139, 149, 150
  hysteresis, 137
  pre-renetting treatment, 136
  substitutes, 133
  syneresis and, 144–7
Rhamnose, 204
Rhamsan gum, 203, 204
Rigor mortis, 380–2

St Paulin cheese, 149
Salad dressings, 62, 111, 300, 461
Salmonella, 180
Sarcoplastic proteins, 375–7
Sarugassum fulvellum, 8
'Sashimi', 8, 385

Sausage, 193, 392
Seaweed, 1
  alkali treatment of, 10–11
  brown, 8, 53
  flour, 111
  red, 5, 6–8, 14, 79
Serine, 123, 133–4
Sheep, 131
Sherbet, 107
Silicate, 404
Skim milk powder, 158, 159
Sodium alginate, 36, 53, 56, 57, 59, 60, 61, 62, 63–4, 65, 70, 218, 296, 299, 306, 307
Sodium carbonate, 27, 57
Sodium caseinate, 108, 158, 159, 162, 225, 393
Sodium chloride, 35, 100, 143, 145, 146, 163, 165, 176, 191, 192, 296, 388, 394, 410
Sodium citrate, 271, 351
Sodium dithionite, 27
Sodium dodecyl sulphate, 188–9
Sodium hexametaphosphate, 73
Sodium hydroxide, 26, 27, 294, 297
Sodium hypochlorite, 27
Sodium phosphate, 388, 389
Sophora japonica, 325
Sorbet, 111
Sorbitol, 35, 267
Soufflés, 175, 193
Soya extrusion, 295
Soy-bean milk, 427
Soy protein, 107, 108, 225
'Spa gelatines', 244
Starch, 202, 266, 272, 282, 304, 305, 343
Steaklets, 391
Steaks, 385
Sucrose, 35, 187, 192, 267
Sugarbeet, 403, 405
Sugars, neutral, 403
Sulphur dioxide, 409
Sulphuric acid, 2, 9, 26, 243
Sunflower pectin, 404, 405
Surimi, 192, 369, 394, 455, 461
Sushi, 385
Swiss cheese, 144

Tamarind gum, 219, 335
Tannin, 280, 281
Tartaric acid, 256
Telopeptides, 237
Thiocyanate, 259
Thrombin, 125
$T_m - \Delta T_m$, 31–3
Tobacco pectin, 403, 404
Tokoroten, 42
Tomatoes, 404
Toppings, 108
Trigonella foenum-graecum, 316
Tripolyphosphate, 389
Troponin, 366
Tween, 80, 257

Urea, 259, 296

Valine, 236

Welan gum, 204
Whey
  acid, 442–3
  isolate, 445
  sources, 436
  sweet, 441–2
  types, 441–3
Whey protein
  application, 459–62
  commercial availability, 446–50
  composition, 436–7
  concentrates, 443, 447–50
  denaturization, 137
  gel properties, 454–9
    strength, 455–9
  production, 443–6
  solution properties, 450–4
  sources, 441–2
  structure, 436–41
Whey protein concentrate (WPC), 435, 437, 444, 449, 459–60, 460–2
Whey protein isolate (WPI), 435, 447
WHO, 113, 407, 408
Wine, clearing, 257, 280–1

Xanthan gum, 111, 203, 208, 209, 219, 221, 222, 230, 304, 305, 322–5, 330–1, 332, 333, 335, 340
   galactomannan mixes, 343
   LBG and, 340
Xanthomonas compestris, 322
Xylose, 403
α-D-Xylose, 335
D-Xylose, 3

Yogurt
   additions to, 163–4, 257
   consumption of, 131
   fruit, 426
   gel development, 153
   gel strength, 160
   homogenization of, 164–5
   manufacture, 158–9
   stabilization, 108
   stirring of, 165